Advances on the Influence of Vegetation and Forest on Urban Air Quality and Thermal Comfort—2nd Edition

Advances on the Influence of Vegetation and Forest on Urban Air Quality and Thermal Comfort—2nd Edition

Guest Editors

Jose Luis Santiago
Esther Rivas
Beatriz Sanchez

MDPI

Basel • Beijing • Wuhan • Barcelona • Belgrade • Novi Sad • Cluj • Manchester

Guest Editors
Jose Luis Santiago
Department of Environment
CIEMAT
Madrid
Spain

Esther Rivas
Department of Environment
CIEMAT
Madrid
Spain

Beatriz Sanchez
Department of Environment
CIEMAT
Madrid
Spain

Editorial Office
MDPI AG
Grosspeteranlage 5
4052 Basel, Switzerland

This is a reprint of the Special Issue, published open access by the journal *Forests* (ISSN 1999-4907), freely accessible at: https://www.mdpi.com/journal/forests/special_issues/M488RLLI1L.

For citation purposes, cite each article independently as indicated on the article page online and as indicated below:

Lastname, A.A.; Lastname, B.B. Article Title. *Journal Name* **Year**, *Volume Number*, Page Range.

ISBN 978-3-7258-6764-6 (Hbk)
ISBN 978-3-7258-6765-3 (PDF)
https://doi.org/10.3390/books978-3-7258-6765-3

Contents

forests

MDPI

Editorial

Advances on the Influence of Vegetation and Forest on Urban Air Quality and Thermal Comfort—2nd Edition

Esther Rivas *, José Luis Santiago and Beatriz Sánchez

Atmospheric Modelling Unit, Department of Environment, CIEMAT, 28040 Madrid, Spain; jl.santiago@ciemat.es (J.L.S.); beatriz.sanchez@ciemat.es (B.S.)
* Correspondence: esther.rivas@ciemat.es

Urbanisation has intensified environmental challenges, particularly the deterioration of air quality and the amplification of heat stress, both of which directly affect public health and urban liveability. Vegetation- and forest-based strategies, often framed as nature-based solutions (NBS), have consequently gained relevance as key tools to mitigate these problems by reducing air pollutant concentrations, improving thermal comfort, and delivering a wide range of co-benefits. Their influence, however, is neither uniform nor linear, since aerodynamic, deposition, shading, and evapotranspiration processes interact in complex ways with urban morphology, atmospheric pollutants, and meteorological conditions. As a result, understanding the role of vegetation in shaping urban environments requires interdisciplinary approaches and evidence from both experimental and modelling studies. Building on the first edition of this Special Issue (https://www.mdpi.com/journal/forests/special_issues/urban_air_thermal, accessed on 24 November 2025), the present collection (https://www.mdpi.com/journal/forests/special_issues/M488RLLI1L, accessed on 24 November 2025) offers new insights into the multiple mechanisms through which vegetation affects air quality, microclimate, and thermal comfort, while also highlighting the challenges of designing context-sensitive green infrastructures that carefully account for the specific environmental, urban, and social conditions in which they are embedded.

This Special Issue comprises 11 papers, including the editorial [1], authored by researchers from various regions (China, Bulgaria, Canada, Italy, Australia, and Spain) and diverse scientific backgrounds, reflecting a broad spectrum of approaches to the study of vegetation in urban contexts. Some contributions focused on the health and climate impacts of large-scale revegetation strategies, demonstrating through mesoscale meteorological simulations that increasing green cover can reduce temperature-related mortality in Southern European cities [2]. Other studies examined the interplay between vegetation and atmospheric pollutants, such as the role of canopy structure in shaping PM2.5 variability in subtropical zones, the dust-retention capacity of urban parks, and the pollutant-filtering efficiency of evergreen and deciduous ornamental tree species [3–5]. The broader biofiltering potential of vegetation was further explored by comparing the pollutant accumulation capacities of mosses, lichens, herbs, and trees, underscoring the value of combining different plant types for effective urban air quality management [6]. Wind comfort was also addressed through a campus-scale case study showing how plant configurations influence local airflow and pedestrian comfort [7]. Regarding the thermal environment, the application of the three-dimensional Green Plot Ratio (GPR) demonstrated the effectiveness of urban greenery in reducing heat stress in residential areas [8], while a complementary study revealed the seasonal trade-offs of street canyon greening, where vegetation that improves summer cooling may compromise winter comfort [9]. In addition, a long-term field experiment combined with microenvironment simulations on an individual Ficus concinna tree quantified the relative contributions of shading and transpiration to urban cooling, showing that transpiration accounted for nearly half of total cooling and dominated under high-temperature conditions, while shading effects peaked under strong solar radiation but could increase nighttime and winter temperatures due to longwave radiation reflection [10]. Finally, a study on garden plants in Suzhou (China) showed how different life forms adopt distinctive functional strategies to cope with air pollution, providing criteria for species selection in stressed urban environments [11].

The papers in this collection reaffirm the complexity of vegetation–atmosphere interactions in cities, while also offering concrete evidence of the benefits of NBS for air quality, thermal comfort, and climate resilience. At the same time, previous research has emphasised that local factors, including plant traits, urban design, and meteorological conditions, strongly influence the impact of vegetation. Vegetation can act simultaneously as a sink and a barrier for pollutants, or as a source of biogenic emissions, while also exerting both cooling and insulating effects that vary seasonally. This underscores the importance of designing green infrastructures not merely as aesthetic additions, but as functional systems capable of balancing ecosystem services and potential disservices. By advancing both theoretical understanding and practical methodologies, the contributions gathered here provide valuable knowledge for researchers, policymakers, and urban planners aiming to create greener, healthier, and more sustainable cities. Ultimately, this Special Issue aspires to serve as a reference point for interdisciplinary research and decision-making on the integration of vegetation into urban environments in the context of ongoing climate change.

Funding: This work is part of the MULTIURBAN-I project (PID2023-149246OB-C21) funded by MICIU/AEI/10.13039/501100011033/ and FEDER/UE and AIRTEC2-CM (TEC-2024/ECO 283) project funded by the Regional Government of Madrid.

Conflicts of Interest: The authors declare no conflicts of interest.

References

1. Rivas, R.; Santiago, J.L.; Sanchez, B. Advances on the Influence of Vegetation and Forest on Urban Air Quality and Thermal Comfort—Series II. *Forests* **2023**, *14*, 284. [CrossRef]
2. de Andrés, J.M.; D'Elia, I.; de la Paz, D.; D'Isidoro, M.; Russo, F.; Mircea, M.; Gualtieri, M.; Vardoulakis, S.; Borge, R. City-Scale Revegetation Strategies Impact on the Temperature-Related Long-Term Mortality: A Quantitative Assessment in Three Cities in Southern Europe. *Forests* **2025**, *16*, 1089. [CrossRef]
3. Huang, Y.; Cai, Y.; Jiao, J.; Pan, C.; Wang, G.; Li, C.; Jia, Z.; Chen, Z.; Zhou, Y.; Zhou, G. The Impact of Meteorological Factors and Canopy Structure on PM2.5 Dynamics Under Different Urban Functional Zones in a Subtropical City. *Forests* **2025**, *16*, 479. [CrossRef]
4. Zhang, X.; Wang, C.; Guo, J.; Zhu, Z.; Xi, Z.; Li, X.; Qiu, L.; Gao, T. Factors Affecting Dust Retention in Urban Parks Across Site and Vegetation Community Scales. *Forests* **2024**, *15*, 2136. [CrossRef]
5. Petrova, S. The Added Value of Urban Trees (*Tilia tomentosa* Moench, *Fraxinus excelsior* L. and *Pinus nigra* J.F. Arnold) in Terms of Air Pollutant Removal. *Forests* **2024**, *15*, 1034. [CrossRef]
6. Petrova, S.; Velcheva, I.; Nikolov, B. Nature-Based Solutions to Reduce Air Pollution: A Case Study from Plovdiv, Bulgaria, Using Trees, Herbs, Mosses and Lichens. *Forests* **2024**, *15*, 928. [CrossRef]
7. Chen, J.; Zeng, J.; Huang, T.; Wang, Y.; Yang, H.; Yu, X.; Wang, Z. Impacts of Plant Configuration on the Outdoor Wind Comfort of Subtropical Coastal Campuses: Evidence from a Study of Quanzhou. *Forests* **2025**, *16*, 461. [CrossRef]
8. Zheng, J.; Li, Z.; Zheng, B. A Study on the Effect of Green Plot Ratio (GPR) on Urban Heat Island Intensity and Outdoor Thermal Comfort in Residential Areas. *Forests* **2024**, *15*, 518. [CrossRef]
9. Liu, J.; Zheng, B.; Yang, F. A Simulation Study of the Impact of Urban Street Greening on the Thermal Comfort in Street Canyons on Hot and Cold Days. *Forests* **2023**, *14*, 2256. [CrossRef]

10. Shi, Z.; Yan, C.; Hu, W.; Luo, Z.; Qiu, G.Y. Disentangling the Cooling Effects of Transpiration and Canopy Shading: Case Study of an Individual Tree in a Subtropical City. *Forests* **2025**, *16*, 1564. [CrossRef]
11. Yang, Z.; Zhang, X.; Qu, Y.; Gao, F.; Li, Y. Response of Common Garden Plant Leaf Traits to Air Pollution in Urban Parks of Suzhou City (China). *Forests* **2023**, *14*, 2253. [CrossRef]

forests

MDPI

Editorial

Advances on the Influence of Vegetation and Forest on Urban Air Quality and Thermal Comfort—Series II

Esther Rivas, Jose-Luis Santiago * and Beatriz Sánchez

Atmospheric Modelling Unit, Department of Environment, CIEMAT, 28040 Madrid, Spain
* Correspondence: jl.santiago@ciemat.es

Cities are composed of different types of urban obstacles such as buildings, cars, trees, hedgerows, etc. The interaction of the atmosphere with urban morphology and the different urban surface properties gives rise to complex atmospheric processes on different scales in the urban canopy layer (UCL).

Spatial distributions of atmospheric pollutants observed in urban environments are also related to these processes. The interactions between the atmosphere and buildings induce reduced ventilation in the streets and, this fact, together with high emission rates of pollutants from road traffic (NO_2, PM10, PM2.5, etc.), induce heterogeneous distributions of pollutants and high levels of pollution in the streets [1–4]. It should also be considered that indoor air quality is influenced by outdoor air pollution. Hence, populations are exposed to these high air pollutant concentrations both indoors and outdoors. Nevertheless, estimating population exposure to atmospheric pollution in both scenarios remains a major challenge [5–7].

Measures for improving air quality in cities are necessary and they are usually based on reducing pollutant emissions or implanting passive mitigation air pollution strategies, such as the so-called green infrastructures (GI). However, how do GI influence air quality? The main effects are [8,9]:

- Aerodynamic effects: trees, hedges, bushes, etc.; are porous obstacles, which modify airflow around them and consequently, the pollutant dispersion.
- Deposition effects: vegetation removes a fraction of pollutants from the air through deposition on leaves and absorption through stomata.
- Emission source of biogenic volatile compounds and pollen.

Deposition effects on air quality are always positive; however, aerodynamic effects might be positive or negative depending on vegetation types and their location and arrangement regarding emission sources and pedestrians. For instance, whereas a vegetation barrier in an open road usually contributes to improving air quality behind the barrier [8,10–12], the effects of street trees depend on several factors such as urban morphology, trees layout, or meteorology and are not always positive [13–17]. Therefore, analyzing the impact of GIs on air quality remains a major challenge due to the complex atmospheric processes involved in urban environments.

On the other hand, cities produce air overheating inside UCL. Urban overheating is driven by reduced ventilation within the streets due to building configurations, together with the spectral response of urban surfaces (which absorb much of the solar radiation) and the thermal storage capacity of urban materials, these processes favor the increase of air temperatures affecting human thermal comfort. This urban overheating has a negative impact on health causing general discomfort, respiratory problems, dehydration and fatigue. Even, in certain meteorological situations, for example, the increasing mortality rate from heat stroke during heat wave episodes. These situations, which are also becoming more frequent due to climate change, often trigger energy consumption in cities due to the massive use of air conditioning systems. In this context, urban vegetation, either integrated into the environment as GI or on the buildings as green facades or green roofs, mitigate urban overheating improving thermal comfort and promote energy savings. Thermal comfort is related to air temperature, mean radiant temperature, wind speed and related humidity. However, how does vegetation affect air temperature and thermal comfort? The main effects are [18,19]:

- Shading and/or isolation effects: trees provide shade to urban surfaces (e.g.; building façade, roads, etc.) decreasing their surface temperature. In addition, green facades and green roofs might act as a protection element for buildings since, on the one hand, they reflect a large part of solar radiation and store little heat and, on the other, their substrates act as thermal insulation for buildings.
- Cooling effects: vegetation provides transpiration cooling because the absorbed solar energy causes an increase of latent heat (water from vegetation is evaporated into the atmosphere) and thus cooling the leaf surfaces and the air around them.

Various experiences show how urbanization exacerbates heat stress increasing daytime mean temperature and green spaces counterbalance this effect providing energy savings [20]. However, the effectiveness of nature-based solutions to reduce air temperature, improve thermal comfort and decrease energy consumption in urban environments depends on many factors and is still a challenge.

In addition to microclimate regulation and air quality services, GIs provide more environmental benefits such as noise reduction or rainwater drainage [18–20], and other benefits such as aesthetic, recreational, psychological or improving perceived mental health [21–23]. GI design should be addressed to obtain a trade-off solution between the different ecosystem services and disservices provided [17]. In general, theoretical studies at the microscale, considering all thermal effects and impacts of urban vegetation on air quality, are not very common in the literature and only a few of them include all vegetation performance [9]. A better understanding of the global effects of vegetation in cities is needed to design effective strategies that contribute to improving air quality, thermal comfort and energy efficiency in cities. In this context, original research articles related to the impact of urban vegetation on air quality, local climate and energy saving in urban environments are welcomed to the current Special Issue entitled "Influence of Vegetation and Forest on Urban Air Quality and Thermal Comfort—Series II". In a previous Special Issue of this journal [24], several novel studies were presented. Zheng et al. [25] evaluated, through microscale modeling, the effects of roof greening on the indoor thermal environment finding that increased roof greening coverage improved the indoor thermal perception. Zhang and Zhao [26] investigated the cooling effects of several species of trees. Regarding the air quality effects, Yin et al. [27] computed the effects of urban park green spaces in Beijing on PM2.5 retention at multiple scales. De la Paz et al. [28] studied air quality and meteorological changes induced by future vegetation in Madrid at the city scale. This study highlights the need to combine nature-based solutions with emission-reduction measures. Santiago et al. [17] simulated at the microscale the impact of a wide set of GI scenarios on traffic-related pollutant concentrations at the pedestrian level. It was found that using GI alone is ineffective as a general air quality mitigation measure, but selecting an appropriate layout of GI elements, GI can also help to reduce exposure to air pollution even in a scenario with high buildings. The contributions of the present Special Issue can address modeling studies at micro- and/or mesoscale (e.g.; computational fluid dynamics models, urban canopy models or mesoscale models), new parameterizations on the effects of urban vegetation and/or experimental works from field and laboratory experiments. Different configurations of trees in the streets, urban parks or vegetation barriers and green facades and roofs will be investigated in relation to their effects on pollutant concentration (aerodynamic effects, pollutant deposition, biogenic emissions, etc.) and/or on local climate and energy efficiency (cooling, shading, etc.). In addition, review papers focused on the current knowledge and future studies about this topic are also welcomed to the current Special Issue.

Funding: This research is part of the RETOS-AIRE project (Grant: RTI2018-099138-B-I00) funded by MCIN/AEI/10.13039/501100011033 and by ERDF A way of making Europe and AIRTEC-CM (S2018/EMT-4329) project funded by Regional Government of Madrid.

Data Availability Statement: Not applicable.

Conflicts of Interest: The authors declare no conflict of interest.

References

1. Borge, R.; Narros, A.; Artíñano, B.; Yagüe, C.; Gómez-Moreno, F.J.; de la Paz, D.; Roman-Cascon, C.; Díaz, E.; Maqueda, G.; Sastre, M.; et al. Assessment of microscale spatio-temporal variation of air pollution at an urban hotspot in Madrid (Spain) through an extensive field campaign. *Atmos. Environ.* **2016**, *140*, 432–445. [CrossRef]
2. Santiago, J.L.; Borge, R.; Martin, F.; de la Paz, D.; Martilli, A.; Lumbreras, J.; Sanchez, B. Evaluation of a CFD-based approach to estimate pollutant distribution within a real urban canopy by means of passive samplers. *Sci. Total Environ.* **2017**, *576*, 46–58. [CrossRef] [PubMed]
3. Rivas, E.; Santiago, J.L.; Lechón, Y.; Martín, F.; Ariño, A.; Pons, J.J.; Santamaría, J.M. CFD modelling of air quality in Pamplona City (Spain): Assessment, stations spatial representativeness and health impacts valuation. *Sci. Total Environ.* **2019**, *649*, 1362–1380. [CrossRef] [PubMed]
4. Sanchez, B.; Santiago, J.L.; Martilli, A.; Martin, F.; Borge, R.; Quaassdorff, C.; de la Paz, D. Modelling NOX concentrations through CFD-RANS in an urban hot-spot using high resolution traffic emissions and meteorology from a mesoscale model. *Atmos. Environ.* **2017**, *163*, 155–165. [CrossRef]
5. Santiago, J.L.; Borge, R.; Sanchez, B.; Quaassdorff, C.; De La Paz, D.; Martilli, A.; Rivas, E.; Martín, F. Estimates of pedestrian exposure to atmospheric pollution using high-resolution modelling in a real traffic hot-spot. *Sci. Total Environ.* **2021**, *755*, 142475. [CrossRef]
6. Santiago, J.L.; Rivas, E.; Gamarra, A.R.; Vivanco, M.G.; Buccolieri, R.; Martilli, A.; Lechón, Y.; Martín, F. Estimates of population exposure to atmospheric pollution and health-related externalities in a real city: The impact of spatial resolution on the accuracy of results. *Sci. Total Environ.* **2022**, *819*, 152062. [CrossRef]
7. Santiago, J.L.; Rivas, E.; Buccolieri, R.; Martilli, A.; Vivanco, M.G.; Borge, R.; Carlo, O.S.; Martín, F. Indoor-outdoor pollutant concentration modelling: A comprehensive urban air quality and exposure assessment. *Air Qual. Atmos. Health* **2022**, *15*, 1583–1608. [CrossRef]
8. Abhijith, K.V.; Kumar, P.; Gallagher, J.; McNabola, A.; Baldauf, R.; Pilla, F.; Broderick, B.; Di Sabatino, S.; Pulvirenti, B. Air pollution abatement performances of green infrastructure in open road and built-up street canyon environments—A review. *Atmos. Environ.* **2017**, *162*, 71–86. [CrossRef]
9. Buccolieri, R.; Santiago, J.L.; Rivas, E.; Sanchez, B. Review on urban tree modelling in CFD simulations: Aerodynamic, deposition and thermal effects. *Urban For. Urban Green.* **2018**, *31*, 212–220. [CrossRef]
10. Santiago, J.L.; Buccolieri, R.; Rivas, E.; Calvete-Sogo, H.; Sanchez, B.; Martilli, A.; Alonso, R.; Elustondo, D.; Santamaría, J.M.; Martin, F. CFD modelling of vegetation barrier effects on the reduction of traffic-related pollutant concentration in an avenue of Pamplona, Spain. *Sustain. Cities Soc.* **2019**, *48*, 101559. [CrossRef]
11. Gallagher, J.; Baldauf, R.; Fuller, C.H.; Kumar, P.; Gill, L.W.; McNabola, A. Passive methods for improving air quality in the built environment: A review of porous and solid barriers. *Atmos. Environ.* **2015**, *120*, 61–70. [CrossRef]
12. Wang, H.; Maher, B.A.; Ahmed, I.A.; Davison, B. Efficient removal of ultrafine particles from diesel exhaust by selected tree species: Implications for roadside planting for improving the quality of urban air. *Environ. Sci. Technol.* **2019**, *53*, 6906–6916. [CrossRef]
13. Jeanjean, A.P.; Buccolieri, R.; Eddy, J.; Monks, P.S.; Leigh, R.J. Air quality affected by trees in real street canyons: The case of Marylebone neighbourhood in central London. *Urban For. Urban Green.* **2017**, *22*, 41–53. [CrossRef]
14. Santiago, J.L.; Martilli, A.; Martin, F. On dry deposition modelling of atmospheric pollutants on vegetation at the microscale: Application to the impact of street vegetation on air quality. *Boundary-Layer Meteorol.* **2017**, *162*, 451–474. [CrossRef]
15. Xue, F.; Li, X. The impact of roadside trees on traffic released PM10 in urban street canyon: Aerodynamic and deposition effects. *Sustain. Cities Soc.* **2017**, *30*, 195–204. [CrossRef]
16. Buccolieri, R.; Jeanjean, A.P.R.; Gatto, E.; Leigh, R.J. The impact of trees on street ventilation, NOx and PM2.5 concentrations across heights in Marylebone Rd street canyon, central London. *Sustain. Cities Soc.* **2018**, *41*, 227–241. [CrossRef]
17. Santiago, J.L.; Rivas, E.; Sanchez, B.; Buccolieri, R.; Esposito, A.; Martilli, A.; Vivanco, M.G.; Martin, F. Impact of different combinations of green infrastructure on traffic-related pollutant concentrations in urban areas. *Forests* **2022**, *13*, 1195. [CrossRef]
18. Salmond, J.A.; Tadaki, M.; Vardoulakis, S.; Arbuthnott, K.; Coutts, A.; Demuzere, M.; Dirks, K.N.; Heaviside, C.; Lim, S.; Macintyre, H. Health and climate related ecosystem services provided by street trees in the urban environment. *Environ. Health* **2016**, *15*, S36. [CrossRef]
19. Santamouris, M.; Ban-Weiss, G.; Osmond, P.; Paolini, R.; Synnefa, A.; Cartalis, C.; Muscio, A.; Zinzi, M.; Morakinyo, T.E.; Ng, E.; et al. Progress in urban greenery mitigation science–assessment methodologies advanced technologies and impact on cities. *J. Civ. Eng. Manag.* **2018**, *24*, 638–671. [CrossRef]

20. Livesley, S.J.; McPherson, E.G.; Calfapietra, C. The urban forest and ecosystem services: Impact on urban water, heat, and pollution cycles at the tree, street and city scale. *J. Environ. Qual.* **2016**, *45*, 119–124. [CrossRef]
21. Roy, S.; Byrne, J.; Pickering, C. A systematic quantitative review of urban tree benefits, costs, and assessment methods across cities in different climatic zones. *Urban For. Urban Green.* **2012**, *11*, 351–363. [CrossRef]
22. Haaland, C.; van Den Bosch, C.K. Challenges and strategies for urban green-space planning in cities undergoing densification: A review. *Urban For. Urban Green.* **2015**, *14*, 760–771. [CrossRef]
23. Van den Berg, M.; Wendel-Vos, W.; van Poppel, M.; Kemper, H.; van Mechelen, W.; Maas, J. Health benefits of green spaces in the living environment: A systematic review of epidemiological studies. *Urban For. Urban Green.* **2015**, *14*, 806–816. [CrossRef]
24. Santiago, J.L.; Rivas, E. Advances on the Influence of Vegetation and Forest on Urban Air Quality and Thermal Comfort. *Forests* **2021**, *12*, 1133. [CrossRef]
25. Zheng, B.; Li, J.; Chen, X.; Luo, X. Evaluating the Effects of Roof Greening on the Indoor Thermal Environment throughout the Year in a Chinese City (Chenzhou). *Forests* **2022**, *13*, 304. [CrossRef]
26. Zhang, R.; Zhao, Z. Giant Trees Exhibited Great Cooling Effect in Residential Area Southwest of China. *Forests* **2022**, *13*, 1516. [CrossRef]
27. Yin, Z.; Zhang, Y; Ma, K. Evaluation of PM2.5 Retention Capacity and Structural Optimization of Urban Park Green Spaces in Beijing. *Forests* **2022**, *13*, 415. [CrossRef]
28. De la Paz, D.; de Andrés, J.M.; Narros, A.; Silibello, C.; Finardi, S.; Fares, S.; Tejero, L.; Borge, R.; Mircea, M. Assessment of Air Quality and Meteorological Changes Induced by Future Vegetation in Madrid. *Forests* **2022**, *13*, 690. [CrossRef]

forests

MDPI

Article

City-Scale Revegetation Strategies Impact on the Temperature-Related Long-Term Mortality: A Quantitative Assessment in Three Cities in Southern Europe

Juan Manuel de Andrés [1], Ilaria D'Elia [2], David de la Paz [1], Massimo D'Isidoro [2], Felicita Russo [2], Mihaela Mircea [2], Maurizio Gualtieri [3], Sotiris Vardoulakis [4] and Rafael Borge [1,*]

1 Environmental Modelling Laboratory, Department of Chemical & Environmental Engineering, Universidad Politécnica de Madrid (UPM), C/ José Gutiérrez Abascal 2, 28006 Madrid, Spain; juanmanuel.deandres@upm.es (J.M.d.A.); david.delapaz@upm.es (D.d.l.P.)

2 ENEA—Italian National Agency for New Technologies, Energy and Sustainable Economic Development, Via dei Mille, 21, 40121 Bologna, Italy; ilaria.delia@enea.it (I.D.); massimo.disidoro@enea.it (M.D.); felicita.russo@enea.it (F.R.); mihaela.mircea@enea.it (M.M.)

3 Department of Environmental and Earth Science, Polaris Research Centre, University of Milano-Bicocca, Piazza Della Scienza 1, 20126 Milano, Italy; maurizio.gualtieri@unimib.it

4 HEAL Global Research Centre, Health Research Institute, University of Canberra, Bruce, ACT 2617, Australia; sotiris.vardoulakis@canberra.edu.au

* Correspondence: rafael.borge@upm.es

Abstract: Nature-based solutions (NBS) have attracted increasing attention in local air quality and climate change adaptation plans as suitable measures to reduce health risks. Although several studies have reported health benefits from short-term urban cooling effects of NBS, medium- to long-term health benefits are still poorly understood. In this study, we assess the changes in long-term mortality related to temperature fluctuations induced by city-scale vegetation actuations in three Southern European cities. We performed two annual high-resolution simulations with the Weather Research and Forecasting model to anticipate the impact of future revegetation strategies on temperature in these urban areas. Further, we assessed the impact of temperature changes on health using a country-specific minimum mortality temperature (MMT) reported in scientific literature. It was found that NBS could provide non-negligible reductions of long-term mortality related to temperature regulation (central estimate of 4.1, 1.2, and 3.4 cases avoided per year in Madrid, Milano, and Bologna, respectively). The effect of vegetation is site-dependent, and the cooling effect explains most of the benefits, especially in densely built-up areas of the cities analyzed. Future research should combine short/long-term temperature effects with other indirect implications (air quality, mental health) in the context of climate change.

Keywords: mesoscale meteorological modelling; urban vegetation; temperature; nature-based solutions; health impact; mortality

1. Introduction

Considering that around 68% of the world's population will live in urban areas by 2050 [1], global threats such as climate change and biodiversity loss cannot be tackled independently of urban growth [2]. In this context, nature-based solutions (NBS) may contribute to less vulnerable and more resilient cities to climate change, enhance human well-being, and protect biodiversity [3–5] and are increasingly included in local strategies [6–9]. The concept of NBS refers to "interventions based on nature that are envisaged to address sustainability challenges such as resource shortages, flood and heat risks and ecosystem degradation caused by processes of urbanization and climate change" [10] and include permeable pavements, urban forests and green corridors, vertical farming and green roofs, among others [11–16].

There are several studies addressing the usefulness of NBS in enhancing biodiversity, increasing carbon storage, improving urban rainwater/stormwater management, or mitigating air pollution and urban noise [17–23]. Proximity to green spaces not only fosters human-nature relationships but also interactions among city dwellers [24] and has been reported to improve physical [25–27], mental [28–30], and social health [31,32]. NBS and, specifically, the introduction of vegetation, is often considered as one of the main tools for urban climate regulation [3]. Several studies highlight the short-term urban cooling impact of vegetation [33], and the corresponding benefits in terms of heatwave and urban heat island mitigation [34,35], which are usually accompanied by marked decreases in short-term mortality [36]. Nonetheless, the findings of these studies highlight the need to perform high-resolution, city-specific analyses since the effect of vegetation on temperature and humidity in urban environments is driven by multiple factors [37,38] and largely depends on local features. For instance, de la Paz et al. [39] found that the net effect of afforestation within the same city might differ depending on the species selection or prior land use.

On the other hand, medium- to long-term health impacts of NBS are less understood [40], and given that climate change influences not only temperature extremes but also thermal variability over extended timeframes, it is important to understand how these prolonged exposures may impact public health [41–45]. According to Zanobetti and O'Neill [46], who made a thorough review of published papers on long-term effects of temperature on health, regional and local temperatures (and changing conditions in weather due to climate change) are clearly associated with multiple health outcomes. They also highlighted that the diversity of study designs and outcomes makes it difficult to draw general conclusions, so continued study of this phenomenon would be useful to inform climate change adaptation policies. More recently, Palmeiro-Silva et al. [47] confirmed that further region-specific analyses are needed to better understand long-term, temperature-related health impacts.

Considering these knowledge gaps and the growing need for developing tools for the correct design, installation, operation and maintenance of NBS in the context of sustainable urban development [21], the work here presented aims at providing a first estimate (to our best knowledge) of the impact that city-scale NBS strategies may have on long-term, temperature-related mortality. For this purpose, temperature changes induced by future vegetation scenarios in three cities in Southern Europe (Madrid, Milano, and Bologna) are modelled and translated into attributable long-term mortality changes. This study has been developed within the VEG-GAP Life project (https://www.lifeveggap.eu) that selected these cities to draw light on the potential effect of NBS in urban areas prone to O_3 and PM air quality issues. The impact of future city-scale revegetation strategies was investigated through the application of two state-of-the-art atmospheric modelling systems (AMS), WRF-CMAQ [48] for Madrid and AMS-MINNI [49–51] for the cities of Milano and Bologna. Mircea et al. [52] found that vegetation impacts on air quality were largely driven by the meteorological changes induced.

The present study intends to add complementary information on the health implications of NBS, focusing on temperature changes exclusively for these three cities, contributing to a more comprehensive view of the potential of urban greening strategies to improve public health.

2. Methodology

2.1. Modelling System, Setup and Performance Evaluation

The weather research and forecasting (WRF) limited-area, non-hydrostatic meteorological mesoscale model [53] was used to assess temperature in the target cities with 1 km^2 spatial resolution and 1 h temporal resolution. Changes in vegetation are reflected in the modeling system by the land use in each grid cell and their corresponding physical properties according to the Noah Land-Surface model (LSM) scheme [54]. Relevant physical properties (albedo, canopy height, roughness length, etc.) are computed as weighted averages using these percentages in the properties table used by the LSM (VEGPARM.tbl) for each grid cell. The WRF model included the building energy parametrization (BEP) [55] with minor modifications to allow a more flexible description of vegetation and urban fraction changes induced by NBS strategies through an extended (7 additional classes to the 24-category USGS land use cover) land use classification [39]. A complete view of the physical options and parametrizations used is shown in the Appendix A (Table A1). More information about the setup and the model performance evaluations is provided in [56].

This model has been extensively applied and assessed for the three urban areas of interest [39,57–59]. Nonetheless, a specific model evaluation for the complete year 2015, baseline scenario for the VEG-GAP project, was performed [56]. Regarding temperature (T2), the key variable in this study, model performance was satisfactory considering both low mean biases and root mean squared errors (see [56]). The model consistently yielded a very high Pearson's correlation coefficient between observed and modelled temperature (around 0.97), demonstrating the ability of the model to accurately capture temperature trends [60]. Further details regarding modelling domains, observational datasets, and statistical evaluation can be found in [56].

2.2. Vegetation Scenarios

The future vegetation scenarios elaborated in the VEG-GAP project derive from the municipal urban green development plans (including location and tree species selection), where NBS are seen as measures to improve urban wellness and quality of life. City-specific vegetation scenarios were defined, and the influence of urban vegetation was derived from the comparison of two model simulations corresponding to the following scenarios (all annual runs) for each city:

- Baseline (SVR): intended to represent the current situation (year 2015) (see [56] for further details);
- Future (SVSR): includes the vegetation associated with city-scale revegetation strategies and NBS interventions involving vegetation in each of the three cities (see [56] and [61] for further details).

2.2.1. Madrid

Madrid is the capital and largest city of Spain (3.3 million inhabitants) and the core of a larger metropolitan area. Madrid municipality has a surface of 604.5 km^2, of which 182.3 km^2 are currently (SVR scenario) covered by green areas. The resulting average population density is 5500 inhabitants/km^2. The city's climate is classified as Csa under the Köppen-Geiger system, indicative of a hot-summer Mediterranean climate [62] The future scenario used in this study represents the updated urban plans and strategies for Madrid. The methodology used for the construction of the new scenario is fully described in [39]. The future scenario contains new important NBS interventions, especially those within the so-called "Metropolitan forest". This project aims to create a 75 km peri-urban green ring, covering more than 5000 hectares, to expand, connect, and balance current peri-urban green areas in Madrid [39]. It involves 30 actuation areas and more than 1.51 million new trees of the most suited species, depending on soil type, topography, and ecological features. More than 50 tree species will be planted, but Mediterranean oaks (*Quercus ilex*) and Aleppo pine (*Pinus halepensis*) make up more than 50% of the total. Within the city, two interventions also stand out: "Urban orchards", which are areas designated for the municipal network of urban vegetable gardens (53 distributed all over the city with an average size around 30,000 m^2) and "Madrid New North" which is a major urban development approved by the local and regional governments. The green areas within intend to contribute to a carbon-neutral mix of uses. Lacking more specific information, the average characteristics (in terms of trees and shrub species and density) of existing parks in Madrid were assumed. Finally, and mainly outside the municipality of Madrid, the "Green Arch" intervention consists of linear plantations along 135 km of drovers' roads, trails, and other rural paths to connect already existing peri-urban forests or green areas across 17 municipalities around Madrid, connecting different biodiversity areas (Figure 1). This intervention involves 92,000 tree native species (*Quercus*, *Pinus*, *Ulmus*, *Olea*, etc.) individuals and 184,000 shrubs (*Rosa*, *Crataegus*, *Juniperus*, etc.).

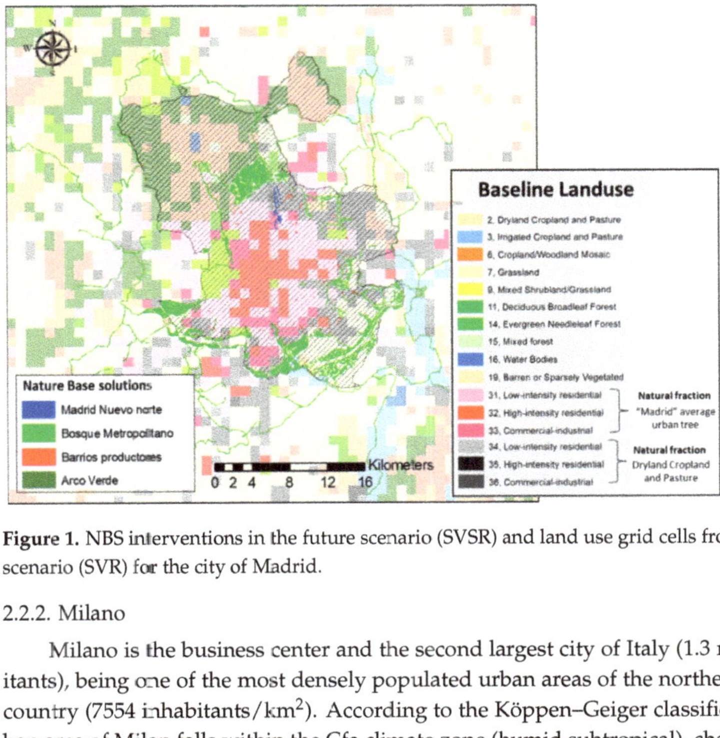

Figure 1. NBS interventions in the future scenario (SVSR) and land use grid cells from the baseline scenario (SVR) for the city of Madrid.

2.2.2. Milano

Milano is the business center and the second largest city of Italy (1.3 million inhabitants), being one of the most densely populated urban areas of the northern part of the country (7554 inhabitants/km^2). According to the Köppen–Geiger classification, the urban area of Milan falls within the Cfa climate zone (humid subtropical), characterized by fully humid conditions and hot summers [63]. Currently, green areas occupy 42.3 km^2 of the total 181.7 km^2 surface of the city. The Municipality of Milan provided, during the project, preliminary information about the green areas planned to be developed. The future vegetation areas are included in four main actions: "Public areas", areas located inside the city where trees are planned to be planted; "20 new parks", areas originally dedicated to different activities (e.g., industrial or railway) to be partially converted into parks; "Ecological Enhancement areas", areas where new green infrastructures are planned to be located; "Blue and Green Infrastructure", roads along which trees are planned to be planted. All these interventions are part of an ongoing strategy to address the fragmentation of formerly contiguous green corridors and the loss of green spaces in the urban expansion process over the last decades, and cover around 750 hectares, mainly in the outskirts of the city (Figure 2). The revegetation plan entails planting around 200,000 individuals. Tree species were selected considering the need for adaptation to climate change and to create multifunctional green spaces, and avoiding infesting and/or aggressive species as well as those outside the phytoclimatic range. Species such as *Quercus robur*, *Quercus pubescens*, *Acer campestre*, *Fraxinus ornus*, or *Carpinus betulus* are the preferred species for this city.

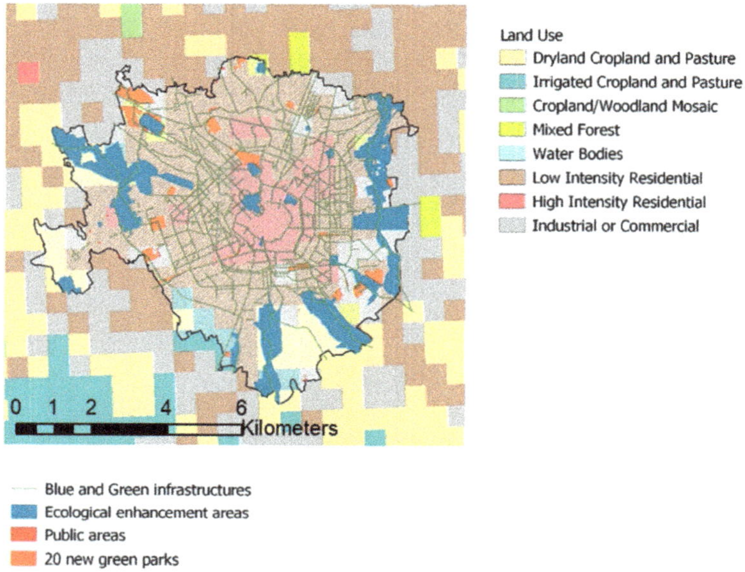

Figure 2. NBS interventions in the future scenario (SVSR) and land use of the baseline scenario (SVR) for the city of Milano.

2.2.3. Bologna

Bologna is the capital and largest city (140.9 km^2) of the Emilia–Romagna region in Northern Italy. It is an important agricultural, industrial, financial, and transport hub that gathers 0.38 million people (2745 inhabitants/km^2). Like Milano, Bologna falls within the Cfa climate zone (humid subtropical) of the Köppen–Geiger classification, characterized by warm-temperate conditions, year-round precipitation, and hot summers [64]. Green areas cover 55.5 km^2 in the SVR scenario. As for the future, the urban green planning for the city of Bologna was still under development during the VEG-GAP project, so a hypothetical vegetation scenario was elaborated using information made available from a previous project. Four types of actions are anticipated: "Historical and green areas" and "greening strategical areas" planned to increase urban vegetation in areas of particular historical and environmental/landscape significance, respectively; "Highway requalification areas" plantations along highway and main roads; "Extraction and manufacturing areas", areas originally dedicated to extraction and manufacturing activities to be potentially converted into parks. All interventions concentrate in the northern part of the city (Figure 3) and are intended to foster ecosystem services to support the new General Urban Plan (PUG). Since this document is not officially enacted, the composition and species selection, focused on maximizing CO_2 capture potential and minimize the emission of reactive VOCs, was based on the outcomes of two previous LIFE projects (GAIA—Green Area Inner City Agreement and BLUEAP—Bologna Local Urban Environment Adaptation Plan for a Resilient City). The combined extension of the four interventions affects an area of 1510 hectares and considers the planting of a variety of trees (*Carpinus*, *Quercus*, *Acer*, and *Populus* among others) and shrub individuals (*Sambucus*, *Cornus*, *Rosa*, etc.).

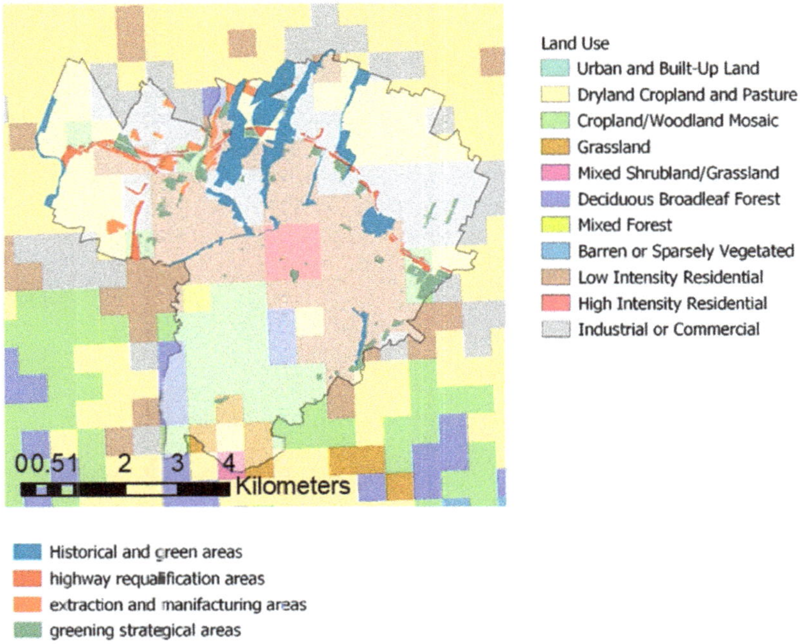

Figure 3. NBS interventions in the Future Scenario (SVSR) and land use of the baseline scenario (SVR) for the city of Bologna.

2.3. Health Impact Assessment

The weather research and forecasting (WRF) limited-area, non-hydrostatic meteorological mesoscale model [53] was used to assess temperature in the target cities with 1 km^2 spatial resolution and 1 h temporal resolution. It was coupled with the building energy parametrization (BEP) [55] with minor modifications to allow a more flexible description of urban vegetation [39]. Physical options and parametrizations used are shown in Table A1 of the Appendix A. More information about the setup and the model performance evaluations is provided in [56].

This study estimates the potential health benefits from NBS interventions at the city scale using the spatial variability of temperature changes at 1 km^2 resolution. It is particularly interesting to investigate long-term associations between temperature and mortality when considering time frames of several decades, as in the case of vegetation scenarios. A variety of methods have been proposed to assess longer-term temperature and health effects that involve multiple temperature metrics and epidemiological approaches [46]. There is a consensus that the impacts of both cold and heat on mortality should be considered [65,66]. Consequently, several authors propose assessment methods based on thermal oscillations, such as the diurnal temperature range (DTR) [67] or deviations from optimum ambient temperature [68] to estimate temperature-related mortality. This approach implicitly takes into account the adaptation of a population to local climatic conditions, highlighting the importance of considering city-specific optimal temperature ranges [69]. The spatial heterogeneity of mortality has also been reported in the few studies that addressed long-term effects [40], so it is important to consider location-specific relative risks. As for temperature data sources, we rely on model outputs. They have high accuracy for predicting this meteorological variable [56] and have been successfully used in epidemiological analyses of temperature-related risks [70], including areas far from weather stations [71].

Considering these factors, the long-term impact on mortality due to temperature changes induced by vegetation was assessed in this study following the methodology proposed by [72]. They analyzed the attribution of heat and cold to mortality in 11 countries, including Spain and Italy. Based on multi-annual datasets over a 40-year period, they identified a country-specific minimum mortality temperature (MMT) that accounts for thermal adaptation of the population in each country and derived percent excess relative risks (ERR) that represent the increments in annual mortality relative risk (%) per 1 °C increase in annual mean degrees above or below the defined MMT (Table 1). They are referred to as heat and cold degrees, respectively, and represent the average departure of daily mean temperatures from MMT [72].

Table 1. Minimum mortality temperature and percent excess relative risks (ERR) per 1 °C change in temperature (from Armstrong et al. [72]).

City	MMT (°C)	Heat ERR% (95% Confidence Interval)	Cold ERR% (95% Confidence Interval)
Madrid	21.4	3.2 (0.6, 5.8)	2.7 (1.5, 3.9)
Milano and Bologna	21.8	8.3 (2.5, 14.4)	−1.1 (−5.2, 3.3)

This methodological approach is consistent with air pollution health impact assessment studies [73,74] and thus, the difference between the number of deaths (ΔM) attributable to changes on heat or cold degrees (ΔT) under any two scenarios of interest can readily be computed in each grid cell of our modelling domains by means of Equation (1).

$$\Delta M = M_0 \cdot \left[1 - e^{\beta \cdot \Delta T}\right] \cdot P \tag{1}$$

where M_0 and P, represent the baseline mortality rate and population for the reference year or baseline scenario, respectively, and β is the fractional increase in RR per °C (ERR%/100).

Of note, the number of locations used by [72] to obtain ERR varied considerably depending on the country (2 locations for Italy and 50 for Spain). This contributes to the larger confidence interval reported in Table 1 for the ERR of Italian cities. In both cases, heat-related mortality risks are larger than those related to cold temperatures. This is consistent with other studies that suggest the vulnerability of the Mediterranean to the effect of high temperatures [75] and contrasts with the findings of other studies for northern European countries, where cold drives most temperature-related deaths [76].

Consistent with the methodology used to derive long-term relative risks [77], the baseline mortality rate represents all-age, all-cause mortality excluding accidents (International Classification of Diseases [ICD]-9 0-799). Baseline natural mortality data were retrieved from the national statistical institutes (INE (www.ine.es) and ISTAT (www.istat.it) for Spain and Italy, respectively) for the year 2015 at the municipality level. The official population census for that year was spatially allocated using the JRC 2018 population density cover (Figure 4) [78]. Spatially disaggregated number of deaths ($M_0 \cdot$P) for the baseline scenario, the data are shown in the Appendix B (Figure A1). The population (P) considered (considering all the grid cells intersected by each municipality) was 3,339,709, 1,442,995, and 416,033 inhabitants for Madrid, Milano, and Bologna municipalities. The corresponding number of deaths (all-age, [ICD]-9 0-799) are 27,557, 18,853, and 5486, respectively, which represent, with an average baseline mortality of 8.3, 12.7, and 13.0 (per 1000 people) in the cities of Madrid, Milano, and Bologna.

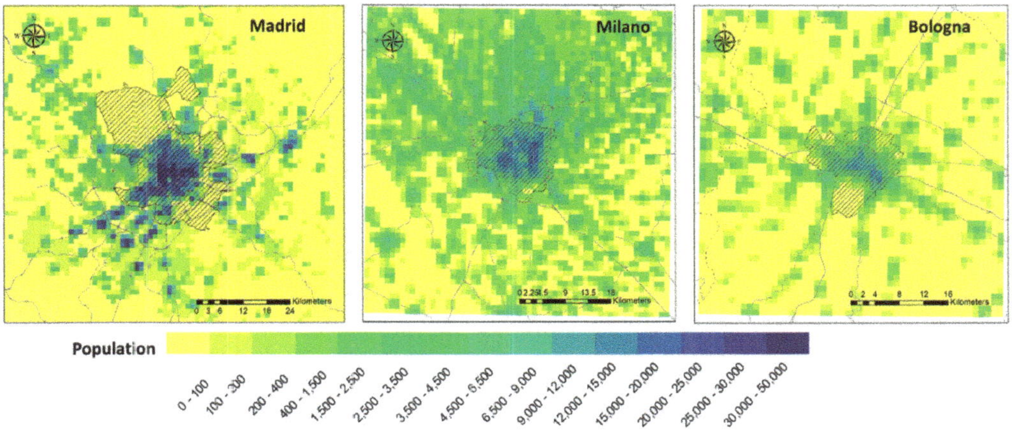

Figure 4. Spatial distribution of the total population in the three selected cities.

3. Results

3.1. Impact of Vegetation on Temperature

Figure 5 shows the near-ground level temperature (T2) annual mean for the three cities for the baseline (SVR) scenario. Madrid (Figure 5a) is the warmest city (15.2 °C annual average) and the area with the highest spatial variability, ranging from 9.1 to 17.1 °C, due to a larger dimension of the modelling domain but also a more complex topography and a greater distance from the sea. Average (minimum, maximum) T2 annual means for the Italian cities are 15.0 (14.4, 16.1) °C for Milano (Figure 5b) and 15.1 (12.0, 16.5) °C for Bologna (Figure 5c). The urban heat island (UHI) effect [79] can be clearly observed in all cases, with maximum annual mean temperatures corresponding to the more densely urbanized areas.

As discussed in D'Isidoro et al. [56], the foreseeable impact of NBS differs depending on local characteristics. As illustrated in Figure 5, the re-vegetation scenario (SVSR) is expected to decrease annual mean temperature downtown in all three cities, with changes up to −2.6 °C in the case of Madrid. The cooling mechanism of vegetation when introduced in densely built areas. According to our results, the cooling effect found in consolidated urban areas is due to increases in latent heat fluxes (up to 17% relative to the SVR scenario) and higher evapotranspiration [80]. On the other hand, vegetation may slightly increase average temperature when new trees are planted outside the urban areas, around +0.1 °C in most of the outskirts of the city. The drivers of these increments are the changes in sensible heat fluxes induced by a decrease in albedo and enhanced energy absorption [39]. To a lesser extent (note the difference on the color scale), this effect is also observed in Milano (Figure 5e) with T2 annual mean changes in the range of −0.1, +0.0 °C, while a net cooling effect dominates in Bologna (Figure 5f), with changes around −0.2 °C in the city center.

The limited impact of the implemented NBS on temperature in Milan may be related to the size and distribution of planned revegetation actions (Figure 2) but may also be attributed to the city's specific urban morphology, characterized by flat topography and high building density. These features pose significant challenges to achieving substantial cooling effects through vegetation alone [38].

Spatially averaged changes over the whole modelling domain are quite subtle, with net changes of less than 0.01 °C in absolute values in all cases.

Changes in annual heat degrees and cold degrees from SVSR (future scenario) relative to SVR (baseline scenario) were computed following the methodology described in Armstrong et al. [72]. The results are presented in Figure 6. Values above 0 in both cases imply an increase in the distance to MMT.

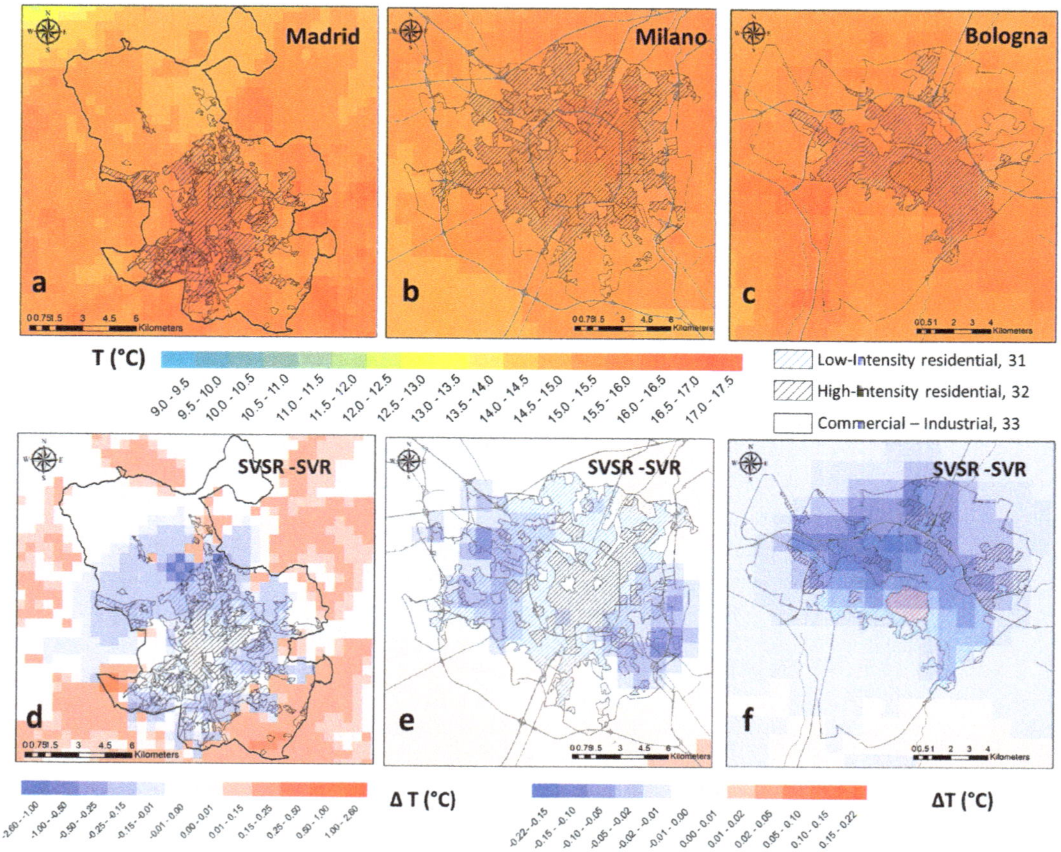

Figure 5. Annual mean temperature (T2) for the baseline scenario (SVR) in (**a**) Madrid, (**b**) Milano and (**c**) Bologna and expected change under the future scenario (SVSR) in (**d**) Madrid, (**e**) Milano and (**f**) Bologna (zoom over the urban areas with indication of the distribution of low and high intensity residential areas as well as commercial-industrial zones, consistently with the WRF-BEP inputs for the SVR scenario).

The results for Madrid indicate a reduction (blue) of heat degrees in most of the city (up to 1.0 °C) and slight increases (red) in the surroundings, most evidently in the areas where future interventions take place. Nonetheless, the increases predicted are below 0.1 °C in all cases, and the aggregated change (sum of changes over the whole city) corresponds to −7.7 °C (Table 2). By contrast, cold degrees are increased in most grid cells, up to 1.6 °C, since the cooling effect of vegetation in consolidated urban areas tends to increase temperature differences below MMT. Only in some areas around the city (negative values, represented by red colors), the average deviation from optimal temperature (MMT) is slightly reduced. Therefore, the implementation of the simulated NBS future scenarios would increase cold degrees in Madrid, with an aggregated change over the city of 6.5 °C (Table 2). Nonetheless, variations at the grid cell level fall in the ±0.2 °C range in all cases.

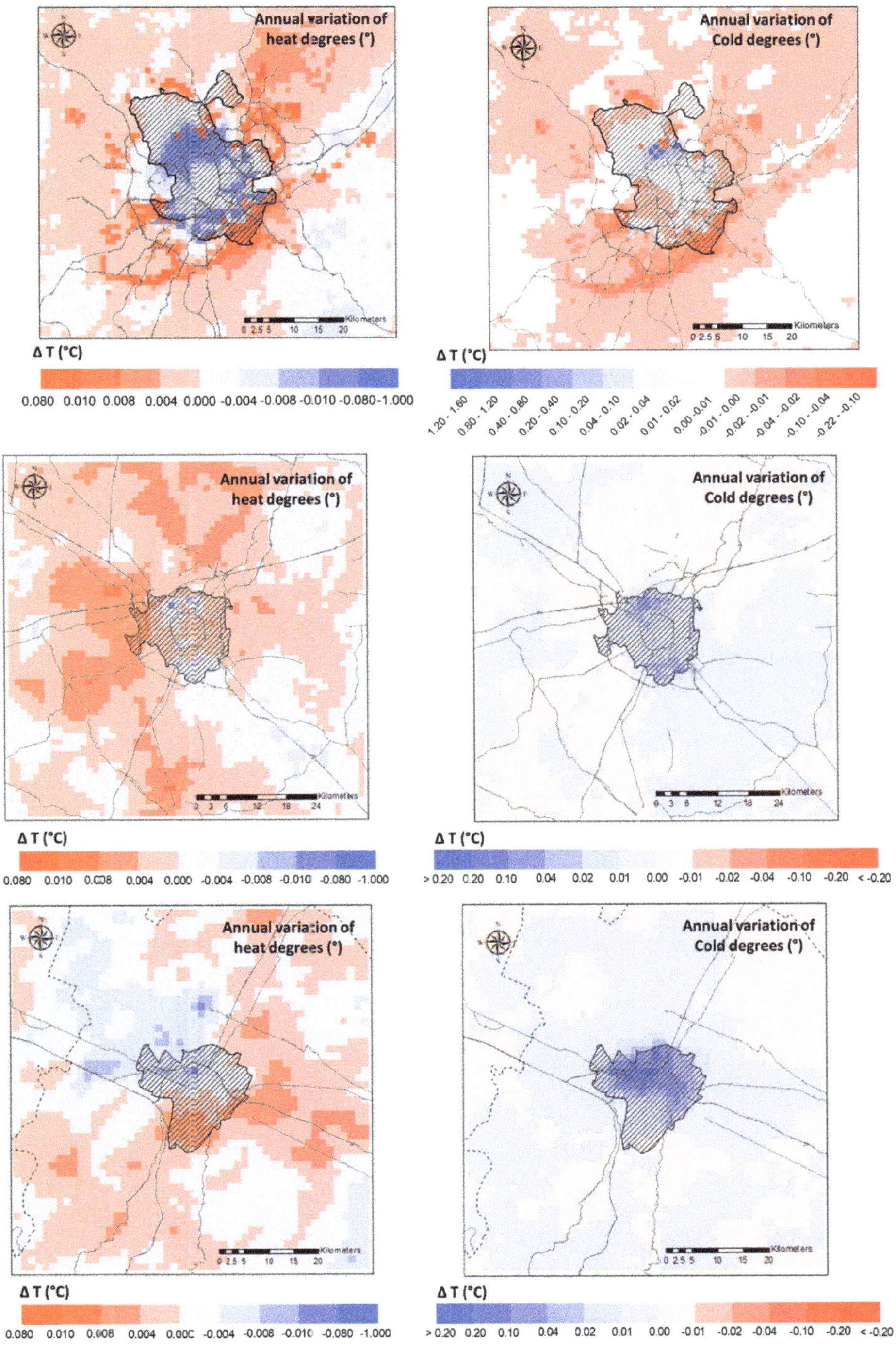

Figure 5. Changes in heat (**left column**) and cold (**right column**) degrees for the city of Madrid (**top**), Milan (**center**), and Bologna (**bottom**). In all cases, they refer to SVSR—SVR.

Table 2. Accumulated mortality (total annual deaths) changes induced by future NBS interventions in the three cities.

City	ΔT (°C)		ΔM (95% Confidence Interval in Brackets)		
	Heat	Cold	Heat Deaths	Cold Deaths	Net Change in Temperature-Related Mortality
Madrid	−7.7	6.5	−7.4 (−1.4, −13.4)	3.2 (1.8, 4.7)	−4.1 (−11.6, 3.3)
Milano	0.1	2.4	0.9 (0.3, 1.6)	−2.1 (−10.0, 6.3)	−1.2 (−9.7, 8.0)
Bologna	−0.7	2.0	−0.4 (−0.1, −0.6)	−3.0 (−14.1, 8.9)	−3.4 (−14.3, 8.8)

Similar results are obtained for Milano and Bologna, although the changes in heat and cold degrees induced by new vegetation are predicted to be smaller. In Milano, both heat and cold degrees would be increased under SVSR (future scenario), although the changes suggest a dominant effect of cooling with an accumulated figure (aggregated change) of 2.4 cold degrees for the whole city (Table 2), in comparison with the nearly negligible increase of heat degrees (Figure 6). Similarly to the Madrid case, Bologna would see a net increase of cold degrees (2.0 accumulation for the whole city) due to the predominantly cooling effect of future vegetation, with changes up to 0.2 °C in the northern part of the city.

That would result in a reduction of annual mean departures above the optimal temperature (MMT) as well, with an aggregated value over the city of −0.7 heat degrees.

Of note, red and blue tones have been used for temperature increases and decreases, respectively, in Figure 6. However, they have different interpretations health-wise for the panels in the left column (variation of heat degrees) and in the right one (variation of cold degrees). Values above 0 in both cases imply an increase in the deviation from MMT and vice versa.

3.2. Impacts on Temperature-Related Mortality

Aggregated changes in annual deaths (ΔM) attributable to changes in heat degrees and cold degrees are summarized in Table 2.

In the city of Madrid, the scenario with new vegetation (SVSR) will result in a net reduction in temperature-related mortality. The expected decrease in heat degrees, along with the higher relative risk (EER 3.2%) associated with heat-related mortality, outweighs the potential increase in cold-related mortality (EER 2.7%). For the Italian cities, heat-related mortality relative risk is considerably higher (EER 8.3%) than that of Spain, so that the impact would be larger even for smaller variations in heat degrees. It should be noted as well that according to Armstrong et al. [72], cold relative risks are negative (EER −1.1%), so an increase of cold degrees would imply reductions in long-term mortality. Consequently, our results suggest a net beneficial long-term effect on the mortality of future vegetation scenarios (SVSR).

Notably, the values reported in Table 2 refer exclusively to the municipal area of the three cities analyzed. However, similar patterns are observed for the whole modeling domains, which include the corresponding metropolitan areas; therefore, the beneficial effects of NBS could be higher than those reported here.

4. Discussion

4.1. Health Impacts

The effects of extreme temperatures on the human body are well documented, and recent literature reports the increasing risk of heat-related mortality associated with climate warming [81,82]. Human adaptation to future meteorological conditions may alleviate, to some extent, the burden on mortality [83], but significant efforts are being devoted to understanding how urban regeneration plans may alleviate the city's overall temperature and the UHI effect [84,85]. The use of NBS in urban areas has been largely proposed as a means to locally reduce temperature [86] and to improve population health [87,88]. Our model simulations quantify the potential long-term temperature-related health benefits associated with plausible NBS in three densely populated southern European cities. We show here that the effects of NBS on urban temperature are site-specific. City-scale aggregated effects on mortality are relatively small in comparison with short-term health impacts related to green urban infrastructure reported elsewhere (e.g., Iungman et al. [89]). Nonetheless, our results point out that a net health benefit (net reduction of the total cases of non-accidental temperature-related mortality) could result from the implementation of NBS in all three cities: −4.1 (−11.6, 3.3) in Madrid, −1.2 (−9.7, 8.0) in Milano and −3.4 (−14.3, 8.8) in Bologna, on an annual basis. Of note, these figures may be explained through different processes. New NBS in Madrid may promote a reduction of temperature above the MMT, but also an increase below the MMT. This will translate into a reduction in heat-related deaths attributable to higher temperatures, but a smaller increase in cold-related deaths associated with cooler temperatures. This is consistent with recent studies that conclude that urban green infrastructure may be particularly beneficial to alleviate the UHI effect in densely urbanized areas [90].

A very limited impact on heat degrees is reported for the city of Milano, although a slight reduction of mortality may be derived from increasing cold degrees according to the ERR defined in [72]. In the domain centered around the city of Bologna, the NBS may determine a reduction in high temperatures above the minimum mortality temperature (MMT) but an increase in the cold ones below the MMT, both results adding to mortality avoidance. As a result, Bologna presents the highest impact of all three cities relative to its population size.

4.2. Policy Implications

While most of the publications available on the effects of NBS are based on surveys collected in the population under investigation or statistical modelling, here we report data obtained from a mesoscale atmospheric modelling system combined with meteorological data. This approach enabled exploration of the potential health benefits of NBS in three EU cities based on plausible scenarios of the positioning of new vegetation. In highly urbanized areas, the selection and regeneration of areas with green infrastructure is a complex task that takes into account the requirements of new service areas and housing, which may reduce the availability of areas suitable for introducing vegetation.

The results we report suggest that the location of new NBS is a key parameter for properly assessing the beneficial effects in terms of local temperature modulation and improvement of public health. Policies to promote NBS should take into account not only space availability but also the distribution of population and green infrastructures, as well as the specific combination of land uses and topography for each city. Our study suggests that reforestation of peri-urban barren areas may slightly increase average temperature, and the results from the three cities point out that the cooling effect of vegetation is stronger in consolidated urban areas that correspond to densely populated areas. That may be of interest in the scope of climate change, especially in Southern Europe, where heat-related mortality is expected to grow in the future [91]. As a result, policy translation of our results would be subject to a detailed analysis of these features in other urban areas, along with the limitations discussed in the following section.

5. Limitations

While this study presents one of the first quantitative analyses of long-term temperature mortality changes induced by urban-scale revegetation strategies, there is a series of limitations that should be considered for the interpretation of the results and to identify further investigations. Our approach relies on relative risks derived at the national level, and ideally, city-specific minimum mortality temperature (MMT) should be derived to more accurately assess health benefits. On the other hand, the uncertainty of our estimates, given by the amplitude of the 95% confidence intervals, is relatively large. In fact, the upper confidence interval regarding net mortality change (Table 2) for the three cities is positive, suggesting that the expected net outcomes may vary considerably. To the greatest extent possible, future research may delve into a comprehensive assessment of public health benefits associated with NBS by scrutinizing multiple health indicators, including mental health [69] or other indirect effects [92].

Regardless of the epidemiological approach followed and health endpoints considered, enhanced methodologies to better reflect exposure may help capture socio-economic variability and the specific impact on vulnerable groups [93]. Dynamic population exposure methods [94] may be explored for a more comprehensive understanding of NBS that not only modulate temperature but may modify population lifestyles and thus, exposure patterns. Projections on demography and population aging may also contribute to a more realistic assessment of long-term health effects [95] and may be included in future research that integrates longitudinal datasets to account for NBS health impacts within a holistic assessment that considers the whole life cycle of the project [96]. Besides health and psychological effects, ecological benefits (e.g., on biodiversity) and potential unintended consequences (e.g., wildfire risk) may be addressed in future studies as well. In addition, our study does not consider indoor effects or the potential interactions with other heat mitigation options such as air conditioning systems. Evolutions of the BEP parameterizations, such as the BEP-BEM (Building Energy Model) scheme [97] or WRF-Comfort [98], may be useful to support future studies in that direction. Of note, future studies may assess the combined effectiveness of a wider range of strategies, such as reflective surfaces, urban form adjustments, or hydrological interventions.

Incorporating advanced mesoscale modelling tools, however, allows further studies to assess the combined effects of temperature and air quality, or to integrate climate change mitigation and adaptation measures in future development projects (climate proofing). Such studies may combine high-resolution dynamic downscaling of climate models [99] with a more detailed description of NBS implementation and vegetation growth modelling. Unlike other methodologies that relay on spatial regression and other statistical models [100] to derive associations between health endpoints and land uses, including urban green space (UGS) [101] the explicit consideration of the physical processes involved in the thermal regulation induced by vegetation may allow more detailed future research where the impact of very specific features such as the type of vegetation or the interactions with previous and surrounding land uses may be explicitly incorporated in the planning process. Nonetheless, models also bring uncertainty, and the representation of the mechanisms through which vegetation modifies momentum, heat, and humidity fluxes (see Section 2.1) is still limited even in advanced mesoscale meteorological models [39].

To assess the combined effects of new vegetation on temperature and air quality at the city scale, it is important to consider that biogenic VOC emissions are likely to enhance ground-level ozone and secondary particulate matter concentrations in urban areas [102,103] that can have a detrimental effect on health [104]. Moreover, urban vegetation may limit the dispersion of fine particulate matter [105], resulting in higher local population exposure and potentially higher health impacts if emission sources such as road traffic are not effectively controlled. Again, that analysis should be city-specific and be based on detailed emission reduction scenarios that reflect local plans and measures intended to meet the new air quality limit values set up in Europe by the Directive (EU) 2024/2881.

6. Conclusions

We present one of the few assessments of long-term health effects associated with temperature changes induced by planned urban revegetation strategies in three cities of southern Europe. This contribution illustrates how mesoscale modelling tools may be integrated into the impact assessment of nature-based solutions (NBS) on public health at the city scale. Our study demonstrates that all-cause mortality excess relative risks (ERR) associated with departures from a minimum mortality temperature (MMT) may be suitable for this kind of health impact assessment. While these EERs are county-specific, our results suggest that city- or even municipality-specific relative risks may be needed to reduce the uncertainty of the estimates. Our results also suggest that the NBS envisaged in Madrid, Milano, and Bologna may have a non-negligible effect on temperature-related long-term mortality and thus, may add to the benefits related to short-term effects from the alleviation of heat and cold temperature extremes. According to our results, the introduction of vegetation in densely populated zones may be particularly beneficial due to the net cooling effect of the vegetation in consolidated urban areas. While this recommendation may hold for other southern Mediterranean cities or others around the world with similar geophysical features, revegetation policies should be specifically designed for each city to take into account the particularities of the urban landscape, population distribution, etc. Despite addressing the technical and scientific limitations highlighted in this study, it is advisable to include social issues for a broader view on how NBS may help improve the sustainability of human agglomerations in the future.

Author Contributions: Conceptualization, S.V. and R.B.; methodology, J.M.d.A., I.D., M.M., M.G., S.V. and R.B.; software, D.d.l.P., M.D. and F.R.; validation, J.M.d.A., I.D., D.d.l.P., M.D., F.R., M.M., M.G. and R.B.; formal analysis and investigation, J.M.d.A., I.D., D.d.l.P., M.D., F.R., M.M., M.G., S.V. and R.B., data curation, I.D., D.d.l.P., M.G. and R.B., writing—original draft preparation, J.M.d.A., I.D., D.d.l.P., M.G. and R.B.; writing—reviewing and editing, J.M.d.A., I.D., D.d.l.P., F.R., M.M., M.G., S.V. and R.B.; visualization, I.D., D.d.l.P. and M.D.; project administration, M.M. and R.B.; funding acquisition, M.M., S.V., and R.B. All authors have read and agreed to the published version of the manuscript.

Funding: This study was developed under the project Life VEG-GAP (https://www.lifeveggap.eu/), accessed on 20 April 2022, which was funded by the European Union Life Program in 2018, Grant Number LIFE18 PRE IT 003. The authors gratefully acknowledge the Universidad Politécnica de Madrid (https://www.upm.es) for providing computing resources on the Magerit Supercomputer. The computing resources and the related technical support used for the AMS-MINNI simulations were provided by the CRESCO/ENEAGRID High Performance Computing infrastructure and its staff. The infrastructure is funded by ENEA, the Italian National Agency for New Technologies, Energy and Sustainable Economic Development, and by Italian and European research programs (http://www.cresco.enea.it/english, (accessed on 25 June 2025)). The authors acknowledge the contribution and collaboration of all three cities in the definition of future scenarios and their involvement in the VEG-GAP project. S.V. acknowledges the HEAL (Healthy Environments And Lives) Network, which receives funding from the National Health and Medical Research Council (NHMRC) Special Initiative in Human Health and Environmental Change (Grant No. 2008937).

Data Availability Statement: The dataset is available on request from the authors.

Conflicts of Interest: The authors declare no conflicts of interest.

Appendix A. Modelling System Setup

The main features of the WRF modelling systems applied in the project are summarized in Table A1.

Table A1. WRF model setup summary.

Model Feature	Madrid	Bologna and Milano
Version	WRFv4.1.2	WRFv3.9 1.1
Nested domains (spatial resolution in km)	4 (27, 9, 3, 1)	3 (12, 4, 1)
Vertical layers (layers within the first km)	38 (17)	39 (12)
Initialization	ERA5 (Copernicus Climate Change Service)	ERA5 (Copernicus Climate Change Service)
Microphysics	WSM6 [106]	WSM6 [106]
Cumulus Parametrization	Off	Off
PBL Scheme	Bougeault-Lacarrère PBL [107]	Mellor Yamada Jancic (MYJ) [108]
Urban Physics	BEP [55]	BEP [55]
Land Surface	Noah LSM [109]	Noah LSM [109]
Longwave Radiation	GFDL [110]	RRTMG [111]
Shortwave Radiation	MM5 [112]	RRTMG [111]

Appendix B. Baseline Mortality

Figure A1 presents the spatially disaggregated number of deaths ($M_0 \cdot P$) for the baseline scenario.

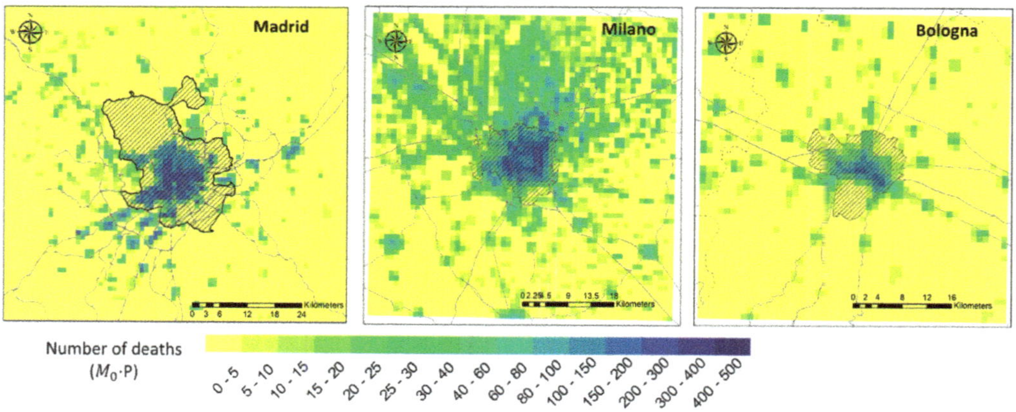

Figure A1. Spatially disaggregated baseline mortality ([ICD]-9 0-799, all ages) is considered in the three selected cities.

References

1. United Nations, Department of Economic and Social Affairs, Population Division. *World Urbanization Prospects: The 2018 Revision (ST/ESA/SER.A/420)*; United Nations: New York, NY, USA, 2019; ISBN 978-92-1-004314-4.
2. Intergovernmental Panel on Climate Change (IPCC). Summary for Policymakers. In *Climate Change 2022—Impacts, Adaptation and Vulnerability: Working Group II Contribution to the Sixth Assessment Report of the Intergovernmental Panel on Climate Change*; Cambridge University Press: Cambridge, UK, 2023; pp. 3–34.
3. Johnson, B.A.; Kumar, P.; Okano, N.; Dasgupta, R.; Shivakoti, B.R. Nature-based solutions for climate change adaptation: A systematic review of systematic reviews. *Nat. Based Solut.* **2022**, *2*, 100042. [CrossRef]
4. Kabisch, N.; Frantzeskaki, N.; Pauleit, S.; Naumann, S.; Davis, M.; Artmann, M.; Haase, D.; Knapp, S.; Korn, H.; Stadler, J.; et al. Nature-based solutions to climate change mitigation and adaptation in urban areas: Perspectives on indicators, knowledge gaps, barriers, and opportunities for action. *Ecol. Soc.* **2016**, *21*, 2. [CrossRef]

5. Xie, L.; Bulkeley, H.; Tozer, L. Mainstreaming sustainable innovation: Unlocking the potential of nature-based solutions for climate change and biodiversity. *Environ. Sci. Policy* **2022**, *132*, 119–130. [CrossRef]
6. Somarakis, G.; Stagakis, S.; Chrysoulakis, N. *ThinkNature Nature-Based Solutions Handbook*; EU Horizon 2020 Research and Innovation: Luxembourg, 2019. [CrossRef]
7. Dumitru, A.; Frantzeskaki, N.; Collier, M. Identifying principles for the design of robust impact evaluation frameworks for nature-based solutions in cities. *Environ. Sci. Policy* **2020**, *112*, 107–116. [CrossRef]
8. Oral, H.V.; Carvalho, P.; Gajewska, M.; Ursino, N.; Masi, F.; van Hullebusch, E.D.; Kazak, J.K.; Exposito, A.; Cipolletta, G.; Andersen, T.R.; et al. A review of nature-based solutions for urban water management in European circular cities: A critical assessment based on case studies and literature. *Blue Green. Syst.* **2020**, *2*, 112–136. [CrossRef]
9. Davies, C.; Wang, B.; Brack, D.; Qin, Q. *Nature-Based Solutions: Lessons from European and Chinese Cases, September 2024, EU–China Cooperation on Environment and Green Economy Project*; European Union Service Contract No. 2017/390-904; EU–China Cooperation on Environment and Green Economy Project: Beijing, China, 2024.
10. Dorst, H.; van der Jagt, A.; Raven, R.; Runhaar, H. Urban greening through nature-based solutions—Key characteristics of an emerging concept. *Sust. Cities Soc.* **2019**, *49*, 101620. [CrossRef]
11. Raimondi, A.; Marrazzo, G.; Sanfilippo, U.; Becciu, G. A probabilistic approach to stormwater runoff control through permeable pavements beneath urban trees. *Sci. Total Environ.* **2023**, *905*, 167196. [CrossRef]
12. Jeong, J.; Park, C.-R. Urban forest indicator assessment for nature-based solutions to connect biodiversity and people. *Ecol. Indic.* **2024**, *169*, 112843. [CrossRef]
13. Zhang, M.; He, J.; Liu, D.; Huang, J.; Yue, Q.; Li, Y. Urban green corridor construction considering daily life circles: A case study of Wuhan city. *China. Ecol. Eng.* **2022**, *184*, 106786. [CrossRef]
14. Kalantari, F.; Mohd Tahir, O.; Mahmoudi Lahijani, A.; Kalantari, S. A Review of Vertical Farming Technology: A Guide for Implementation of Building Integrated Agriculture in Cities. *Adv. Eng. Forum* **2017**, *24*, 76–91. [CrossRef]
15. Vera, S.; Pinto, C.; Tabares-Velasco, P.C.; Bustamante, W. A critical review of heat and mass transfer in vegetative roof models used in building energy and urban environment simulation tools. *Appl. Energy* **2018**, *232*, 752–764. [CrossRef]
16. De Cristo, E.; Evangelisti, L.; Barbaro, L.; De Lieto Vollaro, R.; Asdrubali, F. A Systematic Review of Green Roofs' Thermal and Energy Performance in the Mediterranean Region. *Energies* **2025**, *18*, 2517. [CrossRef]
17. Cohen-Shacham, E.; Walters, G.; Janzen, C.; Maginnis, S. *Nature-Based Solutions to Address Global SOCIETAL Challenges*; IUCN International Union for Conservation of Nature: Gland, Switzerland, 2016. [CrossRef]
18. Manes, S.; Vale, M.M.; Malecha, A.; Pires, A.P.F. Nature-based solutions promote climate change adaptation safeguarding ecosystem services. *Ecosyst. Serv.* **2022**, *55*, 101439. [CrossRef]
19. Le Coent, P.; Graveline, N.; Altamirano, M.A.; Arfaoui, N.; Benitez-Avila, C.; Biffin, T.; Calatrava, J.; Dartee, K.; Douai, A.; Gnonlonfin, A.; et al. Is-it worth investing in NBS aiming at reducing water risks? Insights from the economic assessment of three European case studies. *Nat. Based Solut.* **2021**, *1*, 100002. [CrossRef]
20. Epelde, L.; Mendizabal, M.; Gutiérrez, L.; Artetxe, A.; Garbisu, C.; Feliu, E. Quantification of the environmental effectiveness of nature-based solutions for increasing the resilience of cities under climate change. *Urban. For. Urban. Green.* **2022**, *67*, 127433. [CrossRef]
21. Biswal, B.K.; Bolan, N.; Zhu, Y.G.; Balasubramanian, R. Nature-based Systems (NbS) for mitigation of stormwater and air pollution in urban areas: A review. *Resour. Conserv. Recycl.* **2022**, *186*, 106578. [CrossRef]
22. Chen, H.S.; Lin, Y.C.; Chiueh, P.T. Nexus of ecosystem service-human health-natural resources: The nature-based solutions for urban PM2.5 pollution. *Sust Cities Soc.* **2023**, *91*, 104441. [CrossRef]
23. Gómez-Baggethun, E.; Barton, D.N. Classifying and valuing ecosystem services for urban planning. *Ecol. Econ.* **2023**, *86*, 235–245. [CrossRef]
24. Noe, E.E.; Stolte, O. Dwelling in the city: A qualitative exploration of the human-nature relationship in three types of urban greenspace. *Landsc. Urban. Plann.* **2023**, *230*, 104633. [CrossRef]
25. Coventry, P.A.; Brown, J.V.E.; Pervin, J.; Brabyn, S.; Pateman, R.; Breedvelt, J.; Gilbody, S.; Stancliffe, R.; McEachan, R.; White, P.C.L. Nature-based outdoor activities for mental and physical health: Systematic review and meta-analysis. *SSM Popul. Health* **2021**, *16*, 100934. [CrossRef]
26. Frank, L.D.; Hong, A.; Ngo, V.D. Causal evaluation of urban greenway retrofit: A longitudinal study on physical activity and sedentary behavior. *Prev. Med.* **2019**, *123*, 109–116. [CrossRef]
27. Panter, J.; Guell, C.; Humphreys, D.; Ogilvie, D. Title: Can changing the physical environment promote walking and cycling? A systematic review of what works and how. *Health Place* **2019**, *58*, 102161. [CrossRef]
28. Barron, S.; Rugel, E.J. Tolerant greenspaces: Designing urban nature-based solutions that foster social ties and support mental health among young adults. *Environ. Sci. Policy* **2023**, *139*, 1–10. [CrossRef]
29. Nawrath, M.; Elsey, H.; Dallimer, M. Why cultural ecosystem services matter most: Exploring the pathways linking greenspaces and mental health in a low-income country. *Sci. Total Environ.* **2022**, *806*, 150551. [CrossRef]

30. Subiza-Pérez, M.; Vozmediano, L.; San Juan, C. Green and blue settings as providers of mental health ecosystem services: Comparing urban beaches and parks and building a predictive model of psychological restoration. *Landsc. Urban. Plann.* **2020**, *204*, 103926. [CrossRef]
31. Enssle, F.; Kabisch, N. Urban green spaces for the social interaction, health and well-being of older people—An integrated view of urban ecosystem services and socio-environmental justice. *Environ. Sci. Policy* **2020**, *109*, 36–44. [CrossRef]
32. Lin, D.; Sun, Y.; Yang, Y.; Han, Y.; Xu, C. Urban park use and self-reported physical, mental, and social health during the COVID-19 pandemic: An on-site survey in Beijing, China. *Urban. For. Urban. Green.* **2023**, *79*, 127804. [CrossRef] [PubMed]
33. Zhang, Y.; Zhan, Y.; Yu, T.; Ren, X. Urban green effects on land surface temperature caused by surface characteristics: A case study of summer Beijing metropolitan region. *Infrared Phys. Technol.* **2017**, *86*, 35–43. [CrossRef]
34. Chakraborty, T.; Biswas, T.; Campbell, L.S.; Franklin, B.; Parker, S.S.; Tukman, M. Feasibility of afforestation as an equitable nature-based solution in urban areas. *Sust. Cities Soc.* **2022**, *81*, 103826. [CrossRef]
35. Meerow, S.; Newell, J.P. Spatial planning for multifunctional green infrastructure: Growing resilience in Detroit. *Landsc. Urban. Plann.* **2017**, *159*, 62–75. [CrossRef]
36. McMichael, A.J.; Wilkinson, P.; Kovats, R.S.; Pattenden, S.; Hajat, S.; Armstrong, B.; Vajanapoom, N.; Niciu, E.M.; Mahomed, H.; Kingkeow, C.; et al. International study of temperature, heat and urban mortality: The "ISOTHURM" project. *Int. J. Epidemiol.* **2008**, *37*, 1121–1131. [CrossRef] [PubMed]
37. Yan, M.; Chen, L.; Leng, S.; Sun, R. Effects of local background climate on urban vegetation cooling and humidification: Variations and thresholds. *Urban. For. Urban. Green.* **2023**, *80*, 127840. [CrossRef]
38. Augusto, B.; Roebeling, P.; Rafael, S.; Ferreira, J.; Ascenso, A.; Bodilis, C. Short and medium- to long-term impacts of nature-based solutions on urban heat. *Sust. Cities Soc.* **2020**, *57*, 102122. [CrossRef]
39. de la Paz, D.; de Andrés, J.M.; Narros, A.; Silibello, C.; Finardi, S.; Fares, S.; Tejero, L.; Borge, R.; Mircea, M. Assessment of Air Quality and Meteorological Changes Induced by Future Vegetation in Madrid. *Forests* **2022**, *13*, 690 [CrossRef]
40. Zafeiratou, S.; Stafoggia, M.; Gasparrini, A.; Rao, S.; de Donato, F.; Masselot, P.; Vazquez Fernandez, L.; Diz-Lois Palomares, A.; Aunan, K.; Katsouyanni, K.; et al. Independent effects of long and short-term exposures to non-optimal increased temperature on mortality. *Environ. Pollut.* **2025**, *366*, 125428. [CrossRef] [PubMed]
41. Hu, J.; Zhou, M.; Qin, M.; Tong, S.; Hou, Z.; Xu, Y.; Zhou, C.; Xiao, Y.; Yu, M.; Huang, B.; et al. Long-term exposure to ambient temperature and mortality risk in China: A nationwide study using the difference-in-differences design. *Environ. Pollut.* **2022**, *292*, 118392. [CrossRef]
42. Li, T.; Horton, R.M.; Bader, D.A.; Liu, F.; Sun, Q.; Kinney, P.L. Long-term projections of temperature-related mortality risks for ischemic stroke, hemorrhagic stroke, and acute ischemic heart disease under changing climate in Beijing, China. *Environ. Int.* **2018**, *112*, 1–9. [CrossRef]
43. Zafeiratou, S.; Samoli, E.; Dimakopoulou, K.; Rodopoulou, S.; Analitis, A.; Gasparrini, A.; Stafoggia, M.; De' Donato, F.; Rao, S.; Monteiro, A.; et al. A systematic review on the association between total and cardiopulmonary mortality/morbidity or cardiovascular risk factors with long-term exposure to increased or decreased ambient temperature. *Sci. Total Environ.* **2021**, *772*, 145383. [CrossRef]
44. Yang, C.; Meng, X.; Chen, R.; Cai, J.; Zhao, Z.; Wan, Y.; Kan, H. Long-term variations in the association between ambient temperature and daily cardiovascular mortality in Shanghai, China. *Sci. Total Environ.* **2015**, *538*, 524–530. [CrossRef]
45. Psistaki, K.; Kouis, P.; Michanikou, A.; Yiallouros, P.K.; Papatheodorou, S.I.; Paschalidou, A.K. Temporal trends in temperature-related mortality and evidence for maladaptation to heat and cold in the Eastern Mediterranean region. *Sci. Total Environ.* **2024**, *943*, 173899. [CrossRef]
46. Zanobetti, A.; O'Neill, M.S. Longer-Term Outdoor Temperatures and Health Effects: A Review. *Curr. Epidemiol. Rep.* **2018**, *5*, 125–139. [CrossRef] [PubMed]
47. Palmeiro-Silva, Y.; Chandler, R.E.; Kelman, I. Estimating the effects of temperature on mortality and hospitalisations under RCP2.6 and RCP8.5 scenarios in the short- and long-term by region in Chile: A population-based, modelling study. *Lancet Reg. Health Am.* **2025**, *48*, 101151. [CrossRef]
48. Borge, R.; Santiago, J.L.; de la Paz, D.; Martín, F.; Domingo, J.; Valdés, C.; Sánchez, B.; Rivas, E.; Rozas, M.T.; Lázaro, S.; et al. Application of a short term air quality action plan in Madrid (Spain) under a high-pollution episode—Part II: Assessment from multi-scale modelling. *Sci. Total Environ.* **2018**, *635*, 1574–1584. [CrossRef]
49. D'Isidoro, M.; D'Elia, I.; Vitali, L.; Briganti, G.; Cappelletti, A.; Piersanti, A.; Finardi, S.; Calori, G.; Pepe, N.; Di Giosa, A.; et al. Lessons learnt for air pollution mitigation policies from the COVID-19 pandemic: The Italian perspective. *Atmos. Pollut. Res.* **2022**, *13*, 101620. [CrossRef]
50. D'Elia, I.; Briganti, G.; Vitali, L.; Piersanti, A.; Righini, G.; D'Isidoro, M.; Cappelletti, A.; Mircea, M.; Adani, M.; Zanini, G.; et al. Measured and modelled air quality trends in Italy over the period 2003–2010. *Atmos. Chem. Phys.* **2021**, *21*, 10825–10849. [CrossRef]

51. Piersanti, A.; D'Elia, I.; Gualtieri, M.; Briganti, G.; Cappelletti, A.; Zanini, G.; Ciancarella, L. The Italian national air pollution control programme: Air quality, health impact and cost assessment. *Atmosphere* **2021**, *12*, 196. [CrossRef]
52. Mircea, M.; Borge, R.; Finardi, S.; Briganti, G.; Russo, F.; de la Paz, D.; D'Isidoro, M.; Cremona, G.; Villani, M.G.; Cappelletti, A.; et al. The Role of Vegetation on Urban Atmosphere of Three European Cities. Part 2: Evaluation of Vegetation Impact on Air Pollutant Concentrations and Depositions. *Forests* **2023**, *14*, 1255. [CrossRef]
53. Skamarock, W.C.; Klemp, J.B. A time-split nonhydrostatic atmospheric model for weather research and forecasting applications. *J. Comput. Phys.* **2008**, *227*, 3465–3485. [CrossRef]
54. Liu, Y.; Chen, F.; Warner, T.; Basara, J. Verification of a Mesoscale Data-Assimilation and Forecasting System for the Oklahoma City Area during the Joint Urban 2003 Field Project. *J. Appl. Meteorol. Climatol.* **2006**, *45*, 912–929. [CrossRef]
55. Martilli, A.; Clappier, A.; Rotach, M.W. An Urban Surface Exchange Parameterisation for Mesoscale Models. *Bound. Layer. Meteorol.* **2002**, *104*, 261–304. [CrossRef]
56. D'Isidoro, M.; Mircea, M.; Borge, R.; Finardi, S.; de la Paz, D.; Briganti, G.; Russo, F.; Cremona, G.; Villani, M.G.; Adani, M.; et al. The Role of Vegetation on Urban Atmosphere of Three European Cities—Part 1: Evaluation of Vegetation Impact on Meteorological Conditions. *Forests* **2023**, *14*, 1235. [CrossRef]
57. Adani, M.; Piersanti, A.; Ciancarella, L.; D'Isidoro, M.; Villani, M.G.; Vitali, L. Preliminary tests on the sensitivity of the FORAIR_IT air quality forecasting system to different meteorological drivers. *Atmosphere* **2020**, *11*, 574. [CrossRef]
58. Borge, R.; Alexandrov, V.; del Vas, J.J.; Lumbreras, J.; Rodríguez, E. A comprehensive sensitivity analysis of the WRF model for air quality applications over the Iberian Peninsula. *Atmos. Environ.* **2008**, *42*, 8560–8574. [CrossRef]
59. de la Paz, D.; Borge, R.; Martilli, A. Assessment of a high resolution annual WRF-BEP/CMAQ simulation for the urban area of Madrid (Spain). *Atmos. Environ.* **2016**, *144*, 282–296. [CrossRef]
60. Emery, C.; Tai, E.; Yarwood, G. *Enhanced Meteorological Modeling and Performance Evaluation for Two Texas Ozone Episodes, Final Report to the Texas Natural Resource Conservation Commission under TNRCC Umbrella Contract No. 582-0-31984*; Environ International Corporation: Novato, CA, USA, 2001.
61. Mircea, M.; Briganti, G.; Russo, F.; Finardi, S.; Silibello, C.; Prandi, R.; Carlino, G.; D'Isidoro, M.; Cappelletti, A.; Cremona, G. Vegetation Effects on Air Pollution: A Comprehensive Assessment for Two Italian Cities. *Atmosphere* **2024**, *15*, 1511. [CrossRef]
62. Beck, H.; Zimmermann, N.; McVicar, T.; Vergopolan, N.; Berg, A.; Wood, E.F. Present and future Köppen-Geiger climate classification maps at 1-km resolution. *Sci. Data* **2018**, *5*, 180214. [CrossRef] [PubMed]
63. Puche, M.; Vavassori, A.; Brovelli, M.A. Insights into the Effect of Urban Morphology and Land Cover on Land Surface and Air Temperatures in the Metropolitan City of Milan (Italy) Using Satellite Imagery and In Situ Measurements. *Remote Sens.* **2023**, *15*, 733. [CrossRef]
64. Nardino, M.; Cremonini, L.; Georgiadis, T.; Mandanici, E.; Bitelli, G. Microclimate Classification of Bologna (Italy) as a Support Tool for Urban Services and Regeneration. *Int. J. Environ. Res. Public Health* **2021**, *18*, 4898. [CrossRef]
65. Vardoulakis, S.; Dear, K.; Hajat, S.; Heaviside, C.; Eggen, B.; McMichael, A.J. Comparative assessment of the effects of climate change on heat- and cold-related mortality in the United Kingdom and Australia. *Environ. Health Perspect.* **2014**, *122*, 1285–1292. [CrossRef]
66. Jannis Frasch, L.; König, H.-H.; Konnopka, C. Effects of extreme temperature on morbidity, mortality, and case severity in German emergency care. *Environ. Res.* **2025**, *270*, 121021. [CrossRef]
67. Lee, W.; Bell, M.L.; Gasparrini, A.; Armstrong, B.G.; Sera, F.; Hwang, S.; Lavigne, E.; Zanobetti, A.; Coelho, M.S.Z.S.; Saldiva, P.H.N.; et al. Mortality burden of diurnal temperature range and its temporal changes: A multi-country study. *Environ. Int.* **2018**, *110*, 123–130. [CrossRef]
68. Gasparrini, A.; Guo, Y.; Hashizume, M.; Lavigne, E.; Zanobetti, A.; Schwartz, J.; Tobias, A.; Tong, S.; Rocklöv, J.; Forsberg, B.; et al. Mortality risk attributable to high and low ambient temperature: A multicountry observational study. *Lancet* **2015**, *386*, 369–375. [CrossRef] [PubMed]
69. Carmona, R.; Díaz, J.; Mirón, I.J.; Ortiz, C.; Luna, M.Y.; Linares, C. Mortality attributable to extreme temperatures in Spain: A comparative analysis by city. *Environ. Int.* **2016**, *91*, 22–28. [CrossRef] [PubMed]
70. Mistry, M.N.; Schneider, R.; Masselot, P.; Royé, D.; Armstrong, B.; Kyselý, J.; Orru, H.; Sera, F.; Tong, S.; Lavigne, E.; et al. Comparison of weather station and climate reanalysis data for modelling temperature-related mortality. *Sci. Rep.* **2022**, *12*, 5178. [CrossRef]
71. Royé, D.; Iñiguez, C.; Tobías, A. Comparison of temperature–mortality associations using observed weather station and reanalysis data in 52 Spanish cities. *Environ. Res.* **2020**, *183*, 109237. [CrossRef]
72. Armstrong, B.; Bell, M.L.; Coelho, M.d.S.Z.S.; Guo, Y.L.L.; Guo, Y.; Goodman, P.; Hashizume, M.; Honda, Y.; Kim, H.; Lavigne, E.; et al. Longer-term impact of high and low temperature on mortality: An international study to clarify length of mortality displacement. *Environ. Health Perspect.* **2017**, *125*, 107009. [CrossRef]

73. U.S. EPA. *Environmental Benefits Mapping and Analysis Program: Community Edition (BenMAP-CE) User Manual and Appendices*; U.S. EPA: Research Triangle Park, NC, USA, 2018. Available online: https://www.epa.gov/sites/production/files/2015-04/documents/benmap-ce_user_manual_march_2015.pdf (accessed on 25 June 2025).
74. WHO. AirQ+: Software Tool for Health Risk Assessment of Air Pollution. 2018. Available online: https://www.who.int/europe/tools-and-toolkits/airq---software-tool-for-health-risk-assessment-of-air-pollution (accessed on 25 June 2025).
75. Leone, M.; D'Ippoliti, D.; De Sario, M.; Analitis, A.; Menne, B.; Katsouyanni, K.; de' Donato, F.K.; Basagana, X.; Ben Salah, A.; Casimiro, E.; et al. A time series study on the effects of heat on mortality and evaluation of heterogeneity into European and Eastern-Southern Mediterranean cities: Results of EU CIRCE project. *Environ. Health* **2013**, *12*, 55. [CrossRef]
76. Vázquez Fernández, L.; Diz-Lois Palomares, A.; Vicedo-Cabrera, A.M.; Gasparrini, A.; Freiesleben de Blasio, B.; Di Ruscio, F.; Masselot, P.; Wisløff, T.; Rao-Skirbekk, S. Municipality assessment of temperature-related mortality risks in Norway. *Environ. Res.* **2025**, *266*, 120614. [CrossRef]
77. Gasparrini, A.; Guo, Y.; Hashizume, M. Mortalité attribuable au froid et à la chaleur: Analyse multi-pays. *Environ. Risques Sante* **2015**, *14*, 464–465. Available online: https://www.yearbook-ers.jle.com/download/ers-307077-31575-mortalite_attribuable_au_froid_et_a_la_chaleur_analyse_multi_pays-a.pdf (accessed on 25 June 2025).
78. Gallego Pinilla, F.J. Population density grid of the European Union. *Popul. Environ.* **2010**, *31*, 460–473. [CrossRef]
79. Heaviside, C.; Macintyre, H.; Vardoulakis, S. The Urban Heat Island: Implications for health in a changing environment. *Curr. Environ. Health Rep.* **2017**, *4*, 296–305. [CrossRef] [PubMed]
80. Román-Cascón, C.; Lothon, M.; Lohou, F.; Ojha, N.; Merlin, O.; Aragonés, D.; González-Dugo, M.P.; Andreu, A.; Pellarin, T.; Brut, A.; et al. Can We Use Satellite-Based Soil-Moisture Products at High Resolution to Investigate Land-Use Differences and Land–Atmosphere Interactions? A Case Study in the Savanna. *Remote Sens.* **2020**, *12*, 1701. [CrossRef]
81. Lüthi, S.; Fairless, C.; Fischer, E.M.; Scovronick, N.; Armstrong, B.; Coelho, M.D.S.Z.S.; Guo, Y.L.; Guo, Y.; Honda, Y.; Huber, V.; et al. Rapid increase in the risk of heat-related mortality. *Nat. Commun.* **2023**, *14*, 4894. [CrossRef]
82. Gallo, E.; Quijal-Zamorano, M.; Méndez Turrubiates, R.F.; Tonne, C.; Basagaña, X.; Achebak, H.; Ballester, J. Heat-related mortality in Europe during 2023 and the role of adaptation in protecting health. *Nat. Med.* **2024**, *30*, 3101–3105. [CrossRef]
83. Krummenauer, L.; Costa, L.; Prahl, B.F.; Kropp, J.P. Future heat adaptation and exposure among urban populations and why a prospering economy alone won't save us. *Sci. Rep.* **2021**, *11*, 20309. [CrossRef]
84. Karimi, A.; Mohammad, P.; García-Martínez, A.; Moreno-Rangel, D.; Gachkar, D.; Gachkar, S. New developments and future challenges in reducing and controlling heat island effect in urban areas. *Environ. Dev. Sustain.* **2023**, *25*, 10485–10531. [CrossRef]
85. Yin, Z.; Liu, Z.; Liu, X.; Zheng, W.; Yin, L. Urban heat islands and their effects on thermal comfort in the US: New York and New Jersey. *Ecol. Indic.* **2023**, *154*, 110765. [CrossRef]
86. Wang, C.; Ren, Z.; Dong, Y.; Zhang, P.; Guo, Y.; Wang, W.; Bao, G. Efficient cooling of cities at global scale using urban green space to mitigate urban heat island effects in different climatic regions. *Urban. For. Urban. Green.* **2022**, *74*, 127635. [CrossRef]
87. Pinto, L.V.; Inácio, M.; Ferreira, C.S.S.; Ferreira, A.D.; Pereira, P. Ecosystem services and well-being dimensions related to urban green spaces—A systematic review. *Sust. Cities Soc.* **2022**, *85*, 104072. [CrossRef]
88. Reyes-Riveros, R.; Altamirano, A.; De La Barrera, F.; Rozas-Vásquez, D.; Vieli, L.; Meli, P. Linking public urban green spaces and human well-being: A systematic review. *Urban. For. Urban. Green.* **2021**, *61*, 127105. [CrossRef]
89. Iungman, T.; Ventura Caballé, S.; Segura-Barrero, R.; Cirach, M.; Mueller, N.; Daher, C.; Villalba, G.; Pereira Barboza, E.; Nieuwenhuijsen, M. Co-benefits of nature-based solutions: A health impact assessment of the Barcelona Green Corridor (Eixos Verds) plan. *Environ. Int.* **2025**, *196*, 109313. [CrossRef] [PubMed]
90. Iungman, T.; Cirach, M.; Marando, F.; Pereira Barboza, E.; Khomenko, S.; Masselot, P.; Quijal-Zamorano, M.; Mueller, N.; Gasparrini, A.; Urquiza, J.; et al. Cooling cities through urban green infrastructure: A health impact assessment of European cities. *Lancet* **2023**, *401*, 577–589. [CrossRef]
91. García-León, D.; Masselot, P.; Mistry, M.N.; Gasparrini, A.; Motta, C.; Feyen, L.; Ciscar, J.C. Temperature-related mortality burden and projected change in 1368 European regions: A modelling study. *Lancet Public. Health* **2024**, *9*, 644–653. [CrossRef] [PubMed]
92. Borge, R.; Requia, W.J.; Yagüe, C.; Jhun, I.; Koutrakis, P. Impact of weather changes on air quality and related mortality in Spain over a 25 year period [1993–2017]. *Environ. Int.* **2019**, *133 Part B*, 105272. [CrossRef]
93. WHO Europe. *Heat and Health in the WHO European Region: Updated Evidence for Effective Prevention*; World Health Organization Regional Office for Europe: Copenhagen, Denmark, 2021. Available online: https://www.who.int/europe/publications/i/item/9789289055406 (accessed on 25 June 2025).
94. Picornell, M.; Ruiz, T.; Borge, R.; García-Albertos, P.; de la Paz, D.; Lumbreras, J. Population dynamics based on mobile phone data to improve air pollution exposure assessments. *J. Expo. Sci. Environ. Epidemiol.* **2019**, *29*, 278–291. [CrossRef]
95. Xi, J.Y.; Liang, B.H.; Zhang, W.J.; Yan, B.; Dong, H.; Chen, Y.-Y.; Lin, X.; Gu, J.; Hao, Y.-T. Effects of population aging on quality of life and disease burden: A population-based study. *Glob. Health. Res. Policy* **2025**, *10*, 2. [CrossRef]
96. Chairat, S.; Gheewala, S.H. The conceptual quantitative assessment framework for Nature-based Solutions (NbS). *Nat. Based Solut.* **2024**, *6*, 100152. [CrossRef]

97. Salamanca, F.; Krpo, A.; Martilli, A.; Clappier, A. A new building energy model coupled with an urban canopy parameterization for urban climate simulations—Part I. formulation, verification, and sensitivity analysis of the model. *Theor. Appl. Climatol.* **2010**, *99*, 331–344. [CrossRef]
98. Martilli, A.; Nazarian, N.; Krayenhoff, E.S.; Lachapelle, J.; Lu, J.; Rivas, E.; Rodriguez-Sanchez, A.; Sanchez, B.; Santiago, J.L. WRF-Comfort: Simulating microscale variability in outdoor heat stress at the city scale with a mesoscale model. *Geosci. Model Dev.* **2024**, *17*, 5023–5039. [CrossRef]
99. Lopez-Gomez, I.; Wan, Z.Y.; Zepeda-Núñez, L.; Schneider, T.; Anderson, J.; Sha, F. Dynamical-generative downscaling of climate model ensembles. *Proc. Natl. Acad. Sci. USA* **2025**, *122*, e2420288122. [CrossRef]
100. Su, S.; Zhang, Q.; Pi, J.; Wan, C.; Weng, M. Public health in linkage to land use: Theoretical framework, empirical evidence, and critical implications for reconnecting health promotion to land use policy. *Land. Use Policy* **2016**, *57*, 605–618. [CrossRef]
101. Oosterbroek, B.; de Kraker, J.; Huynen, M.-M.T.E.; Martens, P.; Verhoeven, K. Assessment of green space benefits and burdens for urban health with spatial modeling. *Urban For. Urban. Green.* **2023**, *86*, 128023. [CrossRef]
102. Qiu, Y.; Wu, Z.; Man, R.; Zong, T.; Liu, Y.; Meng, X.; Chen, J.; Chen, S.; Yang, S.; Yuan, B.; et al. Secondary aerosol formation drives atmospheric particulate matter pollution over megacities (Beijing and Seoul) in East Asia. *Atmos. Environ.* **2023**, *301*, 119702. [CrossRef]
103. Sato, K.; Ikemori, F.; Ramasamy, S.; Iijima, A.; Kumagai, K.; Fushimi, A.; Fujitani, Y.; Chatani, S.; Tanabe, K.; Takami, A.; et al. Formation of secondary organic aerosol tracers from anthropogenic and biogenic volatile organic compounds under varied NOx and oxidant conditions. *Atmos. Environ.* X **2022**, *14*, 100169. [CrossRef]
104. Ren, Y.; Qu, Z.; Du, Y.; Xu, R.; Ma, D.; Yang, G.; Shi, Y.; Fan, X.; Tani, A.; Guo, P.; et al. Air quality and health effects of biogenic volatile organic compounds emissions from urban green spaces and the mitigation strategies. *Environ. Pollut.* **2017**, *230*, 849–861. [CrossRef] [PubMed]
105. Morakinyo, T.E.; Lam, Y.F. Study of traffic-related pollutant removal from street canyon with trees: Dispersion and deposition perspective. *Environ. Sci. Pollut. Res.* **2016**, *23*, 21652–21668. [CrossRef]
106. Hong, S.-Y.; Lim, J.-O.J. WRF Single-Moment 6-Class Microphysics Scheme (WSM6). *J. Korean Meteor. Soc.* **2006**, *42*, 129–151.
107. Bougeault, P.; Lacarrere, P. Parameterization of Orography-Induced Turbulence in a Mesobeta--Scale Model. *Mon. Weath. Rev.* **1989**, *117*, 1872–1890. [CrossRef]
108. Janjić, Z.I. The step-mountain eta coordinate model: Further developments of the convection, viscous sublayer, and turbulence closure schemes. *Mon. Weath. Rev.* **1994**, *122*, 927–945. [CrossRef]
109. Chen, F.; Dudhia, J. Coupling an advanced land surface-hydrology model with the Penn State-NCAR MM5 modeling system. Part I: Model implementation and sensitivity. *Mon. Weath. Rev.* **2001**, *129*, 569–585. [CrossRef]
110. Schwarzkopf, M.D.; and Fels, S.B. The simplified exchange method revisited: An accurate, rapid method for computation of infrared cooling rates and fluxes. *J. Geophys. Res. Atmos.* **1991**, *96*, D5. [CrossRef]
111. Iacono, M.J.; Delamere, J.S.; Mlawer, E.J.; Shephard, M.W.; Clough, S.A.; Collins, W.D. Radiative forcing by long-lived greenhouse gases: Calculations with the AER radiative transfer models. *J. Geophys. Res. Atmos.* **2008**, *113*, D13103. [CrossRef]
112. Dudhia, J. Numerical study of convection observed during the winter monsoon experiment using a mesoscale two-dimensional model. *J. Atmos. Sci.* **1989**, *46*, 3077–3107. [CrossRef]

forests

MDPI

Article

The Impact of Meteorological Factors and Canopy Structure on $PM_{2.5}$ Dynamics Under Different Urban Functional Zones in a Subtropical City

Yichen Huang [1,2,3], Yue Cai [4], Jiejie Jiao [5], Chunyu Pan [1,6], Guangyu Wang [6], Chong Li [1,2,3], Zichen Jia [1,2,3], Zhihao Chen [1,2,3], Yufeng Zhou [1,2,3] and Guomo Zhou [1,2,3,*]

1 Key Laboratory of Carbon Sequestration and Emission Reduction in Agriculture and Forestry of Zhejiang Province, Zhejiang A&F University, Hangzhou 311300, China; 2022103022009@stu.zafu.edu.cn (Y.H.); chunyu.pan@ubc.ca (C.P.); chongli@zafu.edu.cn (C.L.); 2022603031009@stu.zafu.edu.cn (Z.J.); 17816899272@163.com (Z.C.); zyffafa@163.com (Y.Z.)

2 Institute of Ecological Civilization and Institute of Carbon Neutrality, Zhejiang A&F University, Hangzhou 311300, China

3 Engineering Research Center of Bamboo Carbon Sequestration for State Forestry and Grassland Administration, Hangzhou 311300, China

4 Ehon Carbon Technologies Company Limited, Hangzhou 311300, China; 2019202012001@stu.zafu.edu.cn

5 Zhejiang Academy of Forestry, Hangzhou 310023, China; jjj029302@hotmail.com

6 Faculty of Forestry, University of British Columbia, 2424 Main Mall, Vancouver, BC V6T 1Z4, Canada; guangyu.wang@ubc.ca

* Correspondence: zhougm@zafu.edu.cn

Abstract: $PM_{2.5}$ pollution has intensified with rapid urbanization and industrialization, raising concerns about its health and environmental impacts. Both meteorological factors and urban forests play crucial roles in influencing $PM_{2.5}$ concentrations. However, limited attention has been given to the direct impact of canopy structure on $PM_{2.5}$ levels at a larger scale. This study analyzes the temporal variation of $PM_{2.5}$, including seasonal and diurnal patterns, across different functional zones (park, traffic, and residential zones) in a subtropical region. It also investigates the seasonal responses of $PM_{2.5}$ to meteorological factors (temperature, humidity, and precipitation) and canopy structure characteristics, including canopy diameter (CD), canopy thickness (CT), canopy area (CA), canopy volume (CV), canopy height ratio (CH), leaf area index (LAI), and tree canopy cover (CO). The results show that among different functional zones, $PM_{2.5}$ concentrations were the highest in park zones, followed by traffic zones. Seasonal variations in $PM_{2.5}$ concentrations were the highest in winter (84.00 ± 45.97 μg/m^3), with greater fluctuations, and the lowest in summer (36.85 ± 17.63 μg/m^3 μg/m^3), with smaller fluctuations. Diurnal variation followed an "N"-shaped curve in spring, summer, and autumn, while a "W"-shaped curve was observed in winter. Correlation analysis indicated significant negative correlations between $PM_{2.5}$ and humidity, temperature, and rainfall, while CD, CA, and CV showed positive correlations with $PM_{2.5}$. Notably, $PM_{2.5}$ exhibited greater sensitivity to changes in canopy structure in winter, followed by autumn. Despite these findings, the influence of canopy structure on $PM_{2.5}$ concentrations was considerably smaller compared to meteorological factors. In particular, every 1 m^2 increase in canopy area could raise $PM_{2.5}$ levels by 0.864 μg/m^3, whereas an average increase of 1 mm in rainfall could raise $PM_{2.5}$ by 13.665 μg/m^3. These findings provide valuable guidance for implementing protective measures, improving air quality, optimizing urban greening strategies, and enhancing public health outcomes.

Keywords: $PM_{2.5}$; functional zones; urban forests; meteorological factors; canopy structure

1. Introduction

With the rapid acceleration of urbanization, air pollution from fine particulate matter ($PM_{2.5}$) has become a major environmental issue that threatens public health in urban areas [1,2]. As an essential ecological infrastructure, urban trees significantly influence the spatial distribution of $PM_{2.5}$ through mechanisms such as dust retention by leaves and aerodynamic effects [3,4]. However, their air purification efficiency is constrained by a variety of complex factors. Recent advancements in remote sensing and LiDAR technology, particularly mobile laser scanning (MLS), which includes handheld, backpack, and vehicle-mounted sensors, have revolutionized the way we collect high-precision three-dimensional canopy structure data. Compared to traditional manual measurement methods, the Backpack Light Detection and Ranging System (BLS) allows for broader spatial coverage and achieves approximately four times the time efficiency, enabling large-scale quantitative analysis of urban tree morphology [5,6].

The regulation of $PM_{2.5}$ by urban forest trees is closely linked to their structural characteristics. Existing studies suggest that among different tree species, conifers, with their higher leaf area index and waxy surfaces, exhibit stronger particulate matter adsorption capacities [7]. Furthermore, trees can influence their $PM_{2.5}$ interception and adsorption capabilities by altering canopy volume and vertical structure [8]. Dense tree canopies are particularly effective at capturing airborne particles, thereby reducing their concentration in the air [9,10]. It has been demonstrated that forests, through their leaves, branches, and complex canopy structures, play a key role in blocking and adsorbing particulate matter, making them an important measure for reducing urban atmospheric pollution. The height and age of trees affect their leaf area index and branch density, thereby influencing their ability to intercept $PM_{2.5}$ and enhancing their efficiency in capturing air pollutants. However, existing research has mainly focused on the particulate adsorption capacity of individual leaves, with limited attention given to the canopy's overall impact on $PM_{2.5}$.

Meteorological factors also influence both $PM_{2.5}$ concentrations and tree growth [11–13]. Increased relative humidity, while beneficial for particle wet deposition, may inhibit stomatal absorption. Rainfall enhances the moisture content of tree leaves, thereby improving their capacity to adsorb $PM_{2.5}$. Additionally, the post-rain moist environment fosters tree growth and increases leaf area, further enhancing $PM_{2.5}$ interception efficiency. Temperature variations affect the growth rate and physiological activity of trees, thus influencing their capacity to process $PM_{2.5}$. Research has shown that higher temperatures may promote tree growth, increasing their leaf area and improving their ability to intercept $PM_{2.5}$. However, excessively high temperatures may lead to rapid water evaporation, impairing trees' physiological functions and affecting their ability to process $PM_{2.5}$ [14,15].

These complex processes are further influenced by urban functional zones (UFZs) [16]. High-intensity emission sources in industrial zones, coupled with line source pollution from major traffic routes, often exceed the threshold that local vegetation can bear [17]. In residential areas, $PM_{2.5}$ concentrations are generally lower, but during the winter heating season, households using coal or wood may contribute to local increases in $PM_{2.5}$ levels [18]. The traffic, commercial activities, and green spaces surrounding functional zones also affect the air quality of these areas [19].

Existing studies are often limited to observations at a single scale: micro-level experiments on individual leaves do not reflect the overall effects of canopy structure, and remote sensing inversions lack precise characterization of three-dimensional structural parameters [20]. This study innovatively integrates BLS with a fixed meteorological station developed by our team, combining high-precision tree canopy morphology data with long-term real-time meteorological and pollution data to form a multi-dimensional observation platform. This approach not only overcomes the spatial scale and data precision limitations of traditional research but also provides a new perspective on the mechanisms by which tree canopies influence $PM_{2.5}$ concentrations, creating a "morphology-meteorology-pollution" multi-source dataset. This study selects three typical urban functional zones (parks, traffic areas, and residential areas) and aims to (1) analyze the temporal variation trends of $PM_{2.5}$ in different functional zones (seasonal and daily variations) and (2) explore the impacts of meteorological factors and canopy structure on $PM_{2.5}$ concentrations across different seasons. The analyses aim to provide a basis for developing differentiated air quality management policies. Moreover, by investigating the interactions between meteorological factors and canopy structure, this study provides theoretical support for predicting $PM_{2.5}$ variation patterns under different climatic conditions. The results will contribute to improving urban greening planning, enhancing air quality, and providing scientific guidance for pedestrian outdoor activities and policy decisions related to public space planning.

2. Materials and Methods

2.1. Study Area

This study was conducted in Tongxiang City, located in Jiaxing City, Zhejiang Province, China (30°28′–30°47′ N, 120°17′–120°39′ E) (Figure 1A). The region experiences a subtropical monsoon climate characterized by abundant sunlight and rainfall. It receives approximately 1702.7 h of sunshine annually, with a frost-free period of about 249.1 days per year. The average annual temperature is 17.1 °C, and the average annual precipitation is 1336 mm. Situated in a plain area with an average elevation of 5.3 m, Tongxiang City covers a total area of 727 square kilometers, with a forest coverage rate of 19.17% (http://www.tx.gov.cn (accessed on 8 March 2024).

In Tongxiang, we selected three representative functional zones—park zone, traffic zone, and residential zones (Figure 1B)—as study sites. As a fixed meteorological station, it can provide PM2.5 concentration monitoring data for local areas, typically within a range of several hundred meters to 2 km [21]. Studies have shown that within smaller plots (such as 20 m × 20 m), the influence of a single environmental variable (e.g., tree or plant coverage) on air quality can be effectively assessed while also reflecting local climatic and environmental characteristics [22]. Therefore, within each functional zone, three 20 × 20 m plots were established. The experimental period was from 1 March 2023 to 29 February 2024. For canopy structure measurements, data were collected on 20 April 2023, 1 July 2023, 26–27 September 2023, and 5 January 2024, representing spring, summer, autumn, and winter, respectively. To minimize within-group variability, Backpack Light Detection and Ranging System (LiBackpack DGC50H) was employed to identify plots with similar characteristics (Table 1). Specifically, in park areas, each plot lacked vehicular roads but included a pedestrian path wide enough for two people to walk side by side. In traffic areas, each plot contained a two-lane bidirectional road. In residential areas, plots were situated within neighborhoods, selecting green spaces adjacent to residential buildings. To reduce between-group variability, all plots were located at least 5 km away from any industrial pollution sources. The study area is situated in a plain, within an urban context, where the primary sources of $PM_{2.5}$ include traffic pollution, industrial emissions, and residential activities [23].

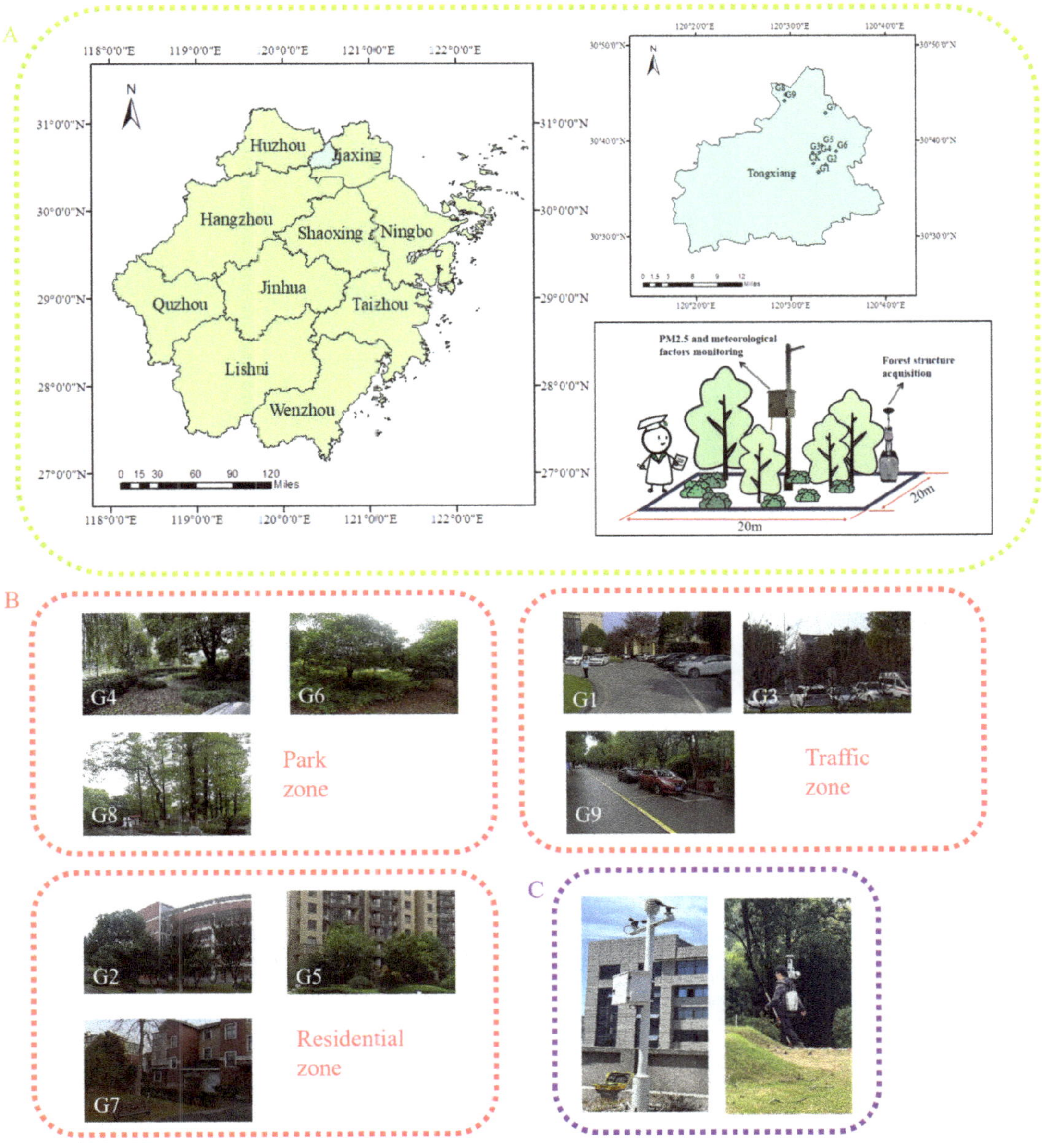

Figure 1. Map of the study area, locations of functional zones, and environmental factor monitoring stations, including weather stations and portable radar devices. (**A**) Locations of functional zones. (**B**) Layout of sample plots. (**C**) Placement of weather stations and utilization of portable radar devices.

Table 1. Basic characteristics of sample plots.

Functional Zones		Species Composition	Tree Count	Average DBH (m)	Average TH (m)	Elevation (m)
Park zone	G4	*Cinnamomum camphora* *Ginkgo biloba* *Pinus parviflora* *Salix babylonica*	11	0.28	10.96	6.0
	G6	*Osmanthus fragrans* *Celtis sinensis* *Hibiscus mutabilis Linn* *Photinia glabra* *Sapindus saponaria* *Cinnamomum camphora* *Acer palmatum Thunb* *Diospyros kaki* *Citrus hystrix*	25	0.25	6.16	6.4
	G8	*Metasequoia glyptostroboides* *Melia azedarach* *Firmiana simplex* *Citrus aurantium*	17	0.29	17.79	5.8
Traffic zone	G1	*Osmanthus fragrans* *Cinnamomum camphora* *Acer palmatum Thunb* *Malus spectabilis* *Photinia glabra* *Pistacia chinensis* *Ginkgo biloba*	10	0.28	8.21	5.3
	G3	*Osmanthus fragrans* *Celtis sinensis* *Acer palmatum Thunb* *Cinnamomum camphora* *Cerasus serrulata* *Acer saccharinum*	14	0.16	5.82	5.5
	G9	*Salix babylonica* *Acer palmatum Thunb* *Elaeocarpus apiculatus*	12	0.24	8.65	5.7
Residential zone	G2	*Cinnamomum camphora* *Osmanthus fragrans* *Cerasus serrulata*	8	0.12	6.73	5.3
	G5	*Sapindus saponaria* *Acer palmatum Thunb* *Osmanthus fragrans*	11	0.19	6.44	5.8
	G7	*Magnolia Grandiflora* *Cinnamomum camphora* *Osmanthus fragrans* *Citrus aurantium* *Sapindus saponaria*	9	0.22	8.00	5.2

2.2. Data Collection

2.2.1. Meteorological Data Collection

$PM_{2.5}$, temperature, relative humidity, and precipitation were measured using a ZX-AQI-L environmental factor online monitoring station developed by our team and manufactured by Shenzhen Zhongxing Environmental Protection Technology (Figure 1C). This station integrates multiple meteorological and environmental pollutants and light intensity, providing high-precision and high-speed monitoring and data output (Table 2). The sensors automatically collect and store data every minute. Before deploying the instruments, calibration of time and readings was conducted. A 3-day pre-experiment was conducted at the same location on the campus. A meteorological station was used as the reference standard, and the $PM_{2.5}$ data for each station were calibrated using linear least squares fitting. Calibration methods for relative humidity and temperature followed the same procedure applied to the $PM_{2.5}$ data. During the experiment, the monitoring station was placed at the center of each sample plot (Figure 1A) at a height of 1.6 m. Daily precipitation data were obtained from the China National Meteorological Science Data Center (http://data.cma.cn, accessed on 8 March 2024). Table 3 shows the mean values of $PM_{2.5}$ and meteorological factors for different functional zones. Additionally, in this study, we did not use an atmospheric dispersion model. Although such models can effectively simulate pollutant dispersion, the primary focus of this study was on real-time monitoring data for localized areas, and due to the scope limitations, we chose to rely on real-time monitoring data to analyze variations in $PM_{2.5}$ concentrations.

Table 2. Specifications of environmental factor monitoring weather stations.

Smart Sensor	Range	Accuracy	Detection Method	Sensor Models	Resolution	Response Time
$PM_{2.5}$ particle concentration sensor	0~1000 μg/m^3	1 μg/m^3	Optical Scattering Method (Laser)	N-PM300-02	1 μg/m^3	60 s
Temperature sensor	−20~85 °C	±0.5 °C	Thermocouple/ Capacitor	SHT30	0.1 °C	60 s
Relative air humidity sensor	0~100% RH	±3% RH	Thermocouple/ Capacitor	SHT30	0.1% RH	60 s

Table 3. Basic information of $PM_{2.5}$ and meteorological factors.

$PM_{2.5}$ and Meteorological Factors	Park Zone	Traffic Zone	Residential Zone
$PM_{2.5}$ concentration	60.35	55.85	53.01
Temperature	17.17	17.88	17.00
Humidness	80.75	79.83	79.88
Precipitation	2.54	2.54	2.54
Wind speed	0.30	0.40	0.24
Illumination	6.38	4.56	5.07

This study utilized meteorological stations to monitor $PM_{2.5}$ concentrations across three urban functional zones. According to the Ambient Air Quality Standards (GB 3095—2012) (Ministry of Environmental Protection, General Administration of Quality Supervision, Inspection and Quarantine of the People's Republic of China, 2012), $PM_{2.5}$ concentrations were classified as follows: a 24 h average concentration of less than 35 μg/m^3 was categorized as Grade I (Good), less than 75 μg/m^3 as Grade II (Moderately Polluted), and greater than 75 μg/m^3 as Grade III (Heavily Polluted). Analysis of variance (Table 4) indicates no significant differences within the same functional zone.

Table 4. $PM_{2.5}$ differences among plots.

Locations of Functional Zones		Mean Value	Standard Deviation	Standard Error	F	p
Park zone	G4	49.97	14.79	7.40	1.40	0.30
	G6	71.94	18.24	9.12	0.55	0.48
	G8	59.81	22.60	11.30	2.19	0.17
	(Intercept)	60.57	19.43	5.61		0.73
Traffic zone	G1	55.92	18.99	9.49	1.61	0.25
	G3	70.8323	27.13	13.56	0.56	0.48
	G9	45.2216	11.75	5.87	2.66	0.14
	(Intercept)	57.3235	21.38	6.17		0.73
Residential zone	G2	56.2548	22.39	11.19	0.34	0.71
	G5	45.7804	15.05	7.58	0.00	0.99
	G7	56.1072	23.35	11.68	0.68	0.43
	(Intercept)	52.7142	19.32	5.58		0.73

2.2.2. Canopy Structure Data Collection

In the preliminary phase of the experiment, we identified tree species and density in nine study plots (Table 1). The plots contained a mixture of evergreen and deciduous trees.

Point cloud data for the plots were then acquired using a Backpack Light Detection and Ranging System, with the specific parameters of the device outlined in Table 3. The acquired data were pre-processed using Cyclone software (Leica Geosystems HDS, San Ramon, CA, USA), which included alignment, merging, and denoising. Subsequently, we analyzed the data to derive several parameters, including canopy diameter (CD), canopy thickness (CT), canopy area (CA), canopy volume (CV), canopy height ratio (CH), leaf area index (LAI), and tree canopy cover (CO), which was calculated by projecting the point cloud of each tree horizontally to generate a canopy distribution image, from which the average diameter in both horizontal and vertical directions was determined. The CV was obtained using an improved stratified slicing method, defined as the total volume of all point cloud slices spaced 1 mm apart within the vertical forest space [24]. The CH was calculated using Formula (1).

$$\mathrm{CH} = \frac{\mathrm{CT}}{\mathrm{TH}} \times 100\% \quad (1)$$

The leaf area index (LAI) was calculated using the algorithm developed by Li et al. [25]. This method involves counting the number of laser point clouds within a three-dimensional grid. The formulas used are given in Equations (2) and (3):

$$l(s) = a(\theta)N(s) \quad (2)$$

where $a(\theta)$ is the leaf inclination correction factor, $N(s)$ is generally set to 1.1 and represents the laser point frequency at layer s, and $l(s)$ is the leaf area index at layer.

$$LAI = \sum_{s=1}^{S} l(s) = 1.1 \times \sum_{s=1}^{S} \frac{n_1(s)}{n_T(s)} \quad (3)$$

where $n_1(s)$ is the total number of grids containing laser points in layer, and $n_T(s)$ is the total number of three-dimensional grids in layer s.

The feature extraction process is illustrated in Figure 2. The stand structure is detailed in Table 5.

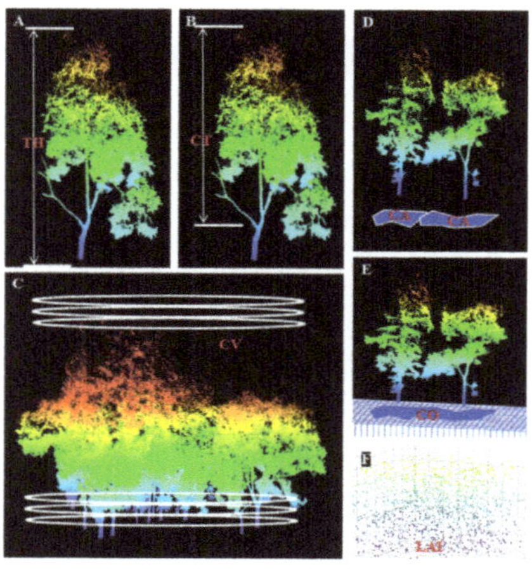

Figure 2. Extraction of structural information using Cyclone software. (**A**) Tree height measurement (TH). (**B**) Canopy thickness measurement (CT). (**C**) Calculation of canopy volume (CV) by vertical slicing at 1 mm intervals. (**D**) Canopy area measurement (CA). (**E**) Canopy cover measurement (CO). (**F**) Leaf area index measurement (LAI).

Table 5. Sensor parameters of the LiBackpack DGC50H.

Parameter	Parameter
Radar sensor	Velodyne Puck VLP-16×2
Accuracy	±3 cm
Detect range	100 m
Dot frequency	600,000(point/s)
Scan patterns	Full color
Resolution ratio	1800 W
Wave length	903 mm
Field angle	Standard 360° Vertical180°

2.3. Statistical Analysis

Statistical analyses were performed using Origin 2024 b and IBM SPSS Statistics 26. To assess seasonal differences among different functional zones and to examine the relationships between $PM_{2.5}$, meteorological factors, and canopy structure, a one-way analysis of variance (ANOVA) was conducted. A significance level of $p < 0.05$ was considered statistically significant. Pearson correlation coefficients were calculated in IBM SPSS Statistics 26 to assess the relationships between $PM_{2.5}$ concentrations, meteorological factors, and canopy structure. Additionally, stepwise multiple linear regression analysis was performed. Finally, the results were visualized using Origin 2024 b.

3. Results

3.1. Temporal Distribution Characteristics of $PM_{2.5}$ in Different Functional Zones

3.1.1. Seasonal Distribution Characteristics of $PM_{2.5}$

As shown in Figure 3, $PM_{2.5}$ concentrations fluctuate significantly across the functional zones, with the overall trend being similar. $PM_{2.5}$ levels decrease starting in March, reach their lowest in summer, and then rise again, peaking in winter. In summer, the minimum $PM_{2.5}$ concentration is 6.42 μg/m^3; in winter, it can reach a maximum of 226.50 μg/m^3. Based on the pollution level proportions (Figure 4), $PM_{2.5}$ concentrations in the three functional zones predominantly fall within Grade II. The proportions of parks, transportation, and residential zones that were determined to be Grade II were 52.6%, 45.2%, and 43.8%, respectively. Park zones have more days with Grade II concentrations (191 days) and more days with Grade III concentrations (90 days) than with Grade I concentrations (82 days). In contrast, the other two functional zones experience more days with good air quality (Grade I) than with severe pollution (Grade III). Seasonally, summer has the highest proportion of Grade I concentrations, while winter has the lowest. Spring shows a higher proportion of Grade II concentrations, while the distribution in the other three seasons is more balanced. Winter has the highest proportion of severe pollution, while summer experiences the least.

The box plots (Figure 5) further reveal differences across functional zones in various seasons. In spring and autumn, $PM_{2.5}$ concentrations in the park (53.54 ± 21.14 μg/m^3 and 60.4 ± 35.1 μg/m^3, respectively) and traffic zones (53.2 ± 21.6 μg/m3 and 58.9 ± 31.2 μg/m^3, respectively) are significantly higher than those in residential zones (47.9 ± 19.5 μg/m^3 and 49.4 ± 28.2 μg/m^3, respectively). In summer, $PM_{2.5}$ concentrations in the park (42.7 ± 21.3 μg/m^3) are significantly higher than those in the traffic zones (38.8 ± 17.2 μg/m^3) and residential zones (32.7 ± 15.5 μg/m^3). In winter, $PM_{2.5}$ concentrations in the traffic zones (93.9 ± 20.6 μg/m^3) are significantly higher than those in the residential zones (81.6 ± 48.9 μg/m^3), while the park zones (85.8 ± 48.2 μg/m^3) show a similar trend to both the traffic and residential zones. Overall, residential zones experience relatively lower $PM_{2.5}$ pollution compared to the other functional zones. Additionally, severe pollution and greater $PM_{2.5}$ fluctuations are observed in autumn and winter, indicating a higher influence of external factors during these seasons.

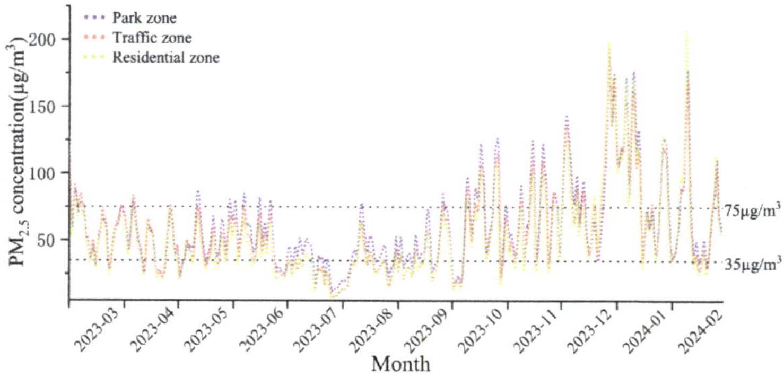

Figure 3. Monthly changes in PM2.5 in different functional zones. The dashed line below the table represents the dividing line between Grade I and Grade II, and the dashed line above represents the dividing line between Grade II and Grade III.

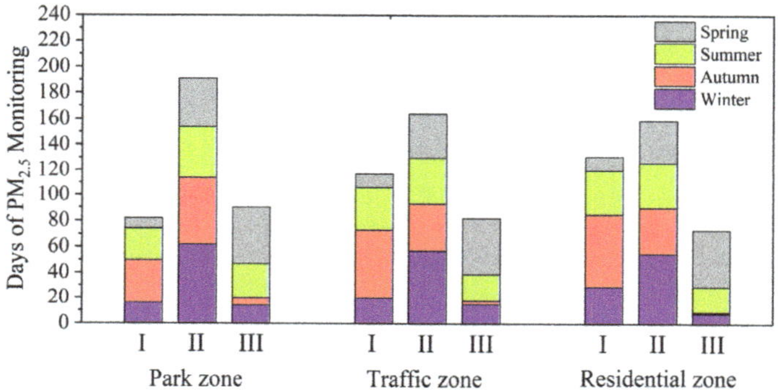

Figure 4. Circular graph of the proportion of pollution degree in different functional zones. I, II, and III represent the Grade I, Grade II, and Grade III, respectively.

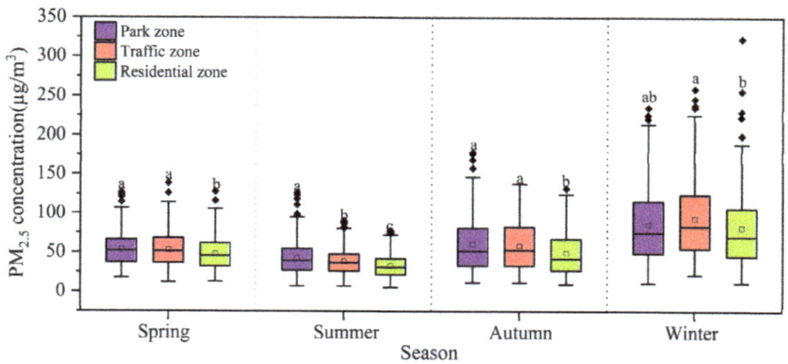

Figure 5. Seasonal boxplots of $PM_{2.5}$ concentrations in the park, traffic, and residential zones; lowercase letters above the boxes indicate significant seasonal differences between functional zones ($p < 0.05$).

3.1.2. Daily Variation of $PM_{2.5}$

As shown in Figure 6, the diurnal variation of $PM_{2.5}$ concentrations follows a similar pattern in spring, summer, and autumn. In these seasons, $PM_{2.5}$ levels rise during the early hours of the morning, decrease during the morning period, and increase again in the evening. However, in winter, $PM_{2.5}$ fluctuations are more frequent. From 10:00 PM to 5:00 AM, $PM_{2.5}$ concentrations generally decrease, followed by an increase starting around 5:00 AM. Between 8:00 AM and 3:00 PM, $PM_{2.5}$ concentrations experience a significant decline, after which there is a sharp rise around 9:00 PM. Summer shows more minor fluctuations than the other three seasons, while autumn and winter exhibit larger variations. Among the different functional zones, residential zones tend to have lower $PM_{2.5}$ concentrations in spring, summer, and autumn, while park zones exhibit higher levels. In winter, traffic zones show the lowest $PM_{2.5}$ concentrations, while park zones have higher concentrations.

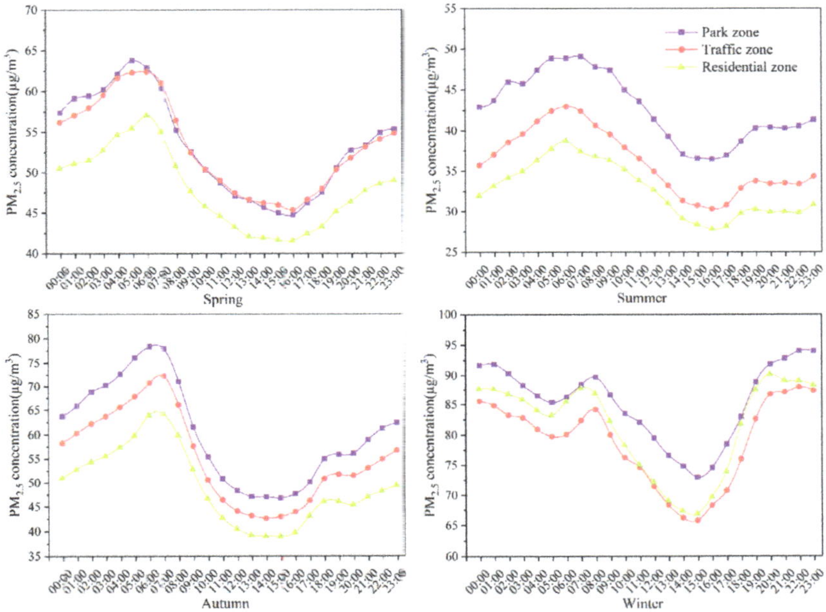

Figure 6. Daily variations of $PM_{2.5}$ in different functional zones across seasons.

3.2. Correlation Between Meteorological Factors and $PM_{2.5}$

This study also collected three key meteorological factors—relative humidity, temperature, and precipitation—and plotted their seasonal variations using box plots (Figure 7). Seasonal differences in these meteorological conditions were observed across the three functional zones. Relative humidity was significantly higher in summer (85.83 $\pm$ 9.79%) compared to spring (76.15 $\pm$ 14.56%), autumn (79.62 $\pm$ 12.86%), and winter (78.64 $\pm$ 17.43%), with no substantial differences between the latter three seasons. Summer exhibited lower variability in relative humidity, while winter showed greater variability with more frequent fluctuations. Temperature exhibited significant seasonal variation, with higher values in summer (26.96 $\pm$ 2.91 °C) and lower values in winter (5.54 $\pm$ 4.81 °C). Temperature followed a general upward trend before declining. Precipitation did not differ significantly between spring (2.50 mm) and autumn (1.71 mm). However, summer precipitation (5.40 mm) was considerably higher than in the other seasons, as indicated by a longer box length, suggesting greater variability and more extreme rainfall events. In contrast, winter precipitation (0.58 mm) was significantly lower and more stable, with fewer extreme rainfall events. Meteorological conditions play a crucial role in the occurrence and development of air pollution.

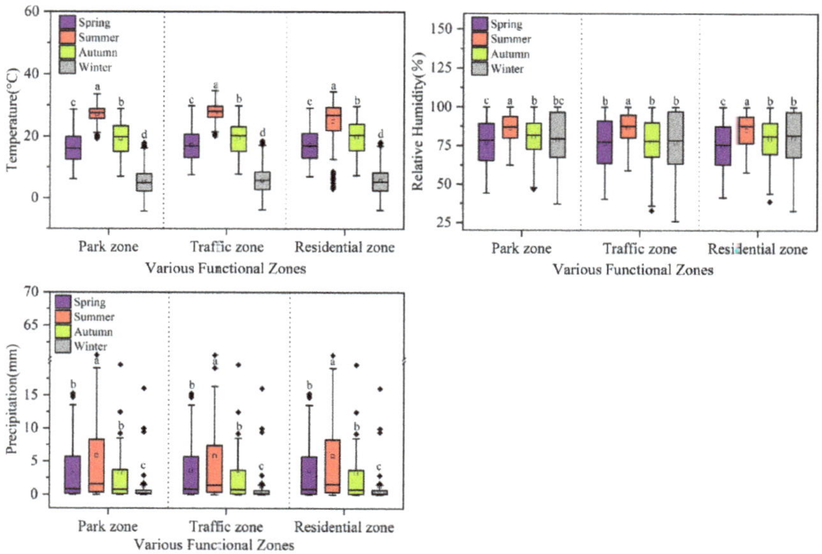

Figure 7. Boxplots of meteorological factors in different functional zones; lowercase letters above the boxes indicate significant seasonal differences within each zone ($p < 0.05$).

We calculated the correlation coefficients between meteorological factors and $PM_{2.5}$ concentrations and analyzed the seasonal differences in these variables at each $PM_{2.5}$ concentration level (Table 6). We found varying degrees of correlation with humidity, temperature, and precipitation. In autumn and winter, average humidity showed a moderate negative correlation with $PM_{2.5}$ concentrations, with winter ($r = -0.51$, $p < 0.01$) exhibiting a stronger correlation than autumn ($r = -0.48$, $p < 0.05$). No significant correlations were observed in the other seasons. Temperature was negatively correlated with $PM_{2.5}$ concentrations in all seasons except summer ($r < -0.48$, $p < 0.05$), with a strong negative correlation observed in spring ($r = -0.69$, $p < 0.01$). The average temperatures for Grade I, II, and III $PM_{2.5}$ concentrations were 21.04 ± 0.57 °C, 17.53 ± 0.34 °C, and 11.30 ± 1.26 °C, respectively. In spring and autumn, the differences between Grade II and III concentrations were greater than those between Grade I and II, while the reverse was true in winter. Precipitation showed a moderate negative correlation with $PM_{2.5}$ concentrations only in autumn ($r = -0.4$, $p < 0.05$). As $PM_{2.5}$ concentrations increased, precipitation tended to decrease.

Table 6. Relationship and correlation between key meteorological variables and $PM_{2.5}$ concentrations.

Statistical Indicators		Spring			Summer			Autumn			Winter		
		Humidity (%)	Temperature (°C)	Precipitation (mm)	Humidity (%)	Temperature (°C)	Precipitation (mm)	Humidity (%)	Temperature (°C)	Precipitation (mm)	Humidity (%)	Temperature (°C)	Precipitation (mm)
$PM_{2.5}$ concentration ($\mu g/m^3$)	<35	75.75	18.33	4.24	87.36	25.74	6.44	77.83	16.45	4.19	76.42	5.93	0.75
	35–75	76.38	17.91	2.09	82.95	24.56	3.55	76.20	16.81	2.19	79.19	3.80	1.06
	≥75	75.08	13.53	1.59	87.66	21.82	6.54	79.44	15.15	1.99	77.07	3.78	0.25
R		−0.26	−0.69 **	−0.22	−0.02	−0.23	0.27	−0.39 *	−0.58 **	−0.4 *	−0.51 **	−0.48 *	−0.05

* $p < 0.05$, ** $p < 0.01$.

3.3. Correlation Between Forest Canopy Structure and $PM_{2.5}$

We calculated the correlation coefficients between $PM_{2.5}$ and canopy structure for the four seasons. As shown in Figure 8, no significant correlation exists between $PM_{2.5}$ and canopy structure in summer. However, in the other three seasons, $PM_{2.5}$ shows a significant and strong positive correlation with canopy diameter (CD), canopy area (CA), and canopy volume (CV) ($r > 0.76$, $p < 0.05$). Specifically, in spring, $PM_{2.5}$ shows a highly significant strong positive correlation with CD and CA ($r > 0.81$, $p < 0.01$) and a significant strong positive correlation with CV ($r = 0.8$, $p < 0.05$). In autumn, $PM_{2.5}$ significantly strongly correlates with CD, CA, and CV ($r > 0.75$, $p < 0.05$). In winter, $PM_{2.5}$ is also significantly positively correlated with all three canopy structure variables, with the strongest correlations observed compared to the other three seasons ($r > 0.89$, $p < 0.01$). Excluding summer, the correlations between $PM_{2.5}$ and CD, CA, and CV are the strongest in winter, with correlation coefficients of $r = 0.86$, 0.93, and 0.87, respectively, and the weakest in autumn ($r = 0.76$, 0.76, and 0.75).

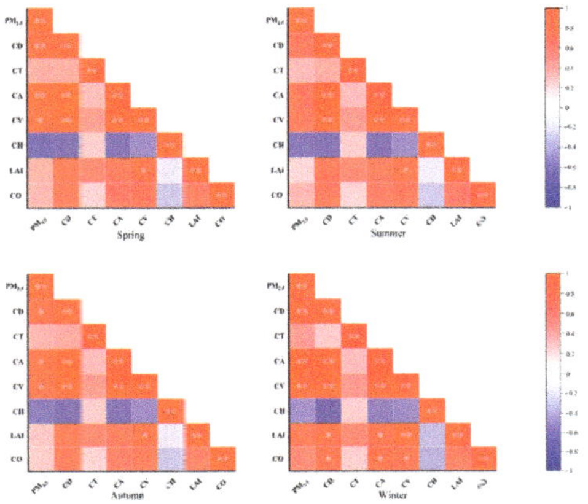

Figure 8. Correlation between $PM_{2.5}$ and canopy structure (CD, CT, CA, CV, CH, LAI, CO). Red indicates a positive correlation; purple indicates a negative correlation. The intensity of the color represents the strength of the correlation. Statistically significant correlations are marked with asterisks (* $p < 0.05$, ** $p < 0.01$).

Based on the results of the correlation analysis ($p < 0.05$), linear regression analysis was conducted for the canopy structure variables (Figure 9). The results indicate that $PM_{2.5}$ concentrations increase with CD, CA, and CV enlargement in all seasons. However, the range of variation in $PM_{2.5}$ is smaller in spring, while it is more pronounced in winter. The best model fits for CD, CA, and CV were observed in winter, with CA showing the best fit ($R^2 = 0.864$). In contrast, the fits for CD, CA, and CV were less optimal in autumn.

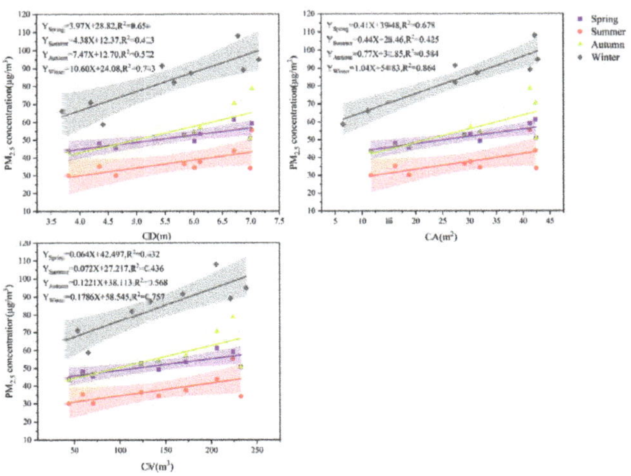

Figure 9. Linear regression of $PM_{2.5}$ with CD, CA, and CV across different seasons (spring, summer, fall, winter).

3.4. Primary Factors Influencing $PM_{2.5}$ Variation Across Seasons

Stepwise regression analysis was employed to identify the key factors influencing the seasonal variation in $PM_{2.5}$ concentrations, considering meteorological factors and canopy structure. The main influencing factors for $PM_{2.5}$ were relative humidity, temperature, precipitation, canopy height, canopy area, and canopy volume (Table 7). In spring, the analysis revealed that environmental temperature, canopy area, and precipitation significantly affected $PM_{2.5}$ levels ($F = 30.869$, df = [2,26], $p < 0.001$, adjusted $R^2 = 0.697$), explaining 69.7% of the variation in $PM_{2.5}$ concentrations. In summer, which exhibited the lowest average $PM_{2.5}$ levels, only canopy volume had a significant influence ($F = 6.545$, df = [1,26], $p < 0.001$, adjusted $R^2 = 0.175$), accounting for 17.57% of the variation. An increase of 1 m^2 in canopy volume resulted in an increase of 0.72 $\mu g/m^3$ in $PM_{2.5}$ concentrations. Both higher $PM_{2.5}$ levels and a greater number of influencing factors were observed in autumn and winter. In autumn, environmental temperature and canopy area had a significant impact on $PM_{2.5}$ ($F = 23.186$, df = [2,26], $p < 0.001$, adjusted $R^2 = 0.6561$), explaining 65.6% of the variation. The canopy area positively affected $PM_{2.5}$ concentrations ($B = 0.72$, $p < 0.001$, $t = 4.774$). In winter, in addition to environmental temperature and canopy area, relative humidity also had a significant effect ($F = 45.36$, df = [3,26], $p < 0.001$, adjusted $R^2 = 0.68$), explaining more than 68% of the variation in $PM_{2.5}$.

Table 7. Stepwise multiple linear regression results with $PM_{2.5}$ concentration as the dependent variable (independent variables include meteorological factors and canopy structure).

Season	Variable	B	SE	p	t	Overall R2	VIF
Spring	(Intercept)	65.27	4.416	<0.001	14.781	0.77	NA
	temperature	−2.077	0.280	<0.001	−7.423		1.702
	CA	0.389	0.076	<0.001	5.121		1.002
	precipitation	4.046	1.374	0.007	2.944		1.7
Summer	(Intercept)	27.298	4.444	<0.001	6.143	0.175	NA
	CV	0.072	0.028	0.017	2.558		1
Autumn	(Intercept)	65.613	8.529	<0.001	7.693	0.631	NA
	CA	0.72	0.151	<0.001	4.774		1.003
	temperature	−1.6	0.349	<0.001	−4.583		1.003
Winter	(Intercept)	263.079	27.084	<0.001	9.713	0.68	NA
	CA	0.864	0.131	<0.001	6.62		1.092
	humidity	−1.615	0.232	<0.001	−6.977		1.032
	temperature	−13.665	2.788	<0.001	−4.901		1.119

4. Discussion

4.1. Temporal Variations in $PM_{2.5}$

The increase in $PM_{2.5}$ concentrations is driven by several factors, including the rise in direct emission sources due to urbanization, the formation of secondary pollutants, the influence of natural conditions, and regional economic and environmental differences [26,27]. This study demonstrates that, from a seasonal perspective, winter $PM_{2.5}$ concentrations are significantly higher than in the other three seasons (Figure 5), exhibiting greater fluctuation (Figure 3), and the proportion of days with severe $PM_{2.5}$ levels is also the highest in winter (Figure 4). This finding is consistent with previous studies. Research by European scholars also indicates higher $PM_{2.5}$ concentrations in winter, with polycyclic aromatic hydrocarbons (PAHs) reaching their highest levels during this season. This phenomenon is likely linked to the increased demand for heating in winter, particularly the rise in coal combustion [20]. We attribute the significantly higher winter $PM_{2.5}$ levels in the study area to anthropogenic emissions and regional transport. The increased heating demand leads to large-scale coal, oil, and biomass usage in northern regions. Moreover, due to centralized heating in winter, China's Huai River policy suggests that northern areas experience significantly higher concentrations of total suspended particles (TSPs) than southern areas [18,28]. The study area, Tongxiang in Zhejiang Province, Eastern China, features low-lying terrain with plains and low hills. Under the influence of a subtropical monsoon climate, winter is characterized by cold, dry northeast winds from the north. We believe that these climatic conditions contribute to the significant increase in pollutant concentrations in winter. Research has shown that in Hangzhou, non-local pollutants contribute up to 70% to $PM_{2.5}$ concentrations. During winter, stable atmospheric conditions inhibit the vertical dispersion of pollutants, thereby enhancing the inter-regional transport of pollutants [29]. Furthermore, meteorological conditions play a key role in winter $PM_{2.5}$ levels. Low temperatures and stable weather conditions during winter hinder air circulation, making it more difficult for pollutants to disperse. Studies also indicate that low wind speeds and high humidity are meteorological factors that promote an increase in $PM_{2.5}$ concentrations [30,31].

$PM_{2.5}$ concentrations exhibit a distinct diurnal variation (Figure 6). We observed that in spring, summer, and autumn, $PM_{2.5}$ concentrations follow an "N"-shaped curve, peaking around 6:00–7:00 AM and reaching a minimum in the evening. In contrast, winter $PM_{2.5}$ concentrations follow a "W"-shaped curve, with a decrease observed at 5:00 AM. Similar findings have been reported in previous studies, where $PM_{2.5}$ concentrations in spring, under sunny and cloudy conditions, typically follow an "N"-shaped curve, with morning concentrations higher than those in the afternoon [32]. This pattern may be related to temperature inversion, a phenomenon in which the temperature near the ground is higher than the temperature of the air above it. This creates a "thermal cap," preventing pollutants from dispersing vertically [33], thus resulting in higher $PM_{2.5}$ concentrations at night compared to during the day. During the summer, stronger solar radiation leads to greater ground energy absorption, with less effective radiation and smaller temperature variations near the surface. This weakens the inversion effect, leading to a smaller diurnal variation in $PM_{2.5}$ concentrations. In winter, with weaker solar radiation and shorter daylight hours, the diurnal variation in $PM_{2.5}$ concentrations becomes more pronounced. The early morning decrease and subsequent rise in $PM_{2.5}$ concentrations during winter may also be related to temperature inversion and atmospheric stability. In the early morning, the ground rapidly cools, forming an inversion layer. Initially, the air is stable, and pollutants settle, causing a decrease in $PM_{2.5}$ concentrations. As time progresses, the ground temperature rises, enhancing air convection and bringing pollutants back to the surface, resulting in an increase in $PM_{2.5}$ levels. In contrast, higher temperatures in other seasons weaken the inversion effect, making it easier for pollutants to accumulate, leading to a continuous upward trend.

Furthermore, traffic flow fluctuations may also play a significant role in the variation of $PM_{2.5}$ concentrations [34]. During the early morning rush hour in all seasons, a clear peak in $PM_{2.5}$ concentrations is typically observed. During summer and autumn, $PM_{2.5}$ concentrations also show significant upward fluctuations in the evening. In comparison, spring and winter show less pronounced upward trends in $PM_{2.5}$ concentrations. This phenomenon may indicate that the influence of temperature inversion on air pollution outweighs the impact of traffic emissions. Therefore, it is crucial to consider not only the seasonal effects on air quality but also the diurnal variation of $PM_{2.5}$ concentrations when planning travel times. Understanding these seasonal variations and their underlying causes is essential for reducing individual exposure to high $PM_{2.5}$ environments and for formulating effective air quality management strategies.

4.2. $PM_{2.5}$ Variations Across Functional Zones

We observe a clear pattern in $PM_{2.5}$ concentrations across different seasons (Figure 5), with concentrations generally higher in the park and traffic zones compared to the residential zones. This may be attributed to traffic emissions and the intensity of outdoor activities. A study in Hong Kong found that diesel vehicles were responsible for approximately 70% to 90% of the vehicle-related $PM_{2.5}$ emissions in roadside environments [35], suggesting that the high traffic flow in the traffic zone contributes to increased particulate matter levels.

As a leisure and recreational area, the park zone is influenced not only by traffic emissions but also by pedestrian activity, which is a critical factor. Previous studies have shown a linear relationship between passenger density and $PM_{2.5}$ concentrations in subway carriages [36]. During the COVID-19 pandemic, reduced human activity led to a noticeable decrease in $PM_{2.5}$ pollution levels in China [37]. Although numerous studies suggest that vegetation can reduce $PM_{2.5}$ levels through processes such as adsorption and deposition [38,39], the higher $PM_{2.5}$ concentrations in the park zone, despite its relatively higher vegetation coverage compared to the residential and traffic zones, warrant further explanation. This phenomenon can be attributed to several factors. In addition to the effects of pedestrian activity and traffic flow, the adsorption capacity of vegetation has a threshold limit [40]. Moreover, forests have been found to reduce wind speed [41]. Research indicates that the canopy coverage provided by large trees can hinder street ventilation and heat dissipation [42]. Lower wind speeds inhibit the dispersion of $PM_{2.5}$, leading to its accumulation. Furthermore, residential zones are typically located farther from major pollution sources, which may partially explain the relatively lower $PM_{2.5}$ levels observed there. Thus, understanding the impact of vegetation on $PM_{2.5}$ concentrations is critical.

Notably, in the winter daily variation (Figure 6), the traffic zone exhibits the lowest $PM_{2.5}$ concentrations during the early morning hours. This phenomenon can be attributed to a reduction in traffic flow during winter due to cold weather, snowfall, and other related factors [43], which leads to a decrease in vehicle emissions. Previous research has demonstrated a declining trend in traffic volume as the severity of cold weather increases. In extremely cold conditions, the daily traffic volume is estimated to decrease by approximately 30% [44]. Additionally, the size of open spaces may influence wind speed, as studies have shown a positive correlation between the street view factor (SVF) and wind speed. Specifically, a 0.1 increase in SVF is associated with a 0.19 m/s increase in wind speed [45]. Traffic zones are generally more open, allowing for higher wind speeds that facilitate the dispersion of pollutants. These factors collectively contribute to the lower PM2.5 concentrations observed in traffic zones compared to park and residential zones during winter.

Contrary to common perceptions, areas with higher vegetation coverage in urban settings do not necessarily exhibit lower $PM_{2.5}$ concentrations. In fact, such areas may display higher concentrations due to the complex interactions between vegetation and urban pollution sources. Multiple factors can influence the relationship between vegetation coverage and $PM_{2.5}$ concentrations. Therefore, a comprehensive understanding of the ecological functions of vegetation and its interactions with urban pollution sources is essential. Further empirical studies are needed to explore this intricate relationship in greater detail.

4.3. The Influence of Meteorological Factors and Canopy Structure on $PM_{2.5}$ Across Different Seasons

Given the seasonal variations in $PM_{2.5}$ concentrations, identifying the specific factors influencing them and their relative magnitudes is crucial. In this study, we analyzed meteorological factors (temperature, humidity, precipitation) and canopy structure (tree height, canopy diameter, canopy thickness, canopy area, canopy volume, leaf area index, and canopy cover). We found seasonal correlations between $PM_{2.5}$ and temperature, relative humidity, and precipitation. Notably, temperature was negatively correlated with $PM_{2.5}$ concentrations in spring, autumn, and winter (Table 6), which contrasts with the findings of many other studies that have observed a positive correlation between temperature and $PM_{2.5}$ levels. These studies suggest that higher temperatures could increase photochemical reactions between atmospheric precursors, resulting in the formation of more particulate matter [46]. However, Jeremy Aw [47] noted that particulate matter tends to peak and then decrease with rising temperatures, as increased temperatures not only promote the formation of particles but also enhance the volatilization loss of these substances. Consequently, in Tongxiang, we observed that $PM_{2.5}$ levels decreased as temperatures increased. A similar trend was observed between humidity and $PM_{2.5}$ levels in autumn and winter, where increasing humidity corresponded with decreasing $PM_{2.5}$ levels. This could be explained by the fact that increased relative humidity leads to the combination of water vapor with aerosol particles, causing them to coagulate and grow, which affects their behavior in the atmosphere and their settling rates. As shown in the box plots, there were no significant differences in humidity between autumn and winter, suggesting that the impact of humidity on $PM_{2.5}$ was consistent across these two seasons. In the correlation analysis, precipitation was found to be negatively correlated with $PM_{2.5}$ only in autumn. Most studies suggest that precipitation significantly reduces $PM_{2.5}$ levels through coagulation and deposition processes [48]. Precipitation may also indirectly reduce $PM_{2.5}$ levels by increasing wind speed and humidity.

Although many studies have demonstrated that trees can reduce $PM_{2.5}$ levels [49], we observed a higher $PM_{2.5}$ concentration in park zones, which have higher tree canopy cover, compared to other zones. Trees primarily reduce $PM_{2.5}$ by adsorbing particles through their leaves, which are a key component of the canopy structure. In this study, we explored the relationship between canopy structure and $PM_{2.5}$ concentrations (Figure 8). Surprisingly, we found a positive correlation between canopy diameter, canopy area, and canopy volume with $PM_{2.5}$ concentrations. Previous research has primarily focused on the effect of individual tree leaves on $PM_{2.5}$ adsorption and deposition or on assessing the reduction in $PM_{2.5}$ through comparisons between study sites and control areas [50,51]. However, there has been limited attention to the complex effects of overall forest structure on $PM_{2.5}$ levels. While trees can remove $PM_{2.5}$, this effect has a threshold [40], and during periods of high pollution, especially in winter, the tree canopy's capacity for purification is often exceeded. Additionally, dense branches and leaves reduce wind speed, slowing the diffusion of air pollutants [52]. Increasing canopy area and volume may also block sunlight, lowering surface temperatures. A decrease in temperature within certain limits can lead to higher $PM_{2.5}$ levels. The canopy can also intercept some precipitation, altering its intensity and potentially increasing $PM_{2.5}$ concentrations [53]. Studies have indicated that leaf IPMC (particulate matter concentration intercepted by tree leaves) decreases as $PM_{2.5}$ concentration increases. This suggests that the higher $PM_{2.5}$ concentrations in park areas may reduce the ability of leaves to capture IPMC, despite the higher pollution levels in these areas [17]. Furthermore, we observed that in winter, $PM_{2.5}$ levels were more sensitive to changes in canopy diameter, canopy area, and canopy volume compared to autumn (Figure 9). This is likely due to higher $PM_{2.5}$ concentrations during autumn and winter, coupled with lower temperatures and atmospheric pressure conditions that slow the settling of pollutants, thus enhancing the canopy's influence on $PM_{2.5}$ concentrations. As a result, the effects of canopy diameter, canopy area, and canopy volume on $PM_{2.5}$ concentrations were more pronounced during these seasons.

The stepwise multiple linear regression analysis results (Table 7) revealed that the seasonal responses of $PM_{2.5}$ to canopy area, canopy volume, temperature, and humidity varied. We found that as temperature and humidity increased, $PM_{2.5}$ concentrations decreased, while increases in canopy area and volume led to higher $PM_{2.5}$ concentrations. Comparing meteorological factors with canopy structure, we found that the impact of canopy structure on $PM_{2.5}$ was smaller than that of meteorological factors, primarily because meteorological variables such as wind speed, humidity, and temperature play more significant roles in air quality. Meteorological conditions directly influence the dispersion, deposition, and chemical reactions of particulate matter [54], while the canopy, though capable of providing some shading and filtration, is often overshadowed by the effects of meteorological variables. Research has shown that during high pollution periods, the dry deposition capacity of plant leaves reaches a saturation point within ten days, which may explain the higher $PM_{2.5}$ levels in winter [55]. Additionally, the more frequent and significant changes in meteorological factors allow them to influence $PM_{2.5}$ concentrations and distribution over larger spatial scales. Thus, meteorological factors and variables such as canopy diameter, area, and volume within the study area must be considered when obtaining and predicting $PM_{2.5}$ data. Furthermore, from the perspective of urban green space management, selecting tree species with smaller canopy volumes or conducting regular pruning can help reduce $PM_{2.5}$ concentrations.

4.4. Limitations

Despite this study revealing the influence of meteorological factors and urban forest structure on $PM_{2.5}$ concentrations, there are several limitations. Due to the high maintenance costs associated with fixed meteorological stations, the number of sample points in this study was limited. Moreover, since $PM_{2.5}$ concentrations are influenced by multiple factors [56], future research should incorporate more variables and environmental factors (e.g., sunlight, traffic flow, tree structure), as well as longer monitoring periods. A more detailed investigation of the forest structure within study sites is also needed to account for the potential impacts of improper or appropriate prunin and the effects of different tree species. To further understand the regulation of $PM_{2.5}$ by urban forests, future studies could consider broader data collection in different types of urban functional zones, particularly in industrial and traffic-dense areas. Additionally, atmospheric diffusion models could be introduced to quantify the dispersion and deposition processes of $PM_{2.5}$ in various geographic settings.

5. Conclusions

This study quantified the temporal variations of $PM_{2.5}$ concentrations across different functional zones and explored the relationships between canopy structure, meteorological factors, and $PM_{2.5}$ levels. The results indicate that $PM_{2.5}$ pollution is most severe in winter, exhibiting large fluctuations (84.00 ± 45.97 μg/m^3), while it is lower and relatively stable in summer (36.85 ± 17.63 μg/m^3). Daily $PM_{2.5}$ variations follow an "N-shaped" curve in spring, summer, and autumn, while a "W-shaped" curve is observed in winter. In different functional zones, $PM_{2.5}$ concentrations in park zones (spring: 53.54 ± 21.14 μg/m^3; summer: 42.7 ± 21.3 μg/m^3; autumn: 60.4 ± 35.1 μg/m^3; winter: 85.8 ± 48.2 μg/m^3) are consistently higher than those in residential zones (spring: 47.9 ± 19.5 μg/m^3; summer: 32.7 ± 15.5 μg/m^3; autumn: 49.4 ± 28.2 μg/m^3; winter: 81.6 ± 48.9 μg/m^3) and traffic zones (spring: 53.21 ± 21.6 μg/m^3; summer: 38.3 ± 17.2 μg/m^3; autumn: 58.9 ± 31.2 μg/m^3; winter: 93.9 ± 20.6 μg/m^3), with traffic zones showing intermediate levels. Correlations were also observed between meteorological factors, canopy structure, and $PM_{2.5}$ concentrations. In spring, autumn, and winter, temperature was moderately negatively correlated with $PM_{2.5}$ ($r < -0.48$, $p < 0.05$), while tree height, canopy area, and canopy volume were strongly positively correlated with $PM_{2.5}$ levels ($r > 0.75$, $p < 0.05$). Humidity was moderately negatively correlated with $PM_{2.5}$ during autumn and winter, and precipitation also showed a moderate negative correlation with $PM_{2.5}$ in winter. $PM_{2.5}$ levels in winter were more sensitive to changes in canopy structure, followed by autumn. Furthermore, the impact of canopy structure on $PM_{2.5}$ was found to be much smaller than that of meteorological factors. In particular, every 1 m^2 increase in canopy area could raise $PM_{2.5}$ levels by 0.864 $\mu g/m^3$, whereas an average increase of 1 mm in rainfall could raise $PM_{2.5}$ by 13.665 $\mu g/m^3$. This study primarily focused on the trends of $PM_{2.5}$ in small urban forests within city ecosystems. Understanding the influence of meteorological conditions and canopy structure on $PM_{2.5}$ can help inform decisions regarding optimal travel times and appropriate protective measures and provide guidance for future efforts to improve air quality, optimize urban green space design, and enhance public health.

Author Contributions: Y.H.: Writing—original draft, Writing—review and editing, Investigation, Conceptualization. Y.C.: Writing—original draft, Writing—review and editing, Investigation, Conceptualization. J.J.: Writing—review and editing. C.P.: Writing—review and editing. G.W.: Writing—review and editing. C.L.: Writing—review and editing. Z.J.: Investigation. Z.C.: Software, Data curation. Y.Z.: Supervision, Project administration. G.Z.: Writing—review and editing, Formal analysis, Supervision, Conceptualization, Methodology, Project administration. All authors have read and agreed to the published version of the manuscript.

Funding: This research was supported by the Key Research and Development Program of Zhejiang Province (Grant Nos. 2022C03039 and 2021C02005) and the National Natural Science Foundation of China (Grant No. 32401573).

Data Availability Statement: The data that support the findings of this study are available from the corresponding author, Guomo Zhou, upon reasonable request.

Acknowledgments: We appreciated the help of Guomo Zhou's team in the canopy structure data collection, and we thank the anonymous reviewers for their constructive comments and suggestions.

Conflicts of Interest: Yue Cai is employed by Ehon Carbon Technologies Company Limited. However, her employer's company was not involved in this study, and there is no relevance between this research and the company.

Abbreviations

The following abbreviations are used in this manuscript:

CD	Canopy diameter
CT	Canopy thickness
CA	Canopy area
CV	Canopy volume
CH	Canopy height ratio
LAI	Leaf area index
CO	Canopy cover

References

1. Pannell, C.W. China's Continuing Urban Transition. *Environ. Plan. A* **2002**, *34*, 1571–1589. [CrossRef]
2. Li, Z.; LeGates, R.; Min, Z.; Jinxin, L. Understanding China's Urbanization: The Great Demographic, Spatial, Economic, and Social Transformation. *China City Plan. Rev.* **2018**, *27*, 79. [CrossRef]
3. Hu, M.; Li, X.; Xu, Y.; Huang, Z.; Chen, C.; Chen, J.; Du, H. Remote sensing monitoring of the spatiotemporal dynamics of urban forest phenology and its response to climate and urbanization. *Urban Clim.* **2024**, *53*, 101810. [CrossRef]
4. Nowak, D.J.; Hirabayashi, S.; Doyle, M.; McGovern, M.; Pasher, J. Air pollution removal by urban forests in Canada and its effect on air quality and human health. *Urban For. Urban Green.* **2018**, *29*, 40–48. [CrossRef]
5. Adhikari, S.; Ma, Q.; Poudel, K.; Renninger, H.J. Aboveground woody biomass estimation of young bioenergy plantations of Populus and its hybrids using mobile (backpack) LiDAR remote sensing. *Trees For. People* **2024**, *18*, 100665. [CrossRef]
6. Ko, C.; Lee, S.; Yim, J.; Kim, D.; Kang, J. Comparison of Forest Inventory Methods at Plot-Level between a Backpack Personal Laser Scanning (BPLS) and Conventional Equipment in Jeju Island, South Korea. *Forests* **2021**, *12*, 308. [CrossRef]

7. Xie, C.; Kan, L.; Guo, J.; Jin, S.; Li, Z.; Chen, D.; Li, X.; Che, S. A dynamic processes study of PM retention by trees under different wind conditions. *Environ. Pollut.* **2018**, *233*, 315–322. [CrossRef]
8. Xie, C.; Guo, J.; Yan, L.; Jiang, R.; Liang, A.; Che, S. The influence of plant morphological structure characteristics on $PM_{2.5}$ retention of leaves under different wind speeds. *Urban For. Urban Green.* **2022**, *71*, 127556. [CrossRef]
9. Han, L.; Zhao, J.; Gao, Y.; Gu, Z.; Xin, K.; Zhang, J. Spatial distribution characteristics of $PM_{2.5}$ and PM_{10} in Xi'an City predicted by land use regression models. *Sustain. Cities Soc.* **2020**, *61*, 102329. [CrossRef]
10. Wu, H.; Yang, C.; Chen, J.; Yang, S.; Lu, T.; Lin, X. Effects of Green space landscape patterns on particulate matter in Zhejiang Province, China. *Atmos. Pollut. Res.* **2018**, *9*, 923–933. [CrossRef]
11. Xing, X.; Chen, Z.; Tian, Q.; Mao, Y.; Liu, W.; Shi, M.; Cheng, C.; Hu, T.; Zhu, G.; Li, Y.; et al. Characterization and source identification of $PM_{2.5}$-bound polycyclic aromatic hydrocarbons in urban, suburban, and rural ambient air, central China during summer harvest. *Ecotoxicol. Environ. Saf.* **2020**, *191*, 110219. [CrossRef] [PubMed]
12. Lu, S.-J.; Wang, D.; Wang, Z.; Li, B.; Peng, Z.-R.; Li, X.-B.; Gao, Y. Investigating the Role of Meteorological Factors in the Vertical Variation in $PM_{2.5}$ by Unmanned Aerial Vehicle Measurement. *Aerosol Air Qual. Res.* **2019**, *19*, 1493–1507. [CrossRef]
13. Santos-Silva, J.C.; Potgieter-Vermaak, S.; Medeiros, S.H.W.; da Silva, L.V.; Ferreira, D.V.; Moreira, C.A.B.; de Souza Zorzenão, P.C.; Pauliquevis, T.; Godoi, A.F.L.; de Souza, R.A.F.; et al. A new strategy for risk assessment of $PM_{2.5}$-bound elements by considering the influence of wind regimes. *Sci. Total Environ.* **2023**, *872*, 162131. [CrossRef]
14. Cai, Y.; Li, C.; Ye, L.; Xiao, L.; Gao, X.; Mo, L.; Du, H.; Zhou, Y.; Zhou, G. Effect of the roadside tree canopy structure and the surrounding on the daytime urban air temperature in summer. *Agric. For. Meteorol.* **2022**, *316*, 108850. [CrossRef]
15. Dong, Y.; Cai, Y.; Li, C.; Wang, H.; Zhou, L.; Sun, J.; Li, C.; Song, B.; Zhou, G. Vertical thermal environment of subtropical broad-leaved urban forests and the influence of canopy structure. *Build. Environ.* **2022**, *224*, 109521. [CrossRef]
16. Hu, T.; Yang, J.; Li, X.; Gong, P. Mapping Urban Land Use by Using Landsat Images and Open Social Data. *Remote Sens.* **2016**, *8*, 151. [CrossRef]
17. Zafra-Mejía, C.; Suárez-López, J.; Rondón-Quintana, H. Analysis of Particulate Matter Concentration Intercepted by Trees of a Latin-American Megacity. *Forests* **2021**, *12*, 723. [CrossRef]
18. Chen, Y.; Ebenstein, A.; Greenstone, M.; Li, H. Evidence on the impact of sustained exposure to air pollution on life expectancy from China's Huai River policy. *Proc. Natl. Acad. Sci. USA* **2013**, *110*, 12936–12941. [CrossRef]
19. Gheshlaghpoor, S.; Amin Zadeh Gohar Rizi, B. Effects of Urban Compactness and Complexity on $PM_{2.5}$ Concentration: The Case Study of Tehran Metropolis. *Int. J. Environ. Res.* **2025**, *19*, 70. [CrossRef]
20. Siudek, P. Seasonal distribution of $PM_{2.5}$-bound polycyclic aromatic hydrocarbons as a critical indicator of air quality and health impact in a coastal-urban region of Poland. *Sci. Total Environ.* **2022**, *827*, 154375. [CrossRef]
21. Rattigan, O.V.; Civerolo, K.L.; Felton, H.D. Trends in wet precipitation, particulate, and gas-phase species in New York State. *Atmos. Pollut. Res.* **2017**, *8*, 1090–1102. [CrossRef]
22. Chen, M.; Jung, J.; Lee, Y.K.; Hur, J. Surface accumulation of low molecular weight dissolved organic matter in surface waters and horizontal off-shelf spreading of nutrients and humic-like fluorescence in the Chukchi Sea of the Arctic Ocean. *Sci. Total Environ.* **2018**, *639*, 624–632. [CrossRef]
23. Chai, L.; Ding, C.; Li, J.; Yang, Z.; Shi, Y. Multi-omics response of Pannonibacter phragmitetus BB to hexavalent chromium. *Environ. Pollut.* **2019**, *249*, 63–73. [CrossRef] [PubMed]
24. Kong, F.; Yan, W.; Zheng, G.; Yin, H.; Cavan, G.; Zhan, W.; Zhang, N.; Cheng, L. Retrieval of three-dimensional tree canopy and shade using terrestrial laser scanning (TLS) data to analyze the cooling effect of vegetation. *Agric. For. Meteorol.* **2016**, *217*, 22–34. [CrossRef]
25. Li, Y.; Guo, Q.; Tao, S.; Zheng, G.; Zhao, K.; Xue, B.; Su, Y. Derivation, Validation, and Sensitivity Analysis of Terrestrial Laser Scanning-Based Leaf Area Index. *Can. J. Remote Sens.* **2016**, *42*, 719–729. [CrossRef]
26. Mahapatra, P.S.; Panda, U.; Mallik, C.; Boopathy, R.; Jain, S.; Sharma, S.K.; Mandal, T.K.; Senapati, S.; Satpathy, P.; Panda, S.; et al. Chemical, microstructural, and biological characterization of wintertime $PM_{2.5}$ during a land campaign study in a coastal city of eastern India. *Atmos. Pollut. Res.* **2021**, *12*, 101164. [CrossRef]
27. Dawidowski, L.; Gelman Constantin, J.; Herrera Murillo, J.; Gómez-Marín, M.; Nogueira, T.; Blanco Jiménez, S.; Díaz-Suárez, V.; Baraldo Victorica, F.; Lichtig, P.; Díaz Resquin, M.; et al. Carbonaceous fraction in $PM_{2.5}$ of six Latin American cities: Seasonal variations, sources and secondary organic carbon contribution. *Sci. Total Environ.* **2024**, *948*, 174630. [CrossRef]
28. Gong, S.Y. Analysis on the Metal Element in the Atmospheric Dustfall at Winter in Linyi. *Adv. Mater. Res.* **2014**, *955–959*, 1176–1178. [CrossRef]
29. Gao, L.; Zhang, R.; Han, Z.; Fu, C.; Yan, P.; Wang, T.; Hong, S.; Jiao, L. A Modeling Study of a Typical Winter $PM_{2.5}$ Pollution Episode in a City in Eastern China. *Aerosol Air Qual. Res.* **2014**, *14*, 311–322. [CrossRef]

30. Chen, L.W.A.; Watson, J.G.; Chow, J.C.; Green, M.C.; Inouye, D.; Dick, K. Wintertime particulate pollution episodes in an urban valley of the Western US: A case study. *Atmos. Chem. Phys.* **2012**, *12*, 10051–10064. [CrossRef]
31. Chuang, M.-T.; Chou, C.C.K.; Lin, C.-Y.; Lee, J.-H.; Lin, W.-C.; Li, M.-H.; Chen, S.-W.; Chang, S.-H. $PM_{2.5}$ episodes in northern Taiwan under southerly winds in late winter. *Atmos. Res.* **2024**, *311*, 107686. [CrossRef]
32. Wang, X.; Wang, C.; Gu, L.; Wang, X.; Wang, Y. Study on the Distribution Characteristics of $PM_{2.5}$ Mass Concentration in Urban Street Green Spaces under Different Weather Conditions in Spring. *Ecol. Environ. Sci.* **2014**, *23*, 972–978. [CrossRef]
33. Yin, P.-Y.; Chang, R.-I.; Day, R.-F.; Lin, Y.-C.; Hu, C.-Y. Improving $PM_{2.5}$ Concentration Forecast with the Identification of Temperature Inversion. *Appl. Sci.* **2021**, *12*, 71. [CrossRef]
34. Khataan, A.; Idris, A.O.; Abo Elazm, M.M.; Kamel, W. Estimating the contribution of traffic flow to pollutant concentrations. *IOP Conf. Ser. Mater. Sci. Eng.* **2021**, *836*, 012002. [CrossRef]
35. Wong, Y.K.; Huang, X.H.H.; Louie, P.K.K.; Yu, A.L.C.; Chan, D.H.L.; Yu, J.Z. Tracking separate contributions of diesel and gasoline vehicles to roadside $PM_{2.5}$ through online monitoring of volatile organic compounds and $PM_{2.5}$ organic and elemental carbon: A 6-year study in Hong Kong. *Atmos. Chem. Phys.* **2020**, *20*, 9871–9882. [CrossRef]
36. Gao, Y.; Chen, F.; Wang, Z. The distribution and influential factors of $PM_{2.5}$ and CO_2 in urban rail carriages. *Indoor Built Environ.* **2019**, *28*, 1383–1395. [CrossRef]
37. Tan, Z.; Li, X.; Gao, M.; Jiang, L. The Environmental Story During the COVID-19 Lockdown: How Human Activities Affect $PM_{2.5}$ Concentration in China? *IEEE Geosci. Remote Sens. Lett.* **2022**, *19*, 1001005. [CrossRef]
38. Kwak, M.J.; Lee, J.; Kim, H.; Park, S.; Lim, Y.; Kim, J.E.; Baek, S.G.; Seo, S.M.; Kim, K.N.; Woo, S.Y. The Removal Efficiencies of Several Temperate Tree Species at Adsorbing Airborne Particulate Matter in Urban Forests and Roadsides. *Forests* **2019**, *10*, 960. [CrossRef]
39. Liu, C.; Dai, A.; Zhang, H.; Sheng, Q.; Zhu, Z. Study on the Correlation Mechanism between the Living Vegetation Volume of Urban Road Plantings and $PM_{2.5}$ Concentrations. *Sustainability* **2023**, *15*, 4653. [CrossRef]
40. Zhang, W.; Zhang, Z.; Meng, H.; Zhang, T. How Does Leaf Surface Micromorphology of Different Trees Impact Their Ability to Capture Particulate Matter? *Forests* **2018**, *9*, 681. [CrossRef]
41. Zhang, J.; Jia, G.; Liu, Z.; Wang, D.; Yu, X. Populus simonii Carr. Reduces Wind Erosion and Improves Soil Properties in Northern China. *Forests* **2019**, *10*, 315. [CrossRef]
42. Zhao, Y.; Li, H.; Bardhan, R.; Kubilay, A.; Li, Q.; Carmeliet, J. The time-evolving impact of tree size on nighttime street canyon microclimate: Wind tunnel modeling of aerodynamic effects and heat removal. *Urban Clim.* **2023**, *49*, 101528. [CrossRef]
43. Datla, S.; Sahu, P.; Roh, H.-J.; Sharma, S. A Comprehensive Analysis of the Association of Highway Traffic with Winter Weather Conditions. *Procedia-Soc. Behav. Sci.* **2013**, *104*, 497–506. [CrossRef]
44. Datla, S.; Sharma, S. Impact of cold and snow on temporal and spatial variations of highway traffic volumes. *J. Transp. Geogr.* **2008**, *16*, 358–372. [CrossRef]
45. Jin, H.; Qiao, L.; Cui, P. Study on the Effect of Streets' Space Forms on Campus Microclimate in the Severe Cold Region of China-Case Study of a University Campus in Daqing City. *Int. J. Environ. Res. Public Health* **2020**, *17*, 8389. [CrossRef]
46. Megaritis, A.G.; Fountoukis, C.; Charalampidis, P.E.; Pilinis, C.; Pandis, S.N. Response of fine particulate matter concentrations to changes of emissions and temperature in Europe. *Atmos. Chem. Phys.* **2013**, *13*, 3423–3443. [CrossRef]
47. Aw, J.; Kleeman, M.J. Evaluating the first-order effect of intraannual temperature variability on urban air pollution. *J. Geophys. Res. Atmos.* **2003**, *108*, 4365–4375. [CrossRef]
48. Speirs, N.B.; Belden, J.L.; Hellum, A.M. The capture of airborne particulates by rain. *J. Fluid Mech.* **2023**, *958*, A40. [CrossRef]
49. Sgrigna, G.; Baldacchini, C.; Dreveck, S.; Cheng, Z.; Calfapietra, C. Relationships between air particulate matter capture efficiency and leaf traits in twelve tree species from an Italian urban-industrial environment. *Sci. Total Environ.* **2020**, *718*, 137310. [CrossRef]
50. Kwak, M.J.; Lee, J.K.; Park, S.; Kim, H.; Lim, Y.J.; Lee, K.-A.; Son, J.-A.; Oh, C.-Y.; Kim, I.; Woo, S.Y. Surface-Based Analysis of Leaf Microstructures for Adsorbing and Retaining Capability of Airborne Particulate Matter in Ten Woody Species. *Forests* **2020**, *11*, 946. [CrossRef]
51. Hong, J.; Lee, M.; Huh, W.; Kim, T.K.; Jeon, J.; Lee, H.; Kim, K.; Byeon, S.; Park, C.; Kim, H.S. Comparisons of $PM_{2.5}$ mitigation with stand characteristics between evergreen Korean pine plantations and deciduous broad-leaved forests in the Republic of Korea. *Environ. Pollut.* **2023**, *334*, 122240. [CrossRef] [PubMed]
52. Liu, Z.; Shen, L.; Yan, C.; Du, J.; Li, Y.; Zhao, H. Analysis of the Influence of Precipitation and Wind on $PM_{2.5}$ and PM_{10} in the Atmosphere. *Adv. Meteorol.* **2020**, *2020*, 1–13. [CrossRef]
53. Tian, X.; Cui, K.; Sheu, H.-L.; Hsieh, Y.-K.; Yu, F. Effects of Rain and Snow on the Air Quality Index, $PM_{2.5}$ Levels, and Dry Deposition Flux of PCDD/Fs. *Aerosol Air Qual. Res.* **2021**, *21*, 210158. [CrossRef]
54. Paraschiv, S.; Paraschiv, L.S. Effects of wind speed, relative humidity, temperature and air pressure on PM_{10} concentration for an urban background area. *IOP Conf. Ser. Mater. Sci. Eng.* **2019**, *595*, 012059. [CrossRef]

55. Liu, R.; Wang, M.; Chen, S.; Zhang, J.; Jin, X.; Ren, Y.; Chen, J. Historical Pollution Exposure Impacts on $PM_{2.5}$ Dry Deposition and Physiological Responses in Urban Trees. *Forests* **2024**, *15*, 1614. [CrossRef]
56. Liu, Y.; Shi, G.; Zhan, Y.; Zhou, L.; Yang, F. Characteristics of $PM_{2.5}$ spatial distribution and influencing meteorological conditions in Sichuan Basin, southwestern China. *Atmos. Environ.* **2021**, *253*, 118364. [CrossRef]

forests

MDPI

Article

Factors Affecting Dust Retention in Urban Parks Across Site and Vegetation Community Scales

Xiang Zhang, Chuanwen Wang, Jiangshuo Guo, Zhongzhen Zhu, Zihan Xi, Xiaohan Li, Ling Qiu and Tian Gao *

College of Landscape Architecture and Art, Northwest A&F University, No. 3 Taicheng Road, Yangling 712100, China; xiangzhang@nwafu.edu.cn (X.Z.); 2022056719@nwafu.edu.cn (C.W.); guojiangshuo@nwafu.edu.cn (J.G.); zhongyuchen@nwafu.edu.cn (Z.Z.); xizihan@nwafu.edu.cn (Z.X.); 2024060911@nwafu.edu.cn (X.L.); qiuling@nwsuaf.edu.cn (L.Q.)

* Correspondence: tian.gao@nwsuaf.edu.cn; Tel.: +86-29-87082997

Abstract: Air pollution poses a significant threat to human health, especially in urban areas. Urban parks function as natural biofilters, and examining the factors influencing dust retention—specifically PM2.5 and PM10 concentrations—across different spatial scales can enhance air quality and resident well-being. This study investigates the factors affecting dust retention in urban parks at both the site and vegetation community scales, focusing on Xi'an Expo Park. Through on-site measurements and a land use regression (LUR) model, the spatial and temporal distributions of PM2.5 and PM10 concentrations were analyzed. The indications of the findings are as follows. (1) The LUR model effectively predicts factors influencing PM2.5 and PM10 concentrations at the site scale, with adjusted R^2 values ranging from 0.482 to 0.888 for PM2.5 and 0.505 to 0.88 for PM10. Significant correlations were found between particulate matter concentrations and factors such as the distance from factories, sampling area size, distance from main roads, presence of green spaces, and extent of hard pavements. (2) At the plant community scale, half-closed (30%–70% canopy cover), single-layered green spaces demonstrated the superior regulation of PM2.5 and PM10 concentrations. Specifically, two vegetation structures—the half-closed single-layered mixed broadleaf-conifer woodland (H1M) and the half-closed single-layered broad-leaved woodland (H1B)—exhibited the highest dust-retention capacities. (3) PM2.5 and PM10 concentrations were highest in winter, followed by spring and autumn, with the lowest levels recorded in summer. Daily particulate matter concentrations peaked between 8:00 and 10:00 a.m. and gradually decreased, reaching a minimum between 4:00 and 6:00 p.m. The objective of this study is to evaluate the impact of urban green spaces on particulate matter (PM) concentrations across multiple scales. By identifying and synthesizing key indicators at these various scales, the research aims to develop effective design strategies for urban green spaces and offer a robust theoretical framework to support the creation of healthier cities. This multi-scale perspective deepens our understanding of how urban planning and landscape architecture can play a critical role in mitigating air pollution and promoting public health.

Keywords: urban park; vegetation structure; green space; land use regression model; PM2.5; PM10

1. Introduction

Air pollution is a significant global health concern, particularly in rapidly urbanization regions. In recent decades, many Chinese cities have experienced severe air pollution, primarily attributed to increased traffic and industrial emissions [1,2]. This issue is especially critical in urban areas, where outdoor air pollution poses serious health risks [3–5]. The International Agency for Research on Cancer, a branch of the World Health Organization (WHO), has classified outdoor air pollution and particulate matter as carcinogenic. The WHO's updated Global Air Quality Guidelines for 2021 reveal that 99% of the world's population breathes air exceeding recommended pollution limits, with approximately 7 million deaths annually attributed to air pollution—over half occurring in developing countries [6].

Particulate matter (PM), particularly that with a diameter less than or equal to 10 μm (PM10) and 2.5 μm (PM2.5), poses a serious threat to public health [7,8]. These airborne particles are linked to an increased risk of lung cancer, cardiovascular and respiratory diseases [9]. PM2.5 remains suspended in the atmosphere for longer periods, adsorbing more harmful substances and penetrating deeper into the respiratory system, affecting the bronchi and alveoli, which can lead to severe health issues [10]. Vulnerable populations, such as the elderly, pregnant women, and infants, are particularly at risk. Exposure to PM2.5 can result in decreased cardiopulmonary function in older adults [11], premature births in pregnant women [12], and increased rates of asthma in children [13]

Controlling air pollution at the source remains challenging. Urban green spaces are vital components of the urban environment and represent the only ecosystems in cities with air-purifying functions [14]. These areas provide healthy outdoor spaces for residents and play a crucial role in regulating the urban ecological environment through thoughtful planning and design [15]. Urban parks, in particular, serve as natural biofilters, absorbing and retaining particulate matter, thereby improving air quality and enhancing resident well-being.

Research has explored the relationship between green space and particulate matter concentrations across various scales, including site, vegetation community, and individual plant levels [16]. At the site-scale, urban air pollution is influenced not only by road traffic and industrial emissions but also by factors such as land-use type, meteorological conditions, and population density [17]. To model air particulate matter distribution more accurately at this scale, methodologies like remote sensing image inversion [18], spatial interpolation [19], machine learning models [20], and land use regression (LUR) models [17] have been developed. Among these, LUR models are particularly effective as they consider both the source and influencing factors of particulate matter and simulate air particulate levels based on geographic variables [21]. This approach effectively elucidates the spatial and temporal distribution characteristics of particulate matter [22], and buffer size in an LUR model can be adjusted to the study scale [23]. Most existing LUR research has focused on large and medium scales. For instance, LUR models have been employed to predict the spatial and temporal trends of PM2.5 in Guangzhou City [24] and to analyze the hourly predictions of NO_2 and PM2.5 concentrations in the Netherlands [25]. Zhu and Zeng [26] examined the impact of lakes on the distribution of PM2.5 and PM10 concentrations in Wuhan, while Gao et al. [27] investigated the relationship between community air pollution and urban form. However, there is a notable scarcity of studies addressing small-scale pollutant distribution using LUR modeling.

At the vegetation community scale, studies on dust retention effects have primarily focused on plant composition and type. Wu et al. [28] found that combinations of trees and shrubs had a PM2.5 sinking capacity twice that of single tree or shrub configurations. Zhu and Zeng [26] reported that tree–shrub–herb combinations were more effective in reducing PM2.5 concentrations compared to other vegetation types. Some researchers argue that broad-leaved forests are superior to coniferous forests in reducing air particulate matter concentrations [29]. This advantage is attributed to the lower stomatal densities of broad-leaved trees, which enhance their ability to regulate PM2.5 concentrations [30]. However, Wu et al. [31] found that conifers exhibit stronger deposition rates for particulate matter due to their morphological and physiological traits, which enhance their capacity to capture airborne pollutants [32]. Conversely, Italian researchers have found that the particulate matter removal capacity follows the following order: broadleaved and coniferous mixed forests > coniferous forests > evergreen forests > deciduous forests. In contrast, Chinese studies report a different ranking, with coniferous forests exhibiting the highest dust retention capacity, followed by evergreen forests, deciduous forests, and finally broadleaved and coniferous mixed forests [33]. These varying conclusions can be attributed to the insufficient quantification of study indicators and a lack of detailed criteria for categorizing vegetation structures.

Our study fills a significant research gap by adopting a multi-scale approach to investigate the effects of urban green spaces on PM concentrations. Most previous research has been conducted at a single scale—either at the site level or the plant community level—due to constraints such as experimental time and spatial limitations. This has often led to inconsistent conclusions and a limited understanding of the integrated role of urban green spaces in air quality regulation.

Comprehensive studies spanning multiple scales are scarce. By studying the effect of green spaces on PM concentrations through a multi-scale approach, we aim to better explain how urban green spaces function as integrated systems in mitigating air pollution. Specifically, our research investigates the factors influencing PM2.5 and PM10 concentrations in urban parks at both the site and plant community scales.

We utilized a combination of on-site measurements and LUR model, focusing on Xi'an Expo Park as our study area. This park was chosen due to its representative features, including diverse vegetation structures, comprehensive functional zoning, and significant pedestrian traffic, making it an ideal location for our research objectives. Specifically, this study (1) employed air quality data collected over all four seasons to conduct LUR modeling and identify significant factors influencing the spatiotemporal distributions of PM2.5 and PM10 concentrations within the park; (2) examined the effects of various vegetation structures on PM2.5 and PM10 levels at the vegetation community scale; and (3) analyzed the seasonal and temporal trends of particulate matter concentrations. By identifying key factors affecting PM2.5 and PM10 concentrations across different spatial scales, our study contributes to improved air quality and the well-being of residents. This multi-scale approach provides novel insights and theoretical support for the planning and development of urban green spaces aimed at constructing healthier cities.

2. Research Area and Methods

2.1. Research Area

Xi'an, located in Shaanxi Province, China (33°42′–34°45′ N, 107°40′–109°49′ E), sits in the central part of the Guanzhong Plain at an average elevation of 550 m above sea level. The city experiences a temperate monsoon climate and serves as a vital political, economic, cultural, scientific, and technological hub in western China. As of the end of 2022, Xi'an's resident population had reached approximately 12.995 million, with around 4.93 million automobiles. This rapid urbanization, along with increased energy consumption from coal, oil and natural gas has led to significant air pollution challenges.

According to the 2022 report from the Ministry of Ecology and Environment on ambient air quality in 168 cities across China, Xi'an ranked 164th, highlighting an urgent need for air quality improvement. The city faces severe air pollution, primarily due to industrial emissions and traffic-related pollutants, including particulate matter such as PM2.5 and PM10, which pose serious health risks to the population.

This study focuses on Xi'an Expo Park, the largest urban park within the third ring expressway of Xi'an, covering an area of 418 hectares. The park is situated in the northeastern part of the city, bordered by the Ba River to the south, with residential areas to its east, west, and north. It is adjacent to an urban expressway and located about 4 km northwest of a thermal power plant (Figure 1). These features make it an ideal location to study the impact of urban green spaces on air quality, particularly in mitigating particulate matter concentrations. Xi'an Expo Park boasts a rich and stable plant community, comprehensive functional zoning, and high pedestrian traffic. Its diverse vegetation structures and strategic location provide a valuable opportunity to investigate the factors influencing dust retention at both the site and vegetation community scales. The natural conditions, park layout, and geographical context make it an excellent site for quantifying and evaluating the experimental models in this study. By focusing on this park, the research aims to provide a scientific foundation for the planning and development of urban green spaces aimed at improving air quality in heavily polluted urban environments like Xi'an.

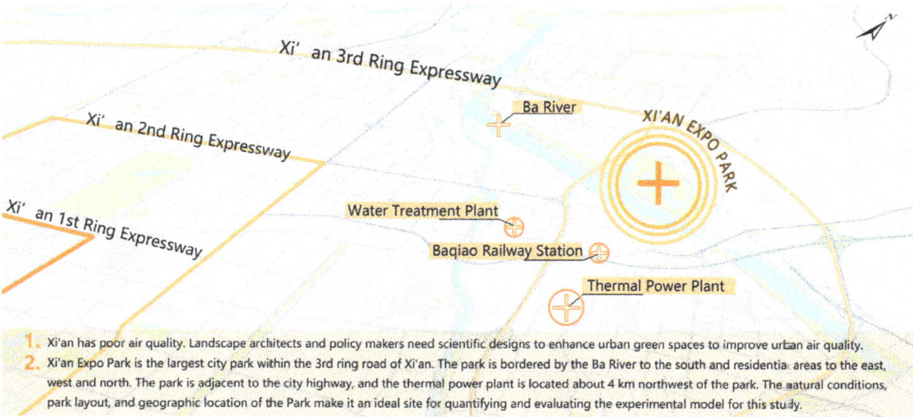

Figure 1. Map of the study area.

2.2. Research Method

2.2.1. LUR Model for Site Scale

The LUR model is a powerful approach used to predict pollutant concentrations by integrating monitoring data with predictor variables that influence pollutant levels within a specified radius around monitoring sites [34,35]. These predictor variables are commonly categorized into three main types: pollution factors, natural factors, and social factors [36]. Pollution factors primarily include transportation and industrial emissions. Natural factors encompass land use, meteorology, altitude and topography. Social factors involve population density, economic factors, and points of interest (POIs).

In this study, the LUR model was employed to simulate the spatial and temporal distributions of PM2.5 and PM10 concentrations at the site scale within Xi'an Expo Park. Data related to land use, pollution sources, meteorological conditions, and natural environmental factors were meticulously selected for analysis. The technical framework of the LUR model is illustrated in Figure 2. The modeling process comprises four main components: variable selection, regression modeling, model validation, and the generation of PM concentration maps.

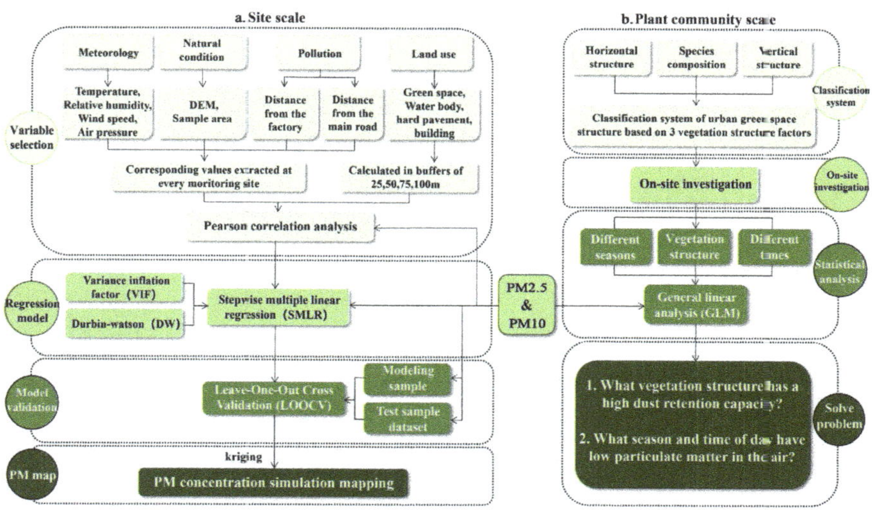

Figure 2. Framework of this study. (**a**) Site scale flowchart. (**b**) Plant community scale flowchart.

LUR Model Data Sources

(1) PM and meteorological data monitoring and sample selection

To ensure accuracy in our outdoor experiments, air pollution and meteorological data were collected under clear, calm, or breezy weather conditions. Meteorological parameters were recorded using a handheld weather station (Kestrel 5500, Advanced Sports International, Philadelphia, PA, USA), while air particulate matter concentrations (PM2.5 and PM10) were measured with a handheld particle counter (Aerocet 831, Zefon International, Ocala, FL, USA). The locations of the monitoring sample points are shown in Figure 3.

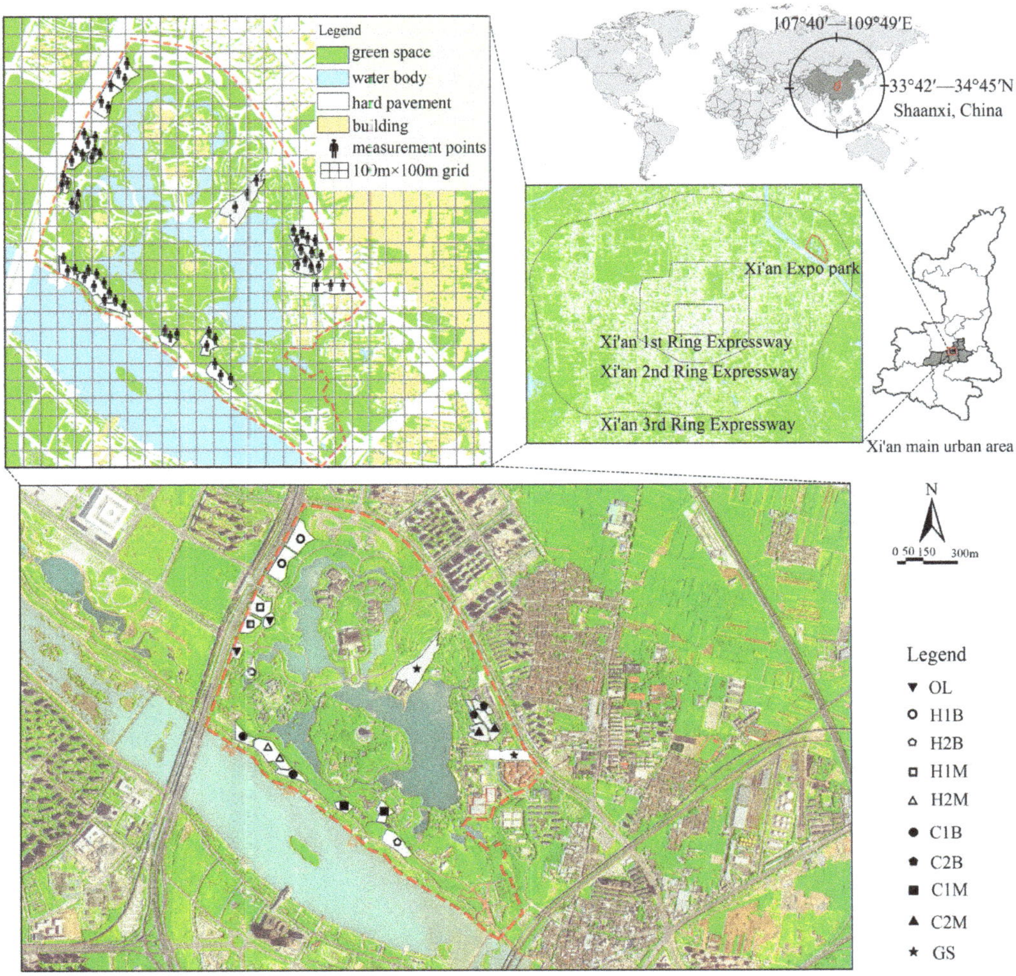

Figure 3. Map illustrating the location of the park, vegetation structure samples plots, land use, and experimental measurement points.

The data collection period spanned from December 2022 to November 2023. Monitoring was conducted two days each month, specifically, one day at the beginning and one day at the end of each month. On each sampling day, measurements were taken during three time periods: 8:00–10:00 a.m., 12:00–2:00 p.m., and 4:00–6:00 p.m. Monitoring was performed sequentially at all sampling points during each time period to ensure consistency and reliability in the data collected.

(2) Land-use data

Land-use data were derived from remote sensing images captured by the Gaofen-1B satellite in August 2023, featuring a spatial resolution of 2 m. Utilizing ENVI 5.3 for interpretation and supervised classification, the land-use in the study area was categorized into four major types: green space, water bodies, hard pavement, and buildings (Figure 3).

Using ArcGIS 10.2, the area of each land-use category was calculated within buffer zones of 0–25, 0–50, 0–75, and 0–100 m around each monitoring point, enabling a comprehensive analysis of land-use factors influencing air quality.

(3) Pollution data and natural information data

In Xi'an, the primary sources of air particulate matter include biomass burning (37.7%), secondary non-fossil sources (30.5%), secondary fossil sources (16.0%), vehicle emissions (9.9%), and coal combustion (5.9%) [37]. A thermal power plant located within the research area was identified as a significant pollution source based on a list of key monitored enterprises published by the Department of Ecology and Environment of Shaanxi Province for 2022–2023. Due to the substantial exhaust gas emissions from the plant during its operation, the distance from the monitoring point to the thermal power plant was utilized as a measure of industrial pollution.

Road traffic data were sourced from a vector road network available on the OpenStreetMap website. The distance from each monitoring point to the center of the nearest major road was employed as an indicator of road-related pollution, analyzed using ArcGIS 10.2. Additionally, elevation data for the study sites were acquired from the Geospatial Data Cloud website, while the area of each sample plot was measured using a handheld GPS device (Garmin GPSmap 629sc, Lenexa, KS, USA).

LUR Model Data Sources

The initial step involved variable selection and regression modeling. Distances from major roads and factories within various buffer zones (0–25 m, 0–50 m, 0–75 m, and 0–100 m) at the monitoring sites were utilized as pollution factors. Different land-use areas served as land-use factors, while air temperature, relative humidity, wind speed, and air pressure were incorporated as meteorological factors. Additionally, elevation and green space area were considered as natural factors. These independent variables were then introduced into the LUR model.

To diagnose multicollinearity among the model variables, the variance inflation factor (VIF) was applied [38]. A lower VIF value indicates a reduced degree of multicollinearity. Specifically, a VIF value of 1 denotes no correlation and no impact on the regression results, while values between 1 and 5 suggest a moderate correlation with negligible effects. Conversely, a VIF value exceeding 10 indicates excessive multicollinearity, which can significantly distort regression outcomes. Consequently, independent variables with a VIF greater than 5 were excluded from the analysis.

Autocorrelation in the model residuals was assessed using the Durbin–Watson (DW) statistic [39]. Stepwise linear regression was employed for all variables that met the conditions; variables with a VIF greater than 5 were removed, and the process was iterated until all remaining variables were significant and free from heterogeneity. Statistical analyses were conducted using ArcGIS 10.2 and SPSS 24.0.

The next step involved model validation. We assessed the model's fitting performance using K-fold cross-validation [40,41], implemented in R version 4.2.2. Finally, we conducted simulation mapping to visualize the spatial distribution of PM concentrations. To clearly visualize the distribution of air particulate matter concentrations within the study area, the kriging method was employed [17]. Initially, a grid of 100 m × 100 m was generated in ArcGIS 10.2, yielding a total of 260 grid points. The air particulate matter concentration values at each grid point were calculated using the PM2.5 and PM10 multiple regression models developed for the Xi'an Expo. Finally, the data from these 260 grid points were utilized to map the spatial distribution of PM2.5 and PM10 concentrations in Xi'an Expo Park through the kriging method.

2.2.2. On-Site Measurement Method for Plant Community Scale

This study employed an on-site measurement method to validate the relationship between various vegetation structures, seasonal variations, and temporal factors with PM2.5 and PM10 concentrations at the plant community scale. The framework for this technical approach is illustrated in Figure 2 and consists of three key components: classification and selection of urban green space systems, on-site measurements, and subsequent statistical analysis.

Classification and Monitoring Sample Selection of Urban Green Spaces

According to Vegetation of China [42], vegetation structure is characterized by the compositional features of plant communities, which exhibit distinct horizontal and vertical arrangements. The horizontal structure refers to the spatial distribution of individual plants or communities, while the vertical structure is determined by the arrangement of different plant types that occupy various elevations, forming distinct canopy layers. For example, a woodland typically consists of tree, shrub, and herb layers. Additionally, the overall appearance of a plant community is influenced by the life forms of the dominant species; for instance, coniferous forests are primarily composed of coniferous trees.

In this study, a three-dimensional classification system for urban green space structure was developed by integrating three elements of vegetation structure: horizontal structure, species composition, and vertical structure (Table 1). The first dimension categorizes the horizontal structure of urban green spaces based on canopy cover into three types: open green space, half-closed green space, and closed green space, utilizing a plant canopy analyzer (LAI-2200, LI-COR Biosciences, Lincoln, NE, USA) (Figure 4). The second dimension further divides open green spaces into two subcategories: lawns and shrubs. For half-closed and closed green spaces, three subcategories are defined: coniferous, broad-leaved, and mixed broadleaf-conifer forests. The third dimension, focusing on vertical structure, further classifies half-closed and closed green spaces into single-layered and multi-layered forests. A single-layered forest consists solely of the tree layer, whereas a multi-layered forest includes a combination of tree, sub-tree, and shrub layers.

Table 1. Classification system of urban green space structure based on three vegetation structure factors.

First Dimension	Second Dimension	Third Dimension	Code
Open green space (<10% canopy cover of trees/shrubs)	Lawn Shrub		OL OS
Half-closed green space (30%–70% canopy cover of trees/shrubs)	Coniferous Broad-leaved Mixed broadleaf-conifer	Single-layered Multi-layered	H1C H2C H1B H2B H1M H2M
Closed green space (>70% canopy cover of trees/shrubs)	Coniferous Broad-leaved Mixed broadleaf-conifer	Single-layered Multi-layered	C1C C2C C1B C2B C1M C2M

Preliminary experiments were conducted through visual interpretation of Google Maps and field visits, taking into account the actual green space characteristics in Xi'an Expo Park. Based on this field research, nine typical vegetation structures were identified within Park (Figure 4). These structures include: open lawn green space (OL), half-closed single-layered mixed broadleaf-conifer woodland (H1M), half-closed multi-layered mixed broadleaf-conifer woodland (H2M), half-closed single-layered broad-leaved woodland (H1B), half-closed multi-layered broad-leaved woodland (H2B), closed single-layered mixed broadleaf-conifer woodland (C1M), closed multi-layered mixed broadleaf-conifer woodland (C2M), closed single-layered broad-leaved woodland (C1B), and closed multi-layered broad-leaved woodland (C2B). Each vegetation structure was sampled in duplicated within the park, while two gray space (GS) areas located on the eastern side of the park were selected as control sites. In total, 20 sample plots were monitored, with their distribution illustrated in Figure 3.

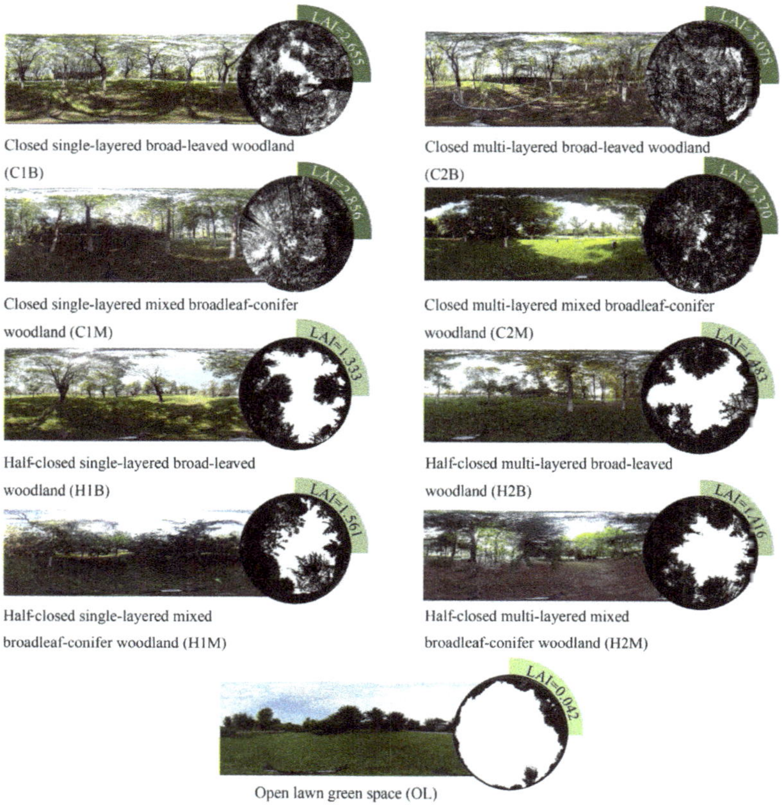

Figure 4. Fisheye camera view of different vegetation structures.

Air Particulate Matter and Meteorological Data Monitoring

For the outdoor experiments, monitoring of air pollution and meteorological data was conducted under clear, breezy, or windless weather conditions. Meteorological parameters were measured using a handheld weather station (Kestrel 5500), while air particulate matter concentrations were recorded using a handheld particle counter (Aerocet 831). Three sampling sites were designated within each sample plot (Figure 3). Care was taken to ensure that the surroundings of each sampling site were free from visible pollution and located at least 20 m away from roads and water edges to minimize potential interference with the experimental data.

The data collection period spanned from December 2022 to November 2023, with measurements taken for two days each month—one day at the beginning and another at the end of the month. For each measurement day, three time periods were selected: 8:00–10:00, 12:00–14:00, and 16:00–18:00. Sequential monitoring was performed simultaneously at all sampling points.

Data Processing and Statistical Analysis

Both PM and meteorological data were collected through the on-site measurement method and managed using Microsoft Office Excel 2019. The General Linear Model (GLM) in SPSS 24.0 was employed to assess the correlations between vegetation structure, season, and different time periods with PM2.5 and PM10 concentrations. Multiple comparisons of means were conducted within the GLM using Tukey's method at a 95% confidence level to determine the grouping of study variables and their variability. In this study, a significance level of $p < 0.05$ was used to indicate significant differences, while $p < 0.01$ was considered highly significant.

3. Results

3.1. Development and Verification of the LUR Model

3.1.1. PM2.5 and PM10 LUR Models for the Park

A total of 24 factors related to pollution, natural conditions, land use, and meteorology were extracted to develop seasonal LUR models for PM2.5 and PM10. The variance inflation factor (VIF) values for both models were below 5, indicating no issues with multicollinearity. Additionally, the Durbin–Watson (DW) values fell within acceptable limits, confirming the absence of spatial autocorrelation. All variables were statistically significant, passing the t- and F-tests with a significance level of $p < 0.05$ (Table 2). Deletion variables are detailed in the Supplementary Table S1.

Table 2. Predictor variables, coefficients, standardized coefficients, t-values, Sig. values, VIF, and validation results for the PM2.5 and PM10 LUR model.

	Variables	Coefficients	Std. Error	Beta	t	Sig.	VIF	Validation Results
Spring PM2.5	Constant	2.039×10^{-15}	0.081		0.000	1.000		F = 23.398; DW = 1.681; Adjusted R^2 = 0.603; K-fold cross validation R^2 = 0.75
	Hard pavement (25 m)	−0.631	0.098	−0.631	−6.444	0.000	1.426	
	Distance from the main road	−0.533	0.108	−0.533	−4.926	0.000	1.740	
	Green space (75 m)	0.256	0.091	0.256	2.815	0.007	1.231	
	Sample area	0.276	0.117	0.276	2.353	0.022	2.041	
Summer PM2.5	Constant	-1.302×10^{-15}	0.057		0.000	1.000		F = 62.184; DW = 1.243; Adjusted R^2 = 0.806; K-fold cross validation R^2 = 0.826
	Distance from the factory	0.538	0.080	0.538	6.742	0.000	1.937	
	Green space (50 m)	0.390	0.060	0.390	6.550	0.000	1.076	
	Building (100 m)	−0.312	0.059	−0.312	−5.279	0.000	1.064	
	Sample area	0.213	0.080	0.213	2.663	0.010	1.944	
Autumn PM2.5	Constant	-6.220×10^{-14}	0.093		0.000	1.000		F = 11.964; DW = 1.349; Adjusted R^2 = 0.482; K-fold cross validation R^2 = 0.396
	Water body (100 m)	−0.507	0.124	−0.507	−4.094	0.000	1.749	
	Autumn pressure	1.024	0.180	1.024	5.696	0.000	3.681	
	Autumn temperature	−1.054	0.205	−1.054	−5.145	0.000	4.772	
	Autumn humidity	0.643	0.196	0.643	3.276	0.002	4.379	
	Distance from the main road	−0.328	0.148	−0.328	−2.222	0.030	2.480	
Winter PM2.5	Constant	2.600×10^{-15}	0.043		0.000	1.000		F = 102.324; DW = 0.726; Adjusted R^2 = 0.888; K-fold cross validation R^2 = 0.780
	Distance from the factory	−0.664	0.070	−0.664	−9.425	0.000	2.610	
	Sample area	−0.377	0.062	−0.377	−6.092	0.000	2.018	
	Green space (100 m)	−0.280	0.047	−0.280	−5.946	0.000	1.164	
	Building (75 m)	0.179	0.047	0.179	3.803	0.000	1.165	
	Distance from the main road	−0.197	0.069	−0.197	−2.879	0.006	2.471	
Spring PM10	Constant	1.749×10^{-15}	0.091		0.000	1.000		F = 16.047; DW = 1.244; Adjusted R^2 = 0.505; K-fold cross validation R^2 = 0.378
	Green space (50 m)	0.210	0.125	0.210	1.673	0.100	1.872	
	Spring humidity	0.518	0.116	0.518	4.471	0.000	1.597	
	Spring temperature	0.389	0.138	0.389	2.823	0.007	2.267	
	Hard pavement (25 m)	−0.247	0.114	−0.247	−2.164	0.035	1.547	
Summer PM10	Constant	-5.782×10^{-15}	0.064		0.000	1.000		F = 46.788; DW = 0.854; Adjusted R^2 = 0.773; K-fold cross validation R^2 = 0.586
	Distance from the factory	0.618	0.069	0.618	8.922	0.000	1.161	
	Water body (75 m)	0.232	0.072	0.232	3.225	0.002	1.251	
	Green space (50 m)	0.271	0.070	0.271	3.886	0.000	1.180	
	Building (75 m)	−0.255	0.069	−0.255	−3.703	0.000	1.148	

Table 2. *Cont.*

	Variables	Coefficients	Std. Error	Beta	t	Sig.	VIF	Validation Results
Autumn PM10	Constant	-1.992×10^{-14}	0.063		0.000	1.000		F = 39.152; DW = 1.976; Adjusted R^2 = 0.764; K-fold cross validation R^2 = 0.726
	Hard pavement (25 m)	0.540	0.077	0.540	7.049	0.000	1.466	
	Sample area	−0.745	0.074	−0.745	−10.077	0.000	1.365	
	Autumn pressure	0.266	0.066	0.266	0.266	0.000	1.092	
	Autumn humidity	0.179	0.073	0.179	0.179	0.017	1.313	
	Water body (50 m)	−0.147	0.067	−0.147	−0.147	0.034	1.13	
Winter PM10	Constant	9.631×10^{-16}	0.45		0.000	1.000		F = 87.551; DW = 0.961; Adjusted R^2 = 0.880; K-fold cross validation R^2 =0.862
	Distance from the factory	−0.642	0.073	−0.642	−8.815	0.000	2.610	
	Sample area	−0.401	0.064	−0.401	−6.252	0.000	2.018	
	Green space (100 m)	−0.266	0.049	−0.266	−5.458	0.000	1.164	
	Building (75 m)	0.166	0.049	0.166	3.407	0.001	1.165	
	Distance from the main road	−0.188	0.071	−0.188	−2.646	0.011	2.471	

For the PM2.5 LUR model, the adjusted R^2 values were 0.603 (spring), 0.806 (summer), 0.482 (autumn), and 0.888 (winter), while the K-fold cross-validation R^2 values were 0.75 (spring), 0.826 (summer), 0.396 (autumn), and 0.780 (winter). In the PM10 LUR model, the adjusted R^2 values were 0.505 (spring), 0.773 (summer), 0.764 (autumn), and 0.880 (winter), with K-fold cross-validation R^2 values of 0.378 (spring), 0.586 (summer), 0.726 (autumn), and 0.862 (winter).

These results confirm the effectiveness of the LUR model for Xi'an Expo Park, with K-fold cross-validation demonstrating strong consistency between the measured and predicted PM2.5 and PM10 concentrations across seasons.

3.1.2. Analysis of the Independent Variable Relationships in the PM2.5 and PM10 LUR Models

The analysis of the 24 independent variable factors revealed their correlations with PM2.5 and PM10 concentrations across all seasons. Based on the magnitude of the correlation coefficients, it was determined that distance from the factory, sample area, distance from the main road, green space, and hard pavement were the most influential factors affecting PM concentration distributions (Figure 5 and Table 2).

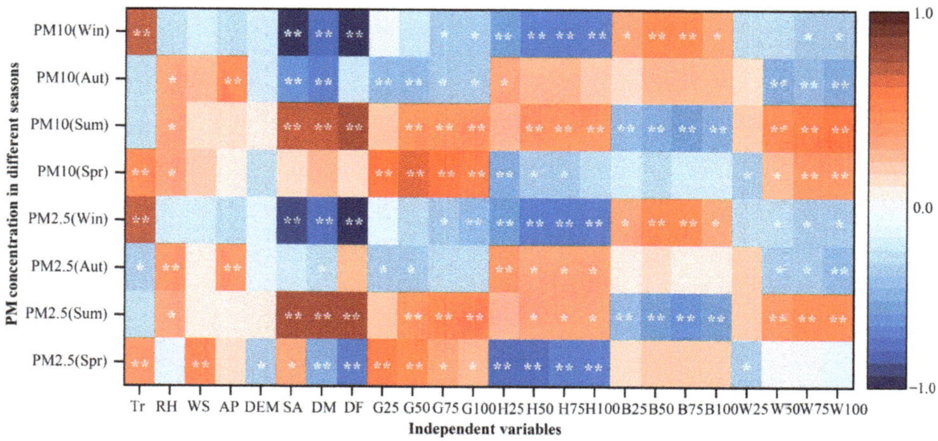

Tr, temperature; RH, relative humidity; WS, wind speed; AP, air pressure; DEM, digital elevation model; SA, sample area; DM, distance from the main road; DF, distance from the factory; G, green space; H, hard pavement; B, building; W, water body

Figure 5. Correlation between PM2.5 and PM10 concentrations and dependent variable factors. * Statistically significant at the $p < 0.05$ level. ** Statistically significant at the $p < 0.01$ level.

In terms of pollution factors, the distance from main roads exhibited a significant positive correlation with PM2.5 and PM10 concentrations during summer, while it showed a notable negative correlation with PM2.5 in spring, autumn, and winter, and a clear negative correlation with PM10 in autumn and winter. The distance from factories demonstrated a statistically significant positive correlation with both PM2.5 and PM10 in summer, alongside significant negative correlations with PM2.5 in spring and winter, and with PM10 in winter.

Regarding land use factors, green space exhibited a significant positive correlation with PM2.5 and PM10 in spring and summer, transitioning to negative correlations in autumn and winter. The hard pavement factor revealed a strong negative correlation with PM2.5 and PM10 in spring and winter, while indicating significant positive correlations in summer and autumn. The building factor showed negative correlations with PM2.5 and PM10 during summer, notable positive correlations in winter, and no correlation in spring and autumn. The water factor presented significant positive correlations with PM2.5 and PM10 in summer, but significant negative correlations in fall and winter.

In terms of natural factors, elevation exhibited a negative correlation in spring and no correlation in other seasons. The sample area showed a significant positive correlation with PM2.5 and PM10 in summer, while indicating significant negative correlations with PM2.5 in winter and notable negative correlations with PM10 in autumn and winter. Temperature among the meteorological factors demonstrated significant positive correlations with PM2.5 and PM10 in spring and winter. Humidity showed significant positive correlations with both PM2.5 and PM10 in summer and autumn. Wind speed exhibited a significant positive correlation with PM2.5 concentration only in spring, while air pressure showed a significant positive correlation with PM2.5 and PM10 in autumn, with no relevance in other seasons (Figure 5).

3.1.3. Mapping PM2.5 and PM10

The distribution of PM2.5 and PM10 concentrations within the built-up area of Xi'an Expo Park is illustrated in Figure 6. The maps reveal significant variations in PM2.5 and PM10 concentrations across the park, influenced by various environmental factors.

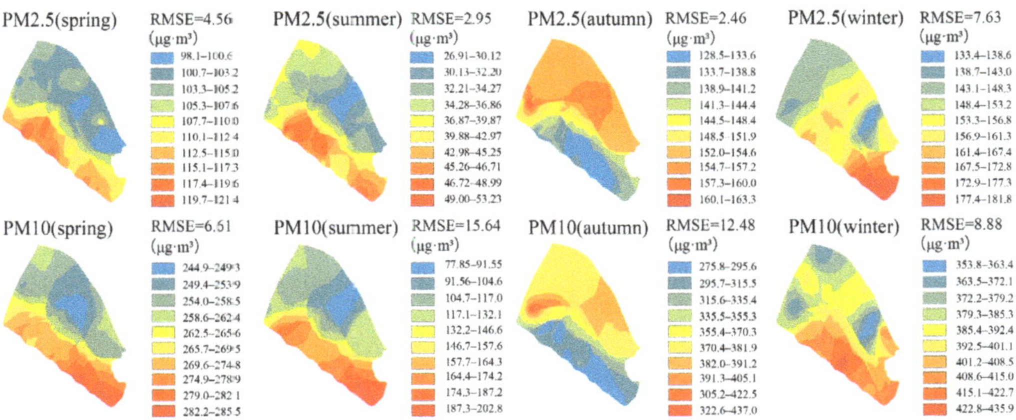

Figure 6. Spatial distribution of PM2.5 and PM10 concentration in the built-up area of Xi'an Expo Park in four seasons.

In spring, PM2.5 concentrations ranged from 98.1 to 121.4 μg/m^3, while PM10 concentrations varied from 244.9 to 285.5 μg/m^3. Areas with low PM2.5 and PM10 concentration were predominantly found in the northern section of the park, characterized by high hard pavement rates, low greening rates, and a considerable distance from water bodies.

During the summer, overall PM2.5 and PM10 concentrations in the park were relatively low, With PM2.5 concentrations ranging from 26.9 to 52.23 μg/m^3 and PM10 concentrations from 77.85 to 202.8 μg/m^3. Notably, areas with elevated PM2.5 and PM10 concentrations were located on the southern side of the park, which, despite being distant from the main road, is close to the abundant water source of the Ba River and features a rich green landscape.

In autumn, PM2.5 concentrations in the Expo Park ranged from 128.5 to 163.3 μg/m^3, and PM10 concentrations varied from 275.8 to 437 μg/m^3. The southern region of the park, noted for its extensive greenery and water resources, exhibited the lowest PM2.5 and PM10 concentrations.

Winter saw the highest concentrations of PM2.5 and PM10 in the Expo Park, with PM2.5 levels ranging from 133.4 to 181.8 μg/m^3 and PM10 level from 353.8 to 435.9 μg/m^3. The elevated concentrations during this season were primarily influenced by emissions from the thermal power plant located on the southern side of the park, leading to higher pollutant levels in this area compared to the northern section.

3.2. Differences in PM2.5 and PM10 Concentration in Green Spaces with Different Vegetation Structures

This study performed a general linear analysis of nine vegetation structures in relation to PM2.5 and PM10 concentrations, while accounting for the influence of meteorological factors. The results indicate highly significant relationships ($p < 0.01$) between vegetation structure type and PM2.5 and PM10 concentrations (Table 3).

Table 3. General linear analysis of vegetation structure and air particle concentration.

PM	Item	DF	Adj MS	F-Value	p-Value	R^2	Adjusted R^2
PM2.5	Vegetation structure	9	121,867.706	21.779	0.000	0.376	0.374
	Temperature	1	4,709,889.739	841.695	0.000		
	Relative humidity	1	1,931,771.44	345.223	0.000		
	Wind speed	1	15,140.93	2.706	0.100		
	Air pressure	1	881,302.932	157.496	0.000		
PM10	Vegetation structure	9	325,253.598	25.672	0.000	0.306	0.304
	Temperature	1	12,532,446.89	989.192	0.000		
	Relative humidity	1	819,527.002	64.686	0.000		
	Wind speed	1	10,053.227	0.794	0.373		
	Air pressure	1	4,483,624.162	353.895	0.005		

Multiple comparisons of means were conducted using the Tukey method at a 95% confidence level (Figure 7). This analysis explored the capacity of different vegetation structures in green spaces to mitigate PM2.5 and PM10 concentrations. The grouping information from the multiple comparisons revealed that the PM2.5 and PM10 concentrations in gray space (GS) were significantly higher than those in all vegetated green spaces.

Among the closed green spaces, the closed single-layered broad-leaved woodland (C1B) exhibited the lowest particulate matter concentrations, demonstrating a high capacity to regulate air particulate matter. Conversely, the closed multi-layered mixed broadleaf-conifer woodland (C2M) had the highest particulate matter concentration.

In half-closed green spaces, the lowest PM2.5 and PM10 concentrations were observed in the half-closed single-layered broad-leaved woodland (H1B) and half-closed single-layered mixed broadleaf-conifer woodland (H1M), both of which showed superior effectiveness in reducing particulate matter levels. The highest air particulate concentrations were recorded in the half-closed multi-layered broad-leaved woodland (H2B), followed by the half-closed multi-layered mixed broadleaf-conifer woodland (H2M) and open lawn green space (OL).

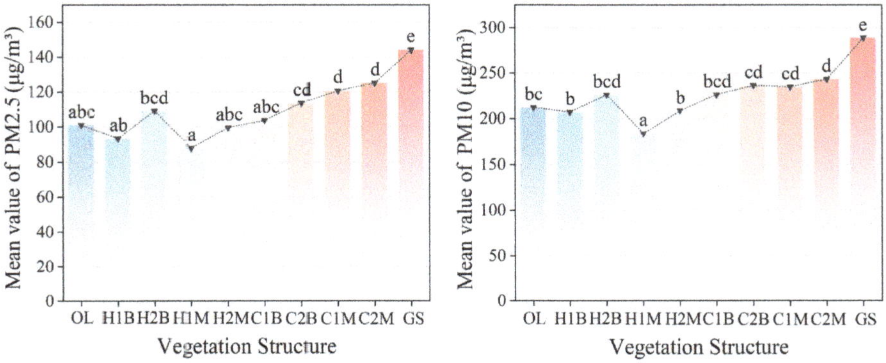

Figure 7. Relationship between vegetation structure and PM2.5 and PM10 concentrations. The lowercase letters in the figure represent the grouping information obtained by multiple comparisons of means using the Tukey method.

3.3. Differences in Air Particle Concentrations in Different Seasons and Time of the Day

This study developed a general linear model to analyze the effects of season, time of day, and meteorological factors on air particulate matter concentrations. The results indicated that both season and time of day had highly significant effects ($p < 0.01$) on particulate matter concentrations (Table 4).

Table 4. General linear analysis of month, time of the day, and air particle matter concentrations.

PM	Item	DF	Adj MS	F-Value	p-Value	R^2	Adjusted R^2
PM2.5	Season	3	626,260.160	115.833	0.000	0.396	0.395
	Temperature	1	1,999,761.289	369.875	0.000		
	Relative humidity	1	857,746.160	158.648	0.000		
	Wind speed	1	1892.141	0.35	0.554		
	Air pressure	1	1,025,237.552	189.627	0.000		
PM10	Season	3	2,489,123.653	214.601	0.000	0.364	0.362
	Temperature	1	3,485,147.940	300.474	0.000		
	Relative humidity	1	1,352,658.747	116.62	0.000		
	Wind speed	1	57,722.926	4.977	0.026		
	Air pressure	1	1,562,455.761	134.708	0.000		
PM2.5	Time of the day	2	500,311.781	89.199	0.000	0.373	0.372
	Temperature	1	5,364,625.559	956.442	0.000		
	Relative humidity	1	2,709,535.056	483.074	0.000		
	Wind speed	1	16.039	0.003	0.957		
	Air pressure	1	1,035,306.493	184.581	0.000		
PM10	Time of the day	2	126,414.467	9.527	0.000	0.272	0.271

Significant differences were observed in PM2.5 and PM10 concentrations across seasons, with consistent overall trends noted (Figure 8). Air particulate matter concentrations were highest in winter, followed by spring and autumn, while summer recorded the lowest levels.

Regarding temporal trends throughout the day, both PM2.5 and PM10 concentrations exhibited approximately linear decreasing patterns. Specifically, particulate matter concentrations were highest in the morning, slightly decreased at noon, and reached their lowest levels in the afternoon (Figure 8). These findings underscore the importance of considering seasonal variations and daily timing when assessing air quality and particulate matter levels.

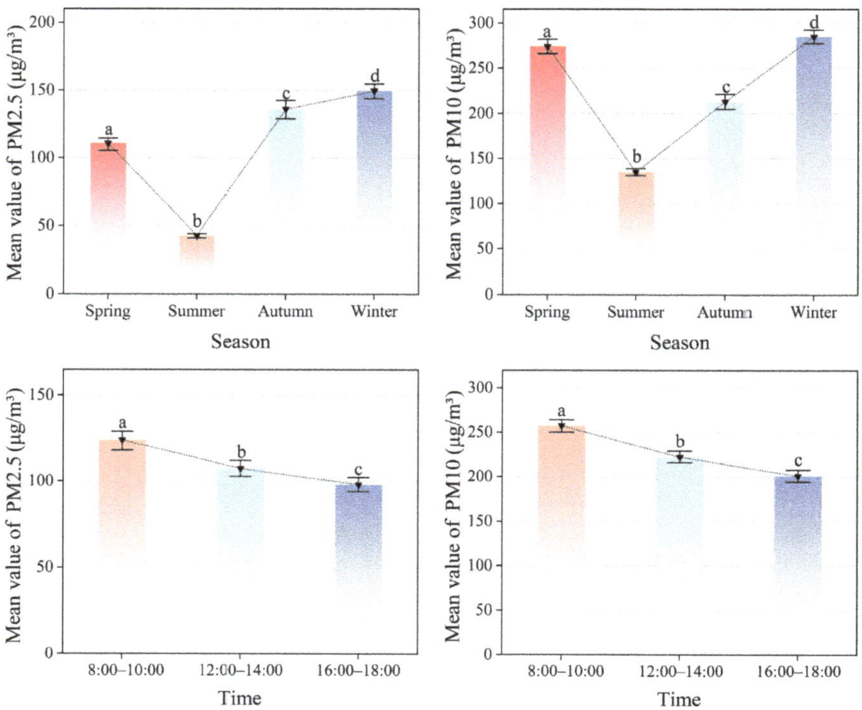

Figure 8. Relationship between different seasons, time of the day and air particle concentration. The lower-case letters in the figure represent the grouping information obtained by multiple comparisons of means using the Tukey method.

4. Discussion

4.1. PM2.5 and PM10 LUR Models in the Xi'an Expo Park

This study assessed the variability of the LUR model coefficients across different seasons. For PM2.5, the adjusted R^2 values ranged from 0.482 to 0.888, with K-fold cross-validation R^2 values ranging from 0.396 to 0.826. For PM10, the adjusted R^2 values varied between 0.505 and 0.880, while the K-fold cross-validation R^2 values ranged from 0.378 to 0.862. Compared to previous studies reporting adjusted R^2 values for PM2.5 between 0.14 and 0.95 and for PM10 between 0.22 and 0.91 [43], our LUR model demonstrates reasonable predictive performance for PM concentrations in Xi'an Expo Park. Notably, 87.5% of our models explained over 50% of the spatial variability in pollutant concentrations. The relatively lower adjusted R^2 and K-fold cross-validation R^2 values in some seasons may be attributed to limited variability in the measured variables and the exclusion of certain influencing factors [44,45]. Factors such as microclimatic variations, transient pollution sources, and unmeasured land-use characteristics might have affected the model's predictive accuracy. Future studies should consider incorporating additional multidimensional influencing factors—such as detailed meteorological data, traffic flow patterns, and finer-scale land-use classifications—to enhance model accuracy and robustness.

Our simulations indicate that the most significant factors affecting PM concentrations in Xi'an Expo Park include distance from factories, sampling area size, distance from main roads, green space coverage, and hard pavement (Table 3). The identified hotspots highlight the predominant influence of traffic and the influence of traffic and industrial emissions on PM concentrations, aligning with other studies that identify factory pollution as a critical contributor to air quality degradation [46]. The combustion of fossil fuels in vehicles, incomplete fuel combustion, and tire-road friction are significant contributors to airborne particulate matter in urban settings [47–50]. Meanwhile, the relationship between the distance from pollution sources and PM concentrations is closely related to seasonal environmental conditions. In summer, higher temperatures and increased solar radiation lead to greater atmospheric turbulence and stronger vertical mixing. These conditions enhance the dispersion of atmospheric particles over longer distances. As a result, pollutants emitted from factories can travel farther from their sources, and the impact of proximity to the pollution source on PM concentrations diminishes. This leads to a positive correlation between distance from factories and PM concentrations in summer—areas farther from the factories may experience higher PM levels due to the widespread dispersion. In contrast, during winter, lower temperatures and more stable atmospheric conditions reduce atmospheric turbulence and vertical mixing. Additionally, phenomena like temperature inversions are more common in winter, trapping pollutants near the ground and preventing their dispersion. The cold, dense air holds pollutants closer to their sources, resulting in higher PM concentrations in areas near factories. Consequently, there is a negative correlation between distance from factories and PM concentrations in winter—PM levels decrease as the distance from the pollution source increases [51]. Seasonal variations in emissions also contribute to these patterns. For instance, increased energy demands for heating in winter may lead to higher emissions from factories and power plants. The combination of reduced pollutant dispersion and increased emissions amplifies PM concentrations near the sources during winter months.

Our findings emphasize that green spaces play a crucial role in regulating PM concentrations. As essential components of urban infrastructure, green spaces mitigate PM pollution through natural processes such as deposition, filtration, and dispersion. Recent studies have demonstrated that urban green infrastructure improves the urban microclimate by enhancing air circulation, reducing ambient temperatures, and facilitating the dispersion of pollutants [52,53]. Additionally, green spaces increase biodiversity by providing habitats for various plant and animal species, which contribute to pollutant removal through bioaccumulation and other ecological processes [54]. They also offer recreational spaces that enhance the well-being of citizens, promoting physical activity and mental health benefits [55,56]. Given the global trend toward increasing urban green infrastructure, it is important for landscape architects and urban planners to balance environmental, social, and economic benefits within limited urban spaces through informed and scientifically grounded design approaches [57]. By aligning our findings with contemporary research, we provide a more comprehensive understanding of the role of green infrastructure in air quality improvement and public health enhancement. In particular, this study found seasonal variations in how green spaces regulate PM concentrations. In spring and summer, green spaces exhibited a positive correlation with PM levels, possibly due to increased pollen and spores from growing plants contributing to particulate matter [58]. Conversely, in winter, evergreen broadleaf and coniferous plants appear to effectively adsorb particulate matter, resulting in a negative correlation with PM concentrations.

Additionally, the area of green space was significantly correlated with PM concentrations, supporting the findings from Douglas et al. [59]. While the sample area correlated positively with PM levels in spring and summer, it correlated negatively in winter. This reflects the unique characteristics of the plants present in these environments. To mitigate PM concentrations, urban planners should strive to reduce the fragmentation of green spaces and apply scientific approaches to vegetation management [60,61].

Lastly, our study found a significant correlation between PM concentrations and hard surfaces. The correlation between hard pavement and PM was negative in spring and winter, but positive in summer and autumn. This indicates that meteorological conditions significantly influence PM distribution in open spaces [43,62]. For instance, in summer, increased humidity correlates with higher PM concentrations, while cooler temperatures in spring and winter have the opposite effect.

Overall, air pollution modeling is a crucial tool for urban planning and public health assessment [63]. By integrating these models with epidemiological data, we can gain a deeper understanding of the health implications associated with varying pollutant levels, leading to more effective management policies [64]. Future studies should aim to explore a broader range of pollutants through LUR modeling. Shi et al. [65] extended LUR modeling to include multiple pollutants such as CO, NO_x, O_3, SO_x, PM2.5, and PM10, enhancing the comprehensiveness of air quality assessments. Additionally, expanding research across different scales—regional, urban, and global—will provide more comprehensive insights [36].

4.2. Effect of Vegetation Structure on PM2.5 and PM10 Concentrations

This study highlights the significant role that green space factors at the site scale play in regulating air particulate matter concentrations in parks. The findings provide valuable insights for landscape architecture, particularly regarding vegetation structure selection.

Our results indicate that green spaces are more effective than open squares in reducing air particulate matter concentrations. This finding aligns with Gómez-Moreno et al. [66], who reported higher PM2.5 concentrations in park squares compared to other types of green spaces. The lowest concentrations of airborne particulate matter were observed in the closed single-layered broad-leaved woodland (C1B), which demonstrates a high capacity of regulating air quality among closed green spaces. In contrast, the highest concentrations were found in the closed multi-layered mixed broadleaf-conifer woodland (C2M).

Among semi-enclosed green spaces, the half-closed single-layered broad-leaved woodland (H1B) and the half-closed single-layered mixed broadleaf-conifer woodland (H1M) exhibited the lowest PM2.5 and PM10 concentration. These two vegetation structures displayed the strongest ability to mitigate air particulate matter levels. Conversely, the half-closed multi-layered mixed broadleaf-conifer woodland (H2M) and open lawns (OL) demonstrated a weaker capacity for regulating particulate matter concentrations compared to H1B and H1M. Therefore, we have discussed the multiple factors influencing the impact of vegetation on particulate matter, including:

1. Leaf Structure: Leaves with complex surfaces, such as rough textures or fine hairs, enhance the capture of airborne particles. Broadleaf species often have larger surface areas and microstructures that facilitate dust retention.
2. Canopy Density: A moderate canopy density (30%–70%) balances air flow and particle deposition. Too dense a canopy can hinder air circulation, reducing the contact between pollutants and leaves, while too sparse a canopy offers less surface area for particle capture [67,68].
3. Plant Type: Trees have a significantly higher dust retention capacity than lawns and shrubs due to their greater biomass, larger leaf area, and more complex branching structures. This finding aligns with the results of Cai et al. [61].

Our results indicate that the capacity of vegetation structures to regulate PM2.5 and PM10 concentrations tends to decrease with increasing structural complexity. Simpler structures allow for better airflow and more effective interception of particulate matter. Single-layered vegetation structures have shown positive impacts in both half-closed and closed environments, supporting the findings of Qiu et al. [69].

Based on our findings, we recommend prioritizing the two vegetation structures with the highest dust retention capacities for park planting design: the half-closed single-layered mixed broadleaf-conifer woodland (H1M) and the half-closed single-layered broad-leaved woodland (H1B). These should be followed by the half-closed multi-layered mixed broadleaf-conifer woodland (H2M) and the open lawns (OL). Conversely, vegetation structures with weaker dust retention capacities included the closed single-layered mixed broadleaf-conifer woodland (C1M), the closed multi-layered mixed broadleaf-conifer woodland (C2M), and the closed multi-layered broad-leaved woodland (C2B). It is noteworthy that these structures predominantly consist of multi-layered vegetation, which challenges the common assumption that trees, shrubs, and grasses inherently possess strong dust retention capacity [70].

4.3. Impact of Different Seasons and Times of the Day on Air Particulate Matter

Air particulate matter concentrations in the Xi'an Expo Park exhibited distinct seasonal variations, with the highest levels recorded in winter, followed by spring and autumn, while summer showed the lowest and most stable concentrations. Located in the Guanzhong Basin, Xi'an experiences significant pollutant accumulation due to its geographical characteristics, which hinder dispersion [71]. The prolonged cold winters in the region, coupled with the prevalent practice of burning coal for heating, contribute substantially to elevated airborne particulate matter concentrations [72]. Additionally, the inversion effect frequently leads to haze conditions [73]. In contrast, summer is characterized by reduced pollutant emissions compared to winter. The high temperatures and humidity levels during this season increase rainfall and wind, which facilitate the diffusion and removal of airborne pollutants [74].

Regarding diurnal patterns, particulate matter concentrations peaked between 8:00 and 10:00 a.m. and gradually declined throughout the day, reaching their minimum levels between 4:00 and 6:00 p.m. This trend can be attributed to the condensation of water vapor in green spaces during the night, which results in higher humidity levels in the morning. Additionally, the lower solar radiation intensity and cooler morning temperatures contribute to increased particulate matter concentrations. As temperatures rise during midday and the afternoon, enhanced plant respiration leads to lower air particulate matter levels.

This finding contrasts with the observations of Gao et al. [75], who reported higher particulate matter concentrations in the morning and evening, with lower levels during the middle of the day.

4.4. Design Recommendations

In this section, we address practical variables that need consideration in urban design, beyond the current government policies that mainly focus on source control measures such as promoting clean energy, factory shutdowns, and vehicular traffic restrictions [76–78]. Our study provides insights into small-scale changes in PM concentrations, which can inform more effective landscape design and policy decisions.

1. For Urban Planners and Policymakers:

Siting Considerations: Whenever possible, situate sensitive land uses—such as residential areas, schools, and hospitals—away from major pollution sources like factories and busy roads to protect public health.

Protective Vegetation Barriers: Establish protective forest belts around factories and main roads to reduce the spread of particulate matter. These green buffers can intercept airborne particles and serve as a barrier between pollution sources and populated areas.

2. For Landscape Architects:

Optimizing Green Space Configurations: Focus on designing green spaces that maximize PM reduction. Our findings indicate that half-closed green spaces with 30%–70% canopy cover and single-layered vegetation structures are most effective in regulating PM concentrations.

Recommended Vegetation Structures: Half-closed single-layered mixed broadleaf-conifer woodland (H1M) combines broadleaf and conifer species to enhance dust retention capacity; and half-closed single-layered broad-leaved woodland (H1B) utilizes broadleaf species known for their larger leaf surface areas, which are effective in capturing particulate matter.

Plant Selection: Minimize the use of pollen-rich plants to reduce allergenic particulate matter. Increase the proportion of evergreen species to provide year-round air quality benefits due to their persistent foliage.

Balancing Green and Hard Surfaces: Appropriately balance the proportions of green spaces and hard surfaces according to the site's size to reduce fragmentation and enhance the effectiveness of green spaces in PM reduction.

These recommendations can add detailed guidance to urban green space planning standards, such as those outlined in GB/T 51346-2019.

3. Public Health Guidance:

Our study also identified temporal patterns in PM concentrations:

Seasonal Variations: PM2.5 and PM10 concentrations were highest in winter, followed by spring and autumn, with the lowest levels recorded in summer.

Daily Variations: Particulate matter concentrations peaked between 8:00 and 10:00 a.m. and gradually decreased, reaching their minimum between 4:00 and 6:00 p.m.

Based on these findings, we advise vulnerable populations—such as older adults, children, and pregnant women—to engage in outdoor activities during the afternoons in spring, summer, and autumn when air quality is comparatively better.

4.5. Limitations and Future Research Directions

This study's LUR model utilized a relatively limited number of natural variables. To enhance the model's predictive capacity, incorporating additional natural variables that may influence air particulate matter, such as three-dimensional building morphology [79], point of interest (POI) [80] and landscape index [81], would provide deeper insights into the complex dynamics of air quality.

Furthermore, future studies could benefit from integrating various study scales, such as urban-level analyses or individual leaf-scale assessments, to conduct multi-scale comparisons. This approach is likely to yield more nuanced results and contribute to more effective urban green space planning solutions aimed at reducing atmospheric particulate matter concentrations in the future.

5. Conclusions

Clean air is essential for public health, especially in the context of increasing air pollution that cannot be effectively mitigated solely by controlling pollution sources. This study addresses the urgent need to leverage urban green spaces for regulating air particulate pollution by examining the factors influencing PM2.5 and PM10 concentrations in urban parks through on-site measurements and LUR modeling.

(1) The LUR model demonstrated its efficacy in predicting the factors that influence PM2.5 and PM10 concentrations within Xi'an Expo Park at the site scale. Key variables, including the distance from factories, sample area, distance from main roads, green space presence, and hard pavement, exhibited strong correlations with the spatial distribution of these particulate matter concentrations.

(2) Further analysis of the vegetation structures revealed that green spaces with 30%–70% canopy cover and single-layered woodlands had greater dust-retention capacity. The ability to regulate PM concentrations became weaker when the canopy cover increased above 70% or decreased below 30%. Notably, the half-closed single-layered broad-leaved woodland (H1B) and the half-closed single-layered mixed broadleaf-conifer woodland (H1M) emerged as the most effective configurations for mitigating air particulate matter concentrations.

(3) Seasonal trends indicated that PM2.5 and PM10 concentrations were highest in winter and lowest in summer, with intermediate levels observed in spring and autumn. Furthermore, air particulate concentrations exhibited a decreasing trend throughout the day, from 8:00 AM to 6:00 PM.

This study provides actionable recommendations for urban planners, policymakers, and landscape architects to optimize urban green spaces for improved air quality. Specifically, the research emphasizes the importance of integrating tree-dominated vegetation with half-closed canopy density and single-layered vertical structures into urban planning guidelines. Additionally, spatial configurations, such as maintaining appropriate edge densities and strategic placement relative to pollution sources, should be prioritized. To implement these findings, we recommend updating urban green space planning standards to incorporate nature-based solutions that balance ecological benefits and social applicability. Public engagement is also critical, and we propose disseminating scientific guidelines to encourage outdoor activities during optimal air quality periods. These strategies aim not only to mitigate air pollution but also to enhance public health and resilience in Xi'an and comparable urban settings, contributing to the sustainable development of healthy cities.

Supplementary Materials: The following supporting information can be downloaded at: https://www.mdpi.com/article/10.3390/f15122136/s1, Table S1: Deleted variables, standardized coefficients, t-values, Sig. values, VIF for the PM2.5 and PM10 LUR model.

Author Contributions: Conceptualization, X.Z., L.Q. and T.G.; methodology, X.Z., L.Q. and T.G.; software, X.Z., L.Q. and T.G.; validation, X.Z., L.Q. and T.G.; formal analysis, X.Z., L.Q. and T.G.; investigation, X.Z., C.W., J.G., Z.Z., Z.X. and X.L.; resources, L.Q. and T.G.; data curation, X.Z., L.Q. and T.G.; writing—original draft preparation, X.Z.; writing—review and editing, X.Z., L.Q. and T.G.; visualization, X.Z., L.Q. and T.G.; supervision, L.Q. and T.G.; project administration, L.Q. and T.G.; funding acquisition, L.Q. and T.G. All authors have read and agreed to the published version of the manuscript.

Funding: This research was funded by the National Natural Science Foundation of China (grant number: 31971722), Social Development Project of Xianyang Science and Technology Bureau (grant number: 2021ZDYF-SF-0022), Shaanxi Academy of Forestry Science and Technology Innovation Program Special Project (grant number: SXLK2021-0216), Shaanxi Academy of Forestry Science and Technology Innovation Program Special Project (grant number: SXLK2023-02-18), Shaanxi Province Ecospatial Governance Key Issues (grant number: 2022HZ1838), Shaanxi Province Ecospatial Governance Key Issues (grant number: 2022HZ1762).

Data Availability Statement: The data that support the findings of this study are available from the corresponding author upon reasonable request.

Conflicts of Interest: The authors declare no conflicts of interest.

References

1. Cheng, Z.; Li, L.; Liu, J. Identifying the spatial effects and driving factors of urban PM2.5 pollution in China. *Ecol. Indic.* **2017**, *82*, 61–75. [CrossRef]
2. Fu, H.; Chen, J. Formation, features and controlling strategies of severe haze-fog pollutions in China. *Sci. Total Environ.* **2017**, *578*, 121–138. [CrossRef] [PubMed]
3. Ozgen, S.; Ripamonti, G.; Cernuschi, S.; Giugliano, M. Ultrafine particle emissions for municipal waste-to-energy plants and residential heating boilers. *Rev. Environ. Sci. Bio/Technol.* **2012**, *11*, 407–415. [CrossRef]
4. Singh, N.; Murari, V.; Kumar, M.; Barman, S.C.; Banerjee, T. Fine particulates over south asia: Review and meta-analysis of PM2.5 source apportionment through receptor model. *Environ. Pollut.* **2017**, *223*, 121–136. [CrossRef]
5. Dong, F.; Zhang, S.; Long, R.; Zhang, X.; Sun, Z. Determinants of haze pollution: An analysis from the perspective of spatiotemporal heterogeneity. *J. Clean Prod.* **2019**, *222*, 768–783. [CrossRef]
6. WHO Global Air Quality Guidelines: Particulate Matter (PM2.5 and PM10), Ozone, Nitrogen Dioxide, Sulfur Dioxide and Carbon Monoxide. 2021. Available online: https://www.who.int/publications/i/item/9789240034228 (accessed on 22 September 2021).
7. Freer-Smith, P.H.; Beckett, K.P.; Taylor, G. Deposition velocities to sorbus aria, acer campestre, populus deltoides × trichocarpa 'beaupré', pinus nigra and × cupressocyparis leylandii for coarse, fine and ultra-fine particles in the urban environment. *Environ. Pollut.* **2005**, *133*, 157–167. [CrossRef]
8. Lu, F.; Xu, D.; Cheng, Y.; Dong, S.; Guo, C.; Jiang, X.; Zheng, X. Systematic review and meta-analysis of the adverse health effects of ambient PM2.5 and PM10 pollution in the chinese population. *Environ. Res.* **2015**, *136*, 196–204. [CrossRef]
9. Guan, W.; Zheng, X.; Chung, K.F.; Zhong, N. Impact of air pollution on the burden of chronic respiratory diseases in China: Time for urgent action. *Lancet* **2016**, *388*, 1939–1951. [CrossRef]
10. Schraufnagel, D.E.; Balmes, J.R.; Cowl, C.T.; De Matteis, S.; Jung, S.; Mortimer, K.; Perez-Padilla, R.; Rice, M.B.; Riojas-Rodriguez, H.; Sood, A.; et al. Air pollution and noncommunicable diseases: A review by the forum of international respiratory societies' environmental committee, part 1: The damaging effects of air pollution. *Chest* **2019**, *155*, 409–416. [CrossRef]
11. Peled, R. Air pollution exposure: Who is at high risk? *Atmos. Environ.* **2011**, *45*, 1781–1785. [CrossRef]
12. Luo, D.; Kuang, T.; Chen, Y.; Huang, Y.; Zhang, H.; Xia, Y. Air pollution and pregnancy outcomes based on exposure evaluation using a land use regression model: A systematic review. *Taiwan. J. Obstet. Gynecol.* **2021**, *60*, 193–215. [CrossRef] [PubMed]

13. Alotaibi, R.; Bechle, M.; Marshall, J.D.; Ramani, T.; Zietsman, J.; Nieuwenhuijsen, M.J.; Khreis, H. Traffic related air pollution and the burden of childhood asthma in the contiguous united states in 2000 and 2010. *Environ. Int.* **2019**, *127*, 858–867. [CrossRef] [PubMed]
14. Veerkamp, C.J.; Schipper, A.M.; Hedlund, K.; Lazarova, T.; Nordin, A.; Hanson, H.I. A review of studies assessing ecosystem services provided by urban green and blue infrastructure. *Ecosyst. Serv.* **2021**, *52*, 101367. [CrossRef]
15. Xie, Q.; Yue, Y.; Sun, Q.; Chen, S.; Lee, S.; Kim, S.W. Assessment of ecosystem service values of urban parks in improving air quality: A case study of wuhan, China. *Sustainability* **2019**, *11*, 6519. [CrossRef]
16. Yin, S.; Chen, D.; Zhang, X.; Yan, J. Review on the multi-scale interactions of urban forests and atmospheric particles: Affecting factors are scale-dependent among tree, stand and region. *Urban For. Urban Green.* **2022**, *78*, 127789. [CrossRef]
17. Hoek, G.; Beelen, R.; de Hoogh, K.; Vienneau, D.; Gulliver, J.; Fischer, P.; Briggs, D. A review of land-use regression models to assess spatial variation of outdoor air pollution. *Atmos. Environ.* **2008**, *42*, 7561–7578. [CrossRef]
18. Guo, Y.; Feng, N.; Christopher, S.A.; Kang, P.; Zhan, F.B.; Hong, S. Satellite remote sensing of fine particulate matter (PM2.5) air quality over beijing using modis. *Int. J. Remote Sens.* **2014**, *35*, 6522–6544. [CrossRef]
19. Sówka, I.; Badura, M.; Pawnuk, M.; Szymański, P.; Batog, P. The use of the gis tools in the analysis of air quality on the selected university campus in poland. *Archiwum Ochrony Srodowiska* **2020**, *46*, 100–106.
20. Chai, J.; Song, J.; Xu, Y.; Zhang, L.; Guo, B. Enhancing the applicability of satellite remote sensing for PM2.5 estimation using machine learning models in China. *J. Sens.* **2022**, *2022*, 7148682. [CrossRef]
21. Olvera, H.A.; Garcia, M.; Li, W.; Yang, H.; Amaya, M.A.; Myers, O.; Burchiel, S.W.; Berwick, M.; Pingitore, N.E. Principal component analysis optimization of a PM2.5 land use regression model with small monitoring network. *Sci. Total Environ.* **2012**, *425*, 27–34. [CrossRef]
22. Karroum, K.; Lin, Y.; Chiang, Y.; Ben Maissa, Y.; El Haziti, M.; Sokolov, A.; Delbarre, H. A review of air quality modeling. *MAPAN* **2020**, *35*, 287–300. [CrossRef]
23. Gulliver, J.; de Hoogh, K.; Fecht, D.; Vienneau, D.; Briggs, D. Comparative assessment of gis-based methods and metrics for estimating long-term exposures to air pollution. *Atmos. Environ.* **2011**, *45*, 7072–7080. [CrossRef]
24. Mo, Y.; Booker, D.; Zhao, S.; Tang, J.; Jiang, H.; Shen, J.; Chen, D.; Li, J.; Jones, K.C.; Zhang, G. The application of land use regression model to investigate spatiotemporal variations of PM2.5 in guangzhou, China: Implications for the public health benefits of PM2.5 reduction. *Sci. Total Environ.* **2021**, *778*, 146305. [CrossRef]
25. Ndiaye, A.; Shen, Y.; Kyriakou, K.; Karssenberg, D.; Schmitz, O.; Flückiger, B.; Hoogh, K.D.; Hoek, G. Hourly land-use regression modeling for NO_2 and PM2.5 in the netherlands. *Environ. Res.* **2024**, *256*, 119233. [CrossRef]
26. Zhu, C.; Zeng, Y. Effects of urban lake wetlands on the spatial and temporal distribution of air pm10 and PM2.5 in the spring in wuhan. *Urban For. Urban Green.* **2018**, *31*, 142–156. [CrossRef]
27. Gao, Y.; Wang, Z.; Liu, C.; Peng, Z. Assessing neighborhood air pollution exposure and its relationship with the urban form. *Build. Environ.* **2019**, *155*, 15–24. [CrossRef]
28. Wu, J.; Luo, K.; Wang, Y.; Wang, Z. Urban road greenbelt configuration: The perspective of PM2.5 removal and air quality regulation. *Environ. Int.* **2021**, *157*, 106786. [CrossRef]
29. Deng, S.; Ma, J.; Zhang, L.; Jia, Z.; Ma, L. Microclimate simulation and model optimization of the effect of roadway green space on atmospheric particulate matter. *Environ. Pollut.* **2019**, *246*, 932–944. [CrossRef]
30. Räsänen, J.V.; Holopainen, T.; Joutsensaari, J.; Ndam, C.; Pasanen, P.; Rinnan, Å.; Kivimäenpää, M. Effects of species-specific leaf characteristics and reduced water availability on fine particle capture efficiency of trees. *Environ. Pollut.* **2013**, *183*, 64–70. [CrossRef]
31. Wu, J.; Wang, Y.; Qiu, S.; Peng, J. Using the modified i-tree eco model to quantify air pollution removal by urban vegetation. *Sci. Total Environ.* **2019**, *688*, 673–683. [CrossRef]
32. Chen, L.; Liu, C.; Zhang, L.; Zou, R.; Zhang, Z. Variation in tree species ability to capture and retain airborne fine particulate matter (PM2.5). *Sci. Rep.* **2017**, *7*, 3206. [CrossRef] [PubMed]
33. Han, D.; Shen, H.; Duan, W.; Chen, L. A review on particulate matter removal capacity by urban forests at different scales. *Urban For. Urban Green.* **2020**, *48*, 126565. [CrossRef]
34. Gilliland, F.; Avol, E.; Kinney, P.; Jerrett, M.; Dvonch, T.; Lurmann, F.; Buckley, T.; Breysse, P.; Keeler, G.; de Villiers, T.; et al. Air pollution exposure assessment for epidemiologic studies of pregnant women and children: Lessons learned from the centers for children's environmental health and disease prevention research. *Environ. Health Perspect.* **2005**, *113*, 1447–1454. [CrossRef] [PubMed]
35. Wang, H.; Yilihamu, Q.; Yuan, M.; Bai, H.; Xu, H.; Wu, J. Prediction models of soil heavy metal(loid)s concentration for agricultural land in dongli: A comparison of regression and random forest. *Ecol. Indic.* **2020**, *119*, 106801. [CrossRef]
36. Ma, X.; Zou, B.; Deng, J.; Gao, J.; Longley, I.; Xiao, S.; Guo, B.; Wu, Y.; Xu, T.; Xu, X.; et al. A comprehensive review of the development of land use regression approaches for modeling spatiotemporal variations of ambient air pollution: A perspective from 2011 to 2023. *Environ. Int.* **2024**, *183*, 108430. [CrossRef]
37. Zhao, L.; Li, T.; Przybysz, A.; Liu, H.; Zhang, B.; An, W.; Zhu, C. Effects of urban lakes and neighbouring green spaces on air temperature and humidity and seasonal variabilities. *Sustain. Cities Soc.* **2023**, *91*, 104438. [CrossRef]
38. Sabaliauskas, K.; Jeong, C.; Yao, X.; Reali, C.; Sun, T.; Evans, G.J. Development of a land-use regression model for ultrafine particles in toronto, canada. *Atmos. Environ.* **2015**, *110*, 84–92. [CrossRef]

39. Briggs, D.J.; Collins, S.; Elliott, P.; Fischer, P.; Kingham, S.; Lebret, E.; Pryl, K.; VAN Reeuwijk, H.; Smallbone, K.; VAN DER Veen, A. Mapping urban air pollution using gis: A regression-based approach. *Int. J. Geogr. Inf. Sci.* **1997**, *11*, 699–718. [CrossRef]
40. Wong, T.-T.; Yeh, P.-Y. Reliable accuracy estimates from k-fold cross validation. *IEEE Trans. Knowl. Data Eng.* **2020**, *32*, 1586–1594. [CrossRef]
41. Bivand, R.S.; Pebesma, E.; Gómez-Rubio, V. *Applied Spatial Data Analysis With R*, 2nd ed.; Springer: New York, NY, USA, 2013; pp. 245–250.
42. Fang, J.; Guo, K.; Wang, G.; Tang, Z.; Xie, Z.; Shen, Z.; Wang, R.; Qiang, S.; Liang, C.; Da, L.; et al. Vegetation classification system and classification of vegetation types used for the compilation of vegetation of China. *Chin. J. Plant Ecol.* **2020**, *44*, 96–110. (In Chinese) [CrossRef]
43. Ma, J.; Cao, Y.; Xu, J.; Qu, Y.; Yu, Z. PM2.5 concentration distribution patterns and influencing meteorological factors in the central and eastern China during 1980–2018. *J. Clean. Prod.* **2021**, *311*, 127565. [CrossRef]
44. Eeftens, M.; Beelen, R.; de Hoogh, K.; Bellander, T.; Cesaroni, G.; Cirach, M.; Declercq, C.; Dėdelė, A.; Dons, E.; de Nazelle, A.; et al. Development of land use regression models for PM2.5, PM2.5 absorbance, PM10 and pmcoarse in 20 european study areas; Results of the escape project. *Environ. Sci. Technol.* **2012**, *46*, 11195–11205. [CrossRef] [PubMed]
45. de Hoogh, K.; Wang, M.; Adam, M.; Badaloni, C.; Beelen, R.; Birk, M.; Cesaroni, G.; Cirach, M.; Declercq, C.; Dėdelė, A.; et al. Development of land use regression models for particle composition in twenty study areas in europe. *Environ. Sci. Technol.* **2013**, *47*, 5778–5786. [CrossRef] [PubMed]
46. Singh, N.; Singh, S. Interstitial lung diseases and air pollution: Narrative review of literature. *Pulm. Ther.* **2021**, *7*, 89–100. [CrossRef]
47. Ehrlich, C.; Noll, G.; Kalkoff, W.D.; Baumbach, G.; Dreiseidler, A. PM10, PM2.5 and PM1.0—Emissions from industrial plants—Results from measurement programmes in germany. *Atmos. Environ.* **2007**, *41*, 6236–6254. [CrossRef]
48. Kim, K.H.; Kumar, P.; Szulejko, J.E.; Adelodun, A.A.; Junaid, M.F.; Uchimiya, M.; Chambers, S. Toward a better understanding of the impact of mass transit air pollutants on human health. *Chemosphere* **2017**, *174*, 268–279. [CrossRef]
49. Casotti Rienda, I.; Alves, C.A. Road dust resuspension: A review. *Atmos. Res.* **2021**, *261*, 105740. [CrossRef]
50. Dirgawati, M.; Heyworth, J.S.; Wheeler, A.J.; Mccaul, K.A.; Blake, D.; Boeyen, J.; Cope, M.; Yeap, B.B.; Nieuwenhuijsen, M.; Brunekreef, B.; et al. Development of land use regression models for particulate matter and associated components in a low air pollutant concentration airshed. *Atmos. Environ.* **2016**, *144*, 69–78. [CrossRef]
51. Han, D.; Zhang, T.; Zhang, X.; Tan, Y. Study on spatiotemporal characteristics and influencing factors of pedestrian-level PM2.5 concentrations in outdoor open spaces of harbin in winter, using a generalized additive model (gam). *Urban CLim.* **2022**, *46*, 101313. [CrossRef]
52. Abhijith, K.V.; Kumar, P.; Gallagher, J.; Mcnabola, A.; Baldauf, R.; Pilla, F.; Broderick, B.; Di Sabatino, S.; Pulvirenti, B. Air pollution abatement performances of green infrastructure in open road and built-up street canyon environments—A review. *Atmos. Environ.* **2017**, *162*, 71–86. [CrossRef]
53. Janhäll, S. Review on urban vegetation and particle air pollution–deposition and dispersion. *Atmos. Environ.* **2015**, *105*, 130–137. [CrossRef]
54. Nowak, D.J.; Hirabayashi, S.; Bodine, A.; Hoehn, R. Modeled PM2.5 removal by trees in ten U.S. Cities and associated health effects. *Environ. Pollut.* **2013**, *178*, 395–402. [CrossRef] [PubMed]
55. Kabisch, N.; van den Bosch, M.; Lafortezza, R. The health benefits of nature-based solutions to urbanization challenges for children and the elderly—A systematic review. *Environ. Res.* **2017**, *159*, 362–373. [CrossRef] [PubMed]
56. Jennings, V.; Rigolon, A.; Thompson, J.; Murray, A.; Henderson, A.; Gragg, R.S. The dynamic relationship between social cohesion and urban green space in diverse communities: Opportunities and challenges to public health. *Int. J. Environ. Res. Public Health* **2024**, *21*, 800. [CrossRef] [PubMed]
57. Wolch, J.R.; Byrne, J.; Newell, J.P. Urban green space, public health, and environmental justice: The challenge of making cities 'just green enough'. *Landsc. Urban Plan.* **2014**, *125*, 234–244. [CrossRef]
58. Rahman, A.; Luo, C.; Khan, M.H.R.; Ke, J.; Thilakanayaka, V.; Kumar, S. Influence of atmospheric PM2.5, PM10, O_3, CO, NO_2, SO_2, and meteorological factors on the concentration of airborne pollen in guangzhou, China. *Atmos. Environ.* **2019**, *212*, 290–304. [CrossRef]
59. Douglas, A.N.J.; Irga, P.J.; Torpy, F.R. Determining broad scale associations between air pollutants and urban forestry: A novel multifaceted methodological approach. *Environ. Pollut.* **2019**, *247*, 474–481. [CrossRef]
60. Mcdonald, A.G.; Bealey, W.J.; Fowler, D.; Dragosits, U.; Skiba, U.; Smith, R.I.; Donovan, R.G.; Brett, H.E.; Hewitt, C.N.; Nemitz, E. Quantifying the effect of urban tree planting on concentrations and depositions of pm10 in two uk conurbations. *Atmos. Environ.* **2007**, *41*, 8455–8467. [CrossRef]
61. Cai, L.; Zhuang, M.; Ren, Y. A landscape scale study in southeast China investigating the effects of varied green space types on atmospheric PM2.5 in mid-winter. *Urban For. Urban Green.* **2020**, *49*, 126607. [CrossRef]
62. Cheng, B.; Ma, Y.; Feng, F.; Zhang, Y.; Shen, J.; Wang, H.; Guo, Y.; Cheng, Y. Influence of weather and air pollution on concentration change of PM2.5 using a generalized additive model and gradient boosting machine. *Atmos. Environ.* **2021**, *255*, 118437. [CrossRef]
63. Oliveri Conti, G.; Heibati, B.; Kloog, I.; Fiore, M.; Ferrante, M. A review of airq models and their applications for forecasting the air pollution health outcomes. *Environ. Sci. Pollut. Res.* **2017**, *24*, 6426–6445. [CrossRef] [PubMed]

64. Ma, X.; Longley, I.; Gao, J.; Salmond, J. Evaluating the effect of ambient concentrations, route choices, and environmental (in)justice on students' dose of ambient no2 while walking to school at population scales. *Environ. Sci. Technol.* **2020**, *54*, 12908–12919. [CrossRef] [PubMed]
65. Shi, Y.; Lau, K.K.; Ng, E. Incorporating wind availability into land use regression modelling of air quality in mountainous high-density urban environment. *Environ. Res.* **2017**, *157*, 17–29. [CrossRef] [PubMed]
66. Gómez-Moreno, F.J.; Artíñano, B.; Ramiro, E.D.; Barreiro, M.; Núñez, L.; Coz, E.; Dimitroulopoulou, C.; Vardoulakis, S.; Yagüe, C.; Maqueda, G.; et al. Urban vegetation and particle air pollution: Experimental campaigns in a traffic hotspot. *Environ. Pollut.* **2019**, *247*, 195–205. [CrossRef]
67. Grote, R.; Samson, R.; Alonso, R.; Amorim, J.; Carinanos, P.; Churkina, G.; Fares, S.; Le Thiec, D.; Niinemets, Ü.; Mikkelsen, T.; et al. Functional traits of urban trees: Air pollution mitigation potential. *Front. Ecol. Environ.* **2016**, *14*, 543–550. [CrossRef]
68. Xing, Y.; Brimblecombe, P. Trees and parks as "the lungs of cities". *Urban For. Urban Green.* **2020**, *48*, 126552. [CrossRef]
69. Qiu, L.; Liu, F.; Zhang, X.; Gao, T. Difference of airborne particulate matter concentration in urban space with different green coverage rates in baoji, China. *Int. J. Environ. Res. Public Health* **2019**, *16*, 1465. [CrossRef]
70. Niu, X.; Li, Y.; Li, M.; Zhang, T.; Meng, H.; Zhang, Z.; Wang, B.; Zhang, W. Understanding vegetation structures in green spaces to regulate atmospheric particulate matter and negative air ions. *Atmos. Pollut. Res.* **2022**, *13*, 101534. [CrossRef]
71. Tuheti, A.; Deng, S.; Li, J.; Li, G.; Lu, P.; Lu, Z.; Liu, J.; Du, C.; Wang, W. Spatiotemporal variations and the driving factors of PM2.5 in xi'an, China between 2004 and 2018. *Ecol. Indic.* **2023**, *146*, 109802. [CrossRef]
72. Zhang, Z.; Zhou, Y.; Zhao, N.; Li, H.; Tohniyaz, B.; Mperejekumana, P.; Hong, Q.; Wu, R.; Li, G.; Sultan, M.; et al. Clean heating during winter season in northern China: A review. *Renew. Sustain. Energy Rev.* **2021**, *149*, 111339. [CrossRef]
73. Shao, M.; Xu, X.; Lu, Y.; Dai, Q. Spatio-temporally differentiated impacts of temperature inversion on surface PM2.5 in eastern China. *Sci. Total Environ.* **2023**, *855*, 158785. [CrossRef] [PubMed]
74. Li, Q.; Zhang, H.; Jin, X.; Cai, X.; Song, Y. Mechanism of haze pollution in summer and its difference with winter in the north China plain. *Sci. Total Environ.* **2022**, *806*, 150625. [CrossRef] [PubMed]
75. Gao, G.; Sun, F.; Thao, N.; Lun, X.; Yu, X. Different concentrations of tsp, pm10, PM2.5, and pm1 of several urban forest types in different seasons. *Pol. J. Environ. Stud.* **2015**, *24*, 2387–2395. [CrossRef]
76. Sun, L.; Wang, S.; Liu, S.; Yao, L.; Luo, W.; Shukla, A. A completive research on the feasibility and adaptation of shared transportation in mega-cities-a case study in beijing. *Appl. Energy* **2018**, *230*, 1014–1033. [CrossRef]
77. Wu, J.; Wang, X.; Wang, H.; Gong, D.; Yang, J.; Jiang, B.; Zhang, C.; Deng, X.; Zhou, L.; Wang, B. Rapidly alleviating particulate matter pollution while maintaining high-speed economic development in the "world's factory". *J. Clean Prod.* **2020**, *266*, 121844. [CrossRef]
78. Lin, B.; Zhu, J. Is the implementation of energy saving and emission reduction policy really effective in chinese cities? A policy evaluation perspective. *J. Clean Prod.* **2019**, *220*, 1111–1120. [CrossRef]
79. Shi, T.; Dirienzo, N.; Requia, W.J.; Hatzopoulou, M.; Adams, M.D. Neighbourhood scale nitrogen dioxide land use regression modelling with regression kriging in an urban transportation corridor. *Atmos. Environ.* **2020**, *223*, 117218. [CrossRef]
80. Zeng, L.; Hang, J.; Wang, X.; Shao, M. Influence of urban spatial and socioeconomic parameters on PM2.5 at subdistrict level: A land use regression study in shenzhen, China. *J. Environ. Sci.* **2022**, *114*, 485–502. [CrossRef]
81. Zhao, L.; Li, T.; Przybysz, A.; Guan, Y.; Ji, P.; Ren, B.; Zhu, C. Effect of urban lake wetlands and neighboring urban greenery on air pm10 and PM2.5 mitigation. *Build. Environ.* **2021**, *206*, 108291. [CrossRef]

forests

MDPI

Article

The Added Value of Urban Trees (*Tilia tomentosa* Moench, *Fraxinus excelsior* L. and *Pinus nigra* J.F. Arnold) in Terms of Air Pollutant Removal

Slaveya Petrova [1,2]

[1] Department of Ecology and Environmental Conservation, Faculty of Biology, Plovdiv University "Paisii Hilendarski", 4000 Plovdiv, Bulgaria; slaveya_petrova@uni-plovdiv.bg or sl_petrova@au-plovdiv.bg

[2] Department of Microbiology and Ecological Biotechnologies, Faculty of Plant Protection and Agroecology, Agricultural University, 4000 Plovdiv, Bulgaria

Abstract: The serious densification of human settlements necessitates an increase in the role and importance of green infrastructures in the overall functioning of urban ecosystems. Accordingly, the aim of the present study was to (1) assess the efficiency of air pollutant removal (potentially toxic elements) of three common ornamental trees (*Tilia tomentosa* Moench, *Fraxinus excelsior* L. and *Pinus nigra* J.F. Arnold) and (2) model the air quality regulatory services (removal of PM_{10}, $PM_{2.5}$ and NO_2). Three different approaches were applied—enrichment factor (EF) and metal accumulation factor (MAI) per tree species, as well as simulation modeling for the whole urban forest. The MAI values of the three studied species were found to be very similar, in the range of 22.35 to 23.08, which suggests that these species could be a good choice for planting in urban areas with worsened air quality. The highest EF values were observed for U (3–18), followed by As (1.6–2.56) and Sr (0.87–2.46). The potential of urban forests in countering air pollution was highlighted by three simulated scenarios for PM_{10}, $PM_{2.5}$ and NO_2 removal. The highest removal efficiency was calculated for evergreen species, followed by the mixed composition of deciduous (90%) and evergreen trees (10%), and the scenario with wholly deciduous trees had the lowest one. The contribution of nature-based solutions in meeting air quality standards and enhancing resilience in urban areas was clearly demonstrated. The functional complementarity of the different functional tree groups (coniferous, evergreen and deciduous broad-leaved species) was proven to be crucial for the support of both functional stabilities of the phytocenosis under diverse climatic conditions and during the change of seasonal cycles in the vegetation.

Keywords: PM_{10}; $PM_{2.5}$; NO_2; heavy metals; enrichment factor; metal accumulation index; ecosystem services

1. Introduction

Humanity is a part of nature and depends on it not only for its own existence, but also for providing its needs, resources for the economy, for its development and its well-being. Ecosystems (natural and anthropogenic) provide food, materials, raw materials, water and other resources without which our life is impossible [1,2]. Far more important and invaluable, however, are all the intangible benefits that ecosystems provide us—air, water and soil purification; climate regulation; biodiversity conservation and even spiritual satisfaction, inspiration and recreation in nature [3]. In fact, the term "ecosystem services" includes all those direct or indirect benefits that nature provides to humanity. In addition to the in-depth understanding of human dependence on natural processes at different spatial and temporal scales, there is also the need to assess the monetary value of these ecosystem services according to economic and management criteria and indicators [4–6].

A city is a complex ecological system that consists of social, economic and natural factors. The quality of life in an urbanized environment is a function of the combined impact of both abiotic and biotic factors (air, soil, water, microclimate, biodiversity, vegetation, etc.). Urbanization leads to the continuous change of ecosystems and landscapes towards increasingly affected air, water, soil, plant and animal communities [7–9]. Air pollution is considered a primary threat to human health in Europe due to the permanent inhalation of gases and fine particles, even at low levels [10]. For this reason, the regulating services of green infrastructures are regarded to be crucial for human well-being [11]. In this group, the priorities are urban air quality improvement [12,13], microclimate regulation and urban heat island mitigation [14], as well as precipitation runoff limitation [15].

Green systems in cities are based on the inclusion of the natural environment into the highly urbanized urban structure. An appropriate green system can play an effective role in purifying the air, improving the microclimate, suppressing noise, beautifying the surrounding space and improving the comfort of living in general [11]. There are many other ecosystem services that could synergize for sustainable urban development; hence, green infrastructures need a proper species composition, especially in terms of ornamental trees, for their successful implementation.

The serious densification of settlement territories with an increase in the density and intensity of construction necessitates an increase in the role and importance of green systems in the overall functioning of the urban ecosystem. The construction and management of green infrastructure in settlements should be carried out on the basis of in-depth studies on the mechanisms of interaction between the urban environment and vegetation, which in turn requires an assessment of the species-specific tolerance and adaptation abilities of individual plant species [16], and thus their ability to serve as biomonitors and as "green filters" of the urban environment [17,18]. Numerous studies have discussed the potential of green systems to improve air quality, even as a physical barrier between air pollution and people [19]. Tremper et al. [20,21] planted an ivy green screen around a playground in a primary school in Enfield, UK. The reduction in the daily PM_{10} concentrations on the playground side of the screen reached 38%, while for the hourly PM10 level it was 41% [20]. Daily NO_2 concentrations after the plants' growth decreased by 15 $\mu g\ m^{-3}$ (21.8%), and the reduction in hourly NO_2 concentrations was found to be up to 18.3 $\mu g\ m^{-3}$ (22.5%) [21]. This study clearly demonstrated the effectiveness of green barriers in reducing air pollution, especially during daytime hours, when both emissions and exposure are highest. Another study from England [22] measured and compared different pollutants' concentrations in front of and behind vegetation of three types—only shrubs, trees and shrubs and only trees. At a local level, the authors revealed that the shrub scenario offered better improvement in air quality, at both away-road and close-road sites (up to 63% reduction in black carbon level), followed by the combination of shrubs + trees. However, the capacity of urban forests to provide regulatory services is often limited to small green spaces. When regarding a city's scale, the effect of vegetation is significantly lower. Notably, Baró [23] showed that only 0.47% of the carbon emissions and 0.52% of the NO_2 emissions in Barcelona (Spain) could be absorbed by urban vegetation. However, there are some options to enhance this regulatory service if the selection of plant species with higher retention potential, spatial scale, urban heterogeneity, and the target air pollutants are taken into account [19].

Such problems could be solved by the implementation of a nature-based solution aiming to improve the urban green infrastructure. This could be done by studying trees' tolerance to air pollution and then performing a revision and revaluation of the adaptation ability of the new planted ornamental trees to the worsened urban conditions [24]. The second step should be related to the selection of tree species, among the possible ones, with a higher removal capacity towards air pollutants, and thus with a greater potential for providing such ecosystem services. Accordingly, the aim of the present study was to (1) assess the efficiency of air pollutant removal (potentially toxic elements) of three common ornamental trees (*Tilia tomentosa* Moench, *Fraxinus excelsior* L. and *Pinus nigra* J.F. Arnold) and (2) model the air quality regulatory services (removal of PM_{10}, $PM_{2.5}$ and NO_2) of the urban forest.

2. Materials and Methods

2.1. Study Area

The city of Plovdiv, Bulgaria (42°8′9.9492″ N; 24°44′31.8048″ E), one of the most polluted cities in Europe for more than 15 years [25], was chosen as the study area. The urban conditions in this city are quite harmful for plants' well-being. The main factors contributing to the worsened air quality have been considered to be domestic heating (during the cold period) and motor traffic [26].

For proper identification of the local air quality, we chose four monitoring plots that should reflect different types of anthropogenic impacts (Figure 1). They were situated in the NE, SE, SW and NW directions from the city center, aiming to include areas with different traffic intensities and domestic heating loads (Table 1). Plot 1 was located in the central part, NE direction, adjacent to a busy road junction with heavy traffic and a high level of traffic emissions. Plot 2 was also located in the central part, SE direction, but the area is subject to moderate motor traffic and a medium level of domestic heating. Plot 3 was located in the west suburb, SW direction, near a busy road junction with very heavy railroad and vehicle traffic. Plot 4 was also located in the west suburb, NW direction, but within a big forest, excluding traffic and buildings, so we planned to use it as a conditional control for the urban environment [24,27].

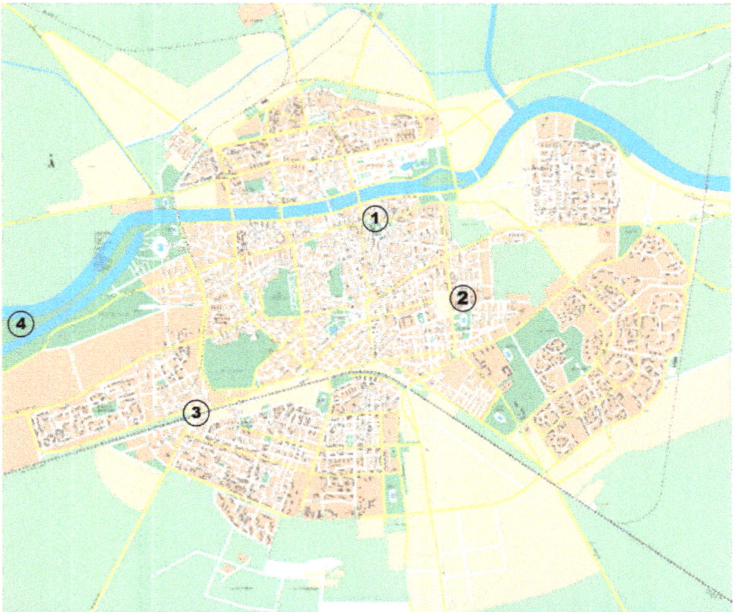

Figure 1. Map of the city of Plovdiv (Bulgaria) and locations of the four monitoring plots.

Table 1. Urbanization intensity evaluated on the basis of built-up area and traffic volume.

Motor Traffic (Car Number per 10 min)	Built-Up Area (%)	Urbanization Intensity	Experimental Plot
0–10	0–10	Low	Plot 4
10–50	10–40	Medium	Plot 2
50–150	40–70	High	Plot 1
>150	>70	Very high	Plot 3

Traffic load was assessed on the basis of the number of cars (or motorcycles) in a period of ten minutes at 8 a.m. on a working day [27]. The percentage of built-up area within a 1 km^2 area was quantified in ArcGis version 11 software using the city general layout plan (Municipality of Plovdiv). As these two features can be quantified in any urban environment, regardless of geographical, cultural and historical variation, they represent an adequate proxy of urbanization intensity [27].

2.2. Experimental Design

During the spring of 2015, 8-year-old seedling material was purchased from a certified nursery and planted at the abovementioned monitoring plots (Figure 2), with 3 individuals per species per plot as a group planting [28]. Periodic observations were made on their development, physiology and health status between 2015 and 2020, and some results from this 6-year period were published in previous studies [16,24,27].

Figure 2. Photos of experimental plots two years after planting.

Leaf sampling was performed in August when the leaves were fully developed. The Expert Tree Pruner with a telescopic handle of 1.5–2.5 m (Draper Tools, Draper Tools Ltd., Hampshire, UK) was used to cut off some branches from the crown. On average, 80–100 fully expanded leaves (needles) per tree were taken, and a representative homogenous sample was prepared for analyses. All the samples were stored in clean, labeled polyethylene bags to avoid contamination during transport [17].

2.3. Chemical Analyses

About 1 g of dried and ground leaves was digested with 5 mL of nitric acid (Merck, Merck KGaA, Darmstadt, Germany) for 24 h at room temperature. After that, the samples were treated for 5 min at maximum power (600 W) in closed cuvettes in a microwave digestion system (Microwave Digestion System CEM MDS 81D, CEM Corporation, Matthews, NC, USA). After cooling at room temperature for 1 h, the cuvettes were opened and 2 mL HNO_3 and 3 mL 30% H_2O_2 were added and allowed to react for another 1 h. The cuvettes were then sealed and placed again for 10 min at 600 W for complete decomposition of the organic matter. The resulting filtrate was diluted with double distilled water to a volume of 50 mL.

The contents of 12 potentially toxic elements (As, Cd, Cr, Cu, Fe, Mn, Ni, Pb, Sr, U, V and Zn) were determined by inductively coupled plasma mass spectrometry (ICP-MS) using an Agilent 7700 ICP-MS (Agilent Technologies Inc., Waldbronn, Germany) instrument (2009), DF 1000. All samples, blanks and standards were diluted with international standardizing agents— final concentrations of 50 ppb Ge and 5 ppb Rh in the solutions. Multy VI (MERCK) calibration standards were prepared immediately prior to the analyses from 1 to 1000 ppb in 0.05 vol% HNO_3. A single standard of 100 ppt Hg was also used in the calibration. Quality control was ensured with a plant standard (NCS DC73348) [29]. The data for each zone are the arithmetic mean of three samples, and the data for each individual sample are the arithmetic mean of three analytical measurements.

2.4. Calculations of Air Pollutant Removal Efficiency

To assess the air pollutant removal efficiency, three different approaches were applied—enrichment factor and metal accumulation factor per tree species, as well as a modeling method for the whole urban forest.

2.4.1. Enrichment Factor (EF)

The enrichment factor (EF) was calculated in order to assess the accumulation of potentially toxic elements in the tree leaves/needles [24]. We used the formula of Mingorance et al. [30]:

$$EF = \frac{C_{urban}}{C_{control}}$$

where C_{urban} and $C_{control}$ are the contents of each element in the plant biomass sampled from the urban (Plots 1–3) and rural/control environments (Plot 4), respectively. We used the following scale for the *EF* to assess the environmental pollution level: (1) EF $\leq$ 1.2—no pollution; (2) 1.2 < EF $\leq$ 2.2—low pollution; (3) 2.2 < EF $\leq$ 3.3—medium pollution; (4) 3.3 < EF $\leq$ 4.3—heavy pollution; (5) EF > 4.3—very heavy pollution [31].

2.4.2. Metal Accumulation Index (MAI)

The metal accumulation index (MAI) [32] gives an option to evaluate the ability of different tree species to accumulate HMs from the ambient air. The MAI values were calculated using the following equation:

$$MAI = \left(\frac{1}{N}\right)\sum_{j=1}^{N} I_j$$

where N is the total number of HMs analyzed, and $I_j = x/\delta x$ is the sub-index for variable j, obtained by dividing the mean value (x) of each metal by its standard deviation (δx).

2.4.3. Simulation Model

The mitigating role of urban vegetation for PM_{10}, $PM_{2.5}$ and NO_2 was estimated by considering the minimum (period without domestic heating, 1 April–30 September) and maximum (period with domestic heating, 1 Ocotber–31 March) concentration values (C) recorded in the Municipality of Plovdiv during the same period (Table 2). These were based on pollutants' yearly time series, obtained by two automatic monitoring stations of air quality in the city. One of them is focused mainly on traffic-related pollution, while the other aims to assess domestic-related pollution. Detailed measurement data have been obtained by the municipal program of air quality management in the city of Plovdiv from 2018 to 2023 [33] and have been averaged into Table 2.

Table 2. Pollutant level in the ambient air (data from automatic measurement system).

Pollutant	Average Concentration per Year (Sum of All Emitters)	Average Concentration during the Period 1 Ocotber–31 March	Average Concentration during the Period 1 April–30 September	Emitter	Annual Emissions per Emitter	Annual Concentration per Emitter
PM_{10}	48 μg/m³	65.67 μg/m³	31.17 μg/m³	Domestic heating	245 t/y	34.4 μg/m³
				Traffic	23 t/y	3.3 μg/m³
				Industry	2 t/y	0.3 μg/m³
				Background	-	10 μg/m³
$PM_{2.5}$	34.2 μg/m³	45.73	22.67 μg/m³	Domestic heating	172.1 t/y	26.2 μg/m³
				Traffic	15.3 t/y	2.5 μg/m³
				Industry	1 t/y	0.2 μg/m³
				Background	-	5.2 μg/m³
NO_2	30 μg/m³	38.67	21.33 μg/m³	Domestic heating	65.1 t/y	2.4 μg/m³
				Traffic	342.9 t/y	12.3 μg/m³
				Industry	144.6 t/y	5.3 μg/m³
				Background	-	10 μg/m³

The mitigating role of urban vegetation for three air pollutants was estimated following Nowak [34], as described in Manes et al. [35]. This method is implemented in the present study using the following equation:

$$Q = F \times L \times T \times 0.5 \times LAI_i$$

where the following variables are used:

Q—the amount of pollutant in the air, removed by trees in a certain period of time;
F—the flow of the pollutant;
L—the total green coverage of the area;
T (in seconds)—the vegetative period of urban trees, which was considered equal to 244 days (8 months, May—November) for deciduous and 365 days for evergreen species;
0.5—the rate of resuspension of particles that return to the atmosphere [36];
LAI_i —a variable used to indicate the removal of 1 m² of soil covered by plants from a functional group (for trees, LAI_i = 4).

For the purposes of physiological modeling, the woody vegetation was divided into two functional groups [11,37]: (1) broad-leaved deciduous trees and (2) evergreen trees (broad-leaved and conifers), allowing an assessment of the ecosystem services in the autumn–winter and spring—summer periods separately. Thus, three different scenarios were generated for ecosystem service simulation modeling purposes: vegetation covers consisting of only evergreen trees (broad-leaved and conifers) (Scenario 1), only broad-leaved deciduous trees (Scenario 2) and a current species composition (dendroflora) (Scenario 3). This approach has the advantage of providing a methodological framework to serve in the formulation of strategies for the development and sustainable management of a green system, as well as to increase its added value in the context of an increased capacity to purify air from pollutants.

The above equation was used for estimating the pollutant removal by simulating as if all the trees belonged to one single functional group. In the three scenarios, one for each group, the total area covered by tree vegetation within the Municipality of Plovdiv was attributed to each one of them. For this simulation, an average between the maximum and minimum deposition values of each pollutant (t ha^{-1}) was considered. Estimates were carried out on a seasonal basis, with and without domestic heating. The data on the area of the green system and the species composition (ratio of deciduous/coniferous species) were according to the Program for the Development, Maintenance and Protection of the Green System of the City of Plovdiv [37]. The data are as follows: 10,187.6 ha total city area, 381.5 ha total green coverage of trees, 90% deciduous trees and 10% coniferous species.

2.5. Statistical Evaluation

The raw data obtained were processed using the SPSS 21 software package. The relationships between the analyzed variables were assessed through Pearson's correlation coefficients. Cluster analysis (complete linkage, Pearson r-distance) was applied for assessing elements' relationships as well as for evaluating accumulation patterns by plant species ($p < 0.05$).

3. Results

3.1. Leaf Concentrations of Potentially Toxic Elements

The concentrations of the studied potentially toxic elements (mainly heavy metals) in the leaves of urban trees are presented in Table 3. Generally, the concentrations of the potentially toxic elements in urban tree leaves were higher at Site 3 and Site 1 (with some exceptions in Tilia), subjected to more intensive traffic, than at Site 2 with a moderate anthropogenic load. The lowest values were measured in samples from Site 4 (with some exceptions in Tilia), subjected to a low anthropogenic impact (very low traffic and low level of air pollution); thus, we used them as a "conditional control" of the urban environment. When comparing the plant species, they showed similar bioaccumulations of As, Cr, U and Zn. Tilia tomentosa was found to be a good accumulator of Fe, Mn, Ni, Pb and Sr; Fraxinus excelsior was a good accumulator of Cu and Sr and Pinus nigra was a good accumulator of Cd, Mn and Pb. The intensity of urban pressure on trees was very well expressed as all three species accumulated high contents of Pb, Sr, U, V and Zn at Site 3 (very heavy railroad and vehicle traffic, resulting in a very high level of air pollution).

Table 3. Concentrations of potentially toxic elements in urban tree leaves, mg/kg (SD)

Species	Site	As	Cd	Cr	Cu	Fe	Mn	Ni	Pb	Sr	U	V	Zn
Tilia tomentosa	1	0.51 (0.03)	0.05 (0.002)	0.76 (0.016)	5.5 (0.26)	151 (7.7)	55 (2.09)	1.03 (0.04)	2 (0.09)	68 (2.7)	0.01 (0.001)	0.21 (0.01)	14 (0.52)
	2	0.43 (0.024)	0.04 (0.002)	0.46 (0.032)	5.8 (0.18)	148 (5.18)	47 (1.79)	1.04 (0.07)	1.5 (0.06)	106 (4.2)	0.01 (0.001)	0.19 (0.01)	18 (0.7)
	3	0.35 (0.018)	0.08 (0.003)	0.53 (0.212)	4.5 (0.19)	155 (6.67)	65 (2.28)	0.78 (0.05)	4.8 (0.17)	123 (4.6)	4.01 (0.148)	0.24 (0.01)	16 (0.51)
	4	0.23 (0.01)	0.09 (0.005)	0.36 (0.019)	5.2 (0.22)	99 (5.15)	54 (2.05)	0.81 (0.05)	2.8 (0.11)	83 (3.6)	0.27 (0.001)	0.23 (0.01)	13 (0.53)
	Av	0.51 (0.03)	0.05 (0.002)	0.76 (0.016)	5.5 (0.26)	151 (7.7)	55 (2.09)	1.03 (0.04)	2.0 (0.09)	68 (2.7)	0.01 (0.001)	0.21 (0.01)	14 (0.52)
Fraxinus excelsior	1	0.38 (0.019)	0.04 (0.003)	0.55 (0.011)	4.5 (0.16)	122 (5.49)	36 (1.13)	0.53 (0.02)	1.3 (0.06)	47 (2.3)	0.02 (0.001)	0.16 (0.01)	14 (0.5)
	2	1.07 (0.055)	0.02 (0.001)	0.35 (0.01)	8.4 (0.29)	105 (3.78)	20 (0.8)	0.52 (0.03)	0.8 (0.03)	76 (3.3)	0.01 (0.001)	0.11 (0.01)	18 (0.68)
	3	0.22 (0.016)	0.06 (0.003)	0.27 (0.011)	29.4 (1.12)	106 (4.66)	31 (1.12)	0.71 (0.04)	2 (0.07)	98 (4.4)	0.64 (0.022)	0.14 (0.01)	15 (0.51)
	4	0.18 (0.011)	0.03 (0.003)	0.21 (0.009)	11.8 (0.44)	60 (2.94)	21 (0.74)	0.45 (0.02)	1 (0.05)	90 (3.9)	0.06 (0.003)	0.08 (0.01)	14 (0.5)
	Av	0.46 (0.024)	0.038 (0.003)	0.345 (0.01)	13.53 (0.5)	98 (4.22)	27 (0.95)	0.55 (0.03)	1.28 (0.05)	78 (3.48)	0.18 (0.007)	0.12 (0.01)	15 (0.55)
Pinus nigra	1	0.16 (0.01)	0.15 (0.007)	0.49 (0.017)	3.0 (0.13)	185 (7.96)	19 (0.87)	0.5 (0.03)	4.4 (0.18)	7.0 (0.59)	0.03 (0.001)	0.29 (0.02)	25 (0.88)
	2	0.17 (0.009)	0.06 (0.003)	0.4 (0.02)	2.0 (0.01)	125 (5.38)	10 (0.43)	0.25 (0.01)	1.3 (0.05)	4.0 (0.44)	0.02 (0.001)	0.13 (0.01)	17 (0.71)
	3	0.18 (0.011)	0.1 (0.005)	0.62 (0.027)	2.7 (0.16)	151 (4.98)	8.0 (0.34)	0.31 (0.02)	4.7 (0.18)	42 (2.02)	3.5 (0.13)	0.31 (0.01)	27 (1.7)
	4	0.09 (0.006)	0.07 (0.005)	0.34 (0.014)	1.5 (0.06)	143 (5.43)	7.0 (0.35)	0.29 (0.02)	4.1 (0.17)	6.0 (0.46)	0.05 (0.002)	0.24 (0.01)	13 (0.44)
	Av	0.15 (0.01)	0.095 (0.005)	0.46 (0.02)	2.3 (0.09)	151 (5.94)	11.0 (0.49)	0.34 (0.02)	3.63 (0.13)	14.75 (0.88)	0.9 (0.03)	0.24 (0.01)	21 (0.93)

Av—average value per studied urban area (Plovdiv, Bulgaria).

Mathematical processing of the data showed 14 positive and 2 negative correlations ($p < 0.05$) (Table 4). The most significant positive correlations were those for V and Pb (r = 0.89), Cd (r = 0.82), Fe (r = 0.79) and Cr (r = 0.76) and between Cd and Pb (r = 0.84), Mn and Ni (r = 0.89). Negative relationships existed between Cu and Fe (r = −0.75) as well as between Cu and V (r = −0.64) ($p < 0.05$). Only for the element As no correlation was found.

Table 4. Values of the correlation coefficient ($p < 0.05$).

	V	Cr	Mn	Fe	Ni	Cu	Zn	As	Sr	Cd	Pb	U
V	1.00											
Cr	0.76	1.00										
Mn	0.06	−0.03	1.00									
Fe	0.79	0.49	0.06	1.00								
Ni	0.03	−0.00	0.89	0.12	1.00							
Cu	−0.64	−0.49	−0.03	−0.75	−0.00	1.00						
Zn	0.53	0.64	−0.43	0.50	−0.30	−0.23	1.00					
As	−0.38	−0.19	0.18	−0.21	0.29	0.04	−0.08	1.00				
Sr	−0.28	−0.12	0.67	−0.50	0.70	0.43	−0.37	0.34	1.00			
Cd	0.82	0.60	−0.09	0.67	−0.17	−0.44	0.59	−0.55	−0.40	1.00		
Pb	0.89	0.53	−0.01	0.66	−0.14	−0.44	0.44	−0.52	−0.31	0.84	1.00	
U	0.43	0.68	0.03	−0.16	−0.08	−0.09	0.24	−0.14	0.29	0.28	0.33	1.00

Significant correlations are marked in red.

The cluster dendrogram shows the separation of two main groups (Figure 3). A general trend could be deduced for the elements Pb, Cr, V, Fe, Cd, Zn and U, which were separated into a separate cluster. Mathematical processing showed that there is a dependence between them and there are elements of synergism in their action, evidence of which is the positive correlation (Table 4) and their falling into the same groups in the cluster analysis. These results give us reason to assume that in the area of our study, the specified elements have a common origin, namely as waste products from road transport. These include emissions from spent fuel, braking systems and tire wear: Fe is released from car braking systems, Pb and Cd from exhaust gases and V from various alloys [38,39]. It was of interest to us to make a comparison between the values of the investigated heavy metals and toxic elements in the leaf samples of Pinus nigra, Tilia tomentosa and Fraxinus excelsior. Greater proximity was confirmed between the two broad-leaved species, and the coniferous species showed specific differences (Figure 4).

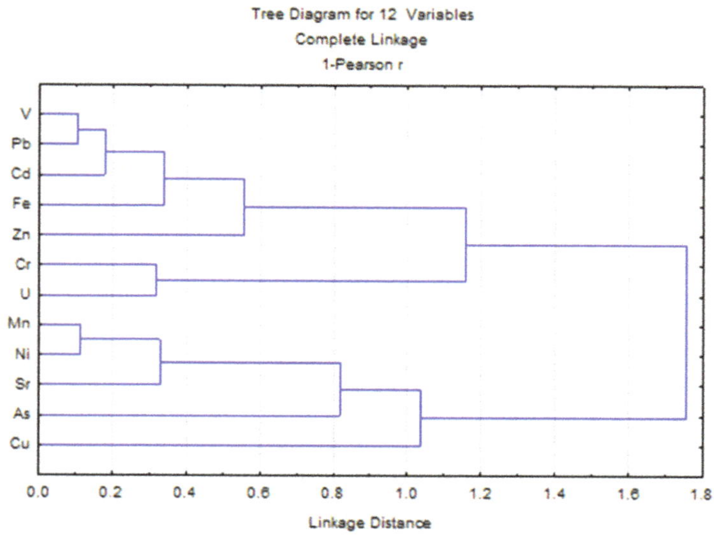

Figure 3. Cluster analysis of the investigated toxic elements in the leaf samples.

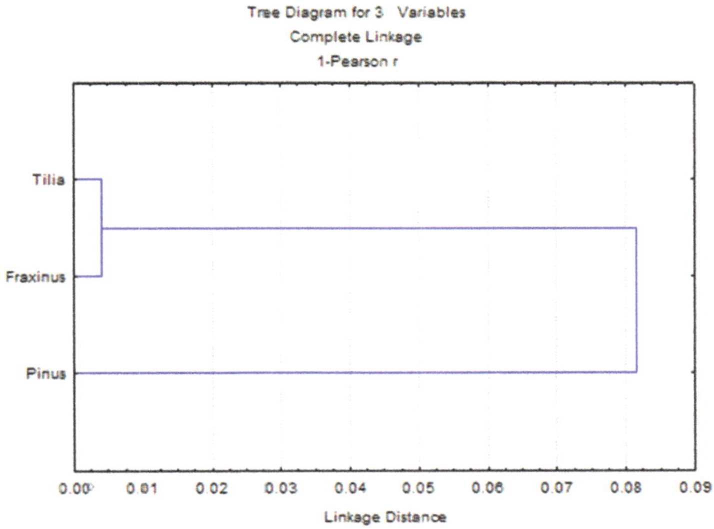

Figure 4. Cluster analysis of similarity between studied tree species based on content of potentially toxic elements in leaves.

3.2. Calculating Efficiency of Air Pollutant Removal (Potentially Toxic Elements)

The values of the enrichment factor (EF) are shown in Table 5. It can reflect the ability of plants to uptake a single HM. The average EF of U was the highest, in the range from 3 to 18, while the rest of the elements varied between 0.72 (Cd) and 2.56 (As).

Table 5. Enrichment factor values of potentially toxic elements in urban tree leaves.

Species	As	Cd	Cr	Cu	Fe	Mn	Ni	Pb	Sr	U	V	Zn
T. tomentosa	1.65	0.72	1.47	1.01	1.39	1.02	1.14	0.99	1.15	4	0.96	1.15
F. excelsior	2.56	1.27	1.64	1.15	1.63	1.29	1.22	1.28	0.87	3	1.5	1.07
P. nigra	1.67	1.36	1.35	1.53	1.06	1.57	1.17	0.89	2.46	18	1	1.62

Tree leaves accumulate different elements simultaneously. Using the data obtained in the present study, we calculated the accumulation index to assess the overall performance of the trees in terms of metal accumulation (MAI). The MAI values found were as follows: *Tilia tomentosa* = 22.37, *F. excelsior* = 23.08 and *P. nigra* = 22.35. The obtained MAI values in the present study are higher than those reported previously by Hu et al. [40], Alahabadi et al. [41] and Liu et al. [32], which suggests that these species could be a good choice for planting in urban areas with worsened air quality.

3.3. Modeling Efficiency of Air Pollutant Removal (PM_{10}, $PM_{2.5}$ and NO_2)

The calculated annual and seasonal quantities of pollutants removed from the composition of atmospheric air on an annual basis, according to the two simulated scenarios and the real distribution of the dendroflora of the city of Plovdiv, are presented in Figure 5. According to the observations made, deciduous trees were active in the period from May to November (6 months without domestic heating and 2 months with emissions from heating), and in this period, if all urban forest is represented by such species, they could accumulate up to 301.1 t of PM_{10}, 228.6 t of $PM_{2.5}$ and 215.1 t of NO_2. The real values, calculated on the basis of actual urban forest (90% deciduous, active for 8 months, and 10% evergreen species, active in the whole year), are significantly higher—377.4 t of PM_{10}, 272.5 t of $PM_{2.5}$

and 254.1 t of NO_2 ($p < 0.05$). The obtained results clearly indicate the role of the green system as a filter of ambient air in settlements.

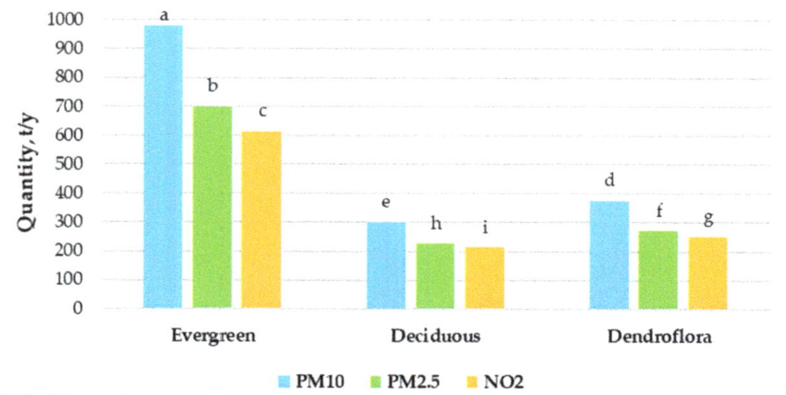

Figure 5. Simulation model of the removed amounts of pollutants from the atmospheric air in the city of Plovdiv according to the three scenarios. Different letters indicate significant differences.

It is noteworthy that with the configuration of only evergreen trees, the obtained effect of purifying the atmosphere from the specified pollutants is three times greater compared to the configuration of only deciduous trees ($p < 0.05$), which also shows the lowest degree of removal of pollutants. The actual dendroflora configuration is slightly better able to provide this ecosystem service—25% more PM_{10}, 19% more $PM_{2.5}$ and 18% more NO_2 removal ($p < 0.05$), highlighting the importance of the green system's species composition to maintain its functional stability.

4. Discussion

Data concerning the EF values (Table 3) pointed out that from the three studied ornamental trees, *Tilia tomentosa* had the lowest pollutant removal efficiency, as an EF value of more than 1.2 for potentially toxic elements was found only for four of them (U—4.0, As—1.65, Cr—1.49, Fe—1.37). *Fraxinus excelsior* was found to be a better accumulator of As (2.56), Cr (1.64), Fe (1.63), V (1.5), Mn (1.29), Pb (1.28) and Cd 91.27), while *Pinus nigra* tended to accumulate U (18), Sr (2.46), As (1.67), Zn (1.62), Mn (1.57), Cu (1.53), Cd (1.36) and Cr (1.35). Similar EF values for Cu, Pb, Zn and Mn in *Tilia* leaves have been reported by Serbula et al. [42] in rural areas of Serbia, while data from an industrial site in the city of Bor were twice higher. When regarding the Serbian data for *Pinus* needles, the EF values of these elements ranged from 0.6 (Mn) to 4 (Pb). A study from Spain (Huelva) has reported some EF values for pine needles (*Pinus pinea*), collected from three heavily industry-influenced locations, ranging from 4 up to 16 [30].

The MAI values obtained for Plovdiv's tree leaves are notably higher (up to 6–7-fold) when compared to some other studies [32,40,41]. For example, the MAI values found by Hu et al. [40] in the city of Yanan (China) were 2.56 (*Sophora japonica*), 2.33 (*Picea asperata*) and 2.16 (*Fraxinus chinensis*), while the MAI value calculated for *Sophora japonica* from Beijing (China) was 9.0 [32]. Data from Yazd city (Iran) [41] were also in the range between 3.31 (*Fraxinus excelsior*) and 6.66 (*Pinus moldarica*). Roy et al. [43] found that *Ficus bengalensis* had higher MAI values from the 10 species studied, with the MAI from the industrial site = 5.15 and the MAI from the commercial site = 6.09, indicating its higher accumulation capacity. These differences between results from the abovementioned studies could be explained by the influence of various factors such as differences in local atmospheric chemistry and meteorology properties, sampling altitude, sampling time, plant characteristics, etc., which affect the removal capacity of air pollutants from urban plants [44,45]. The number of analyzed elements was also different—12 elements in our study and generally lower numbers (4–7) of elements in the others. According to the above MAI values, all studied species have sufficient accumulation properties, and as they can grow in contaminated environments, they can be used as bioindicators or biomonitors for potentially toxic element contamination in the urban environment. Thus, the present study confirms the other authors' recommendations [41] that these plants should be used more frequently as barriers between contaminated and vulnerable areas, such as parks, schools, hospitals and residential areas.

Particulate matter is a dominant air pollutant in urban areas, and both fractions (PM_{10} and $PM_{2.5}$) have negative effects on public health [46]. Many studies have demonstrated the significant contribution of urban forests in improving air quality [17,34]. One of the main parameters affecting this regulatory ecosystem service is the physiology of the main tree functional groups (evergreen broadleaves, deciduous broadleaves and conifers), as well as their abundance and spatial distribution within cities [35,47]. A spatial analysis integrating dynamic green system modeling and geostatistics was applied to estimate the seasonal and annual amounts of pollutants removed by the two functional groups of trees. The main emitter of PM in the city of Plovdiv was considered to be domestic heating; hence, the simulations were set to distinguish the accumulation provided by deciduous and evergreen species between the cold and warm seasons. A similar approach has been proven to be appropriate and effective in other urbanized areas from different continents [34,47–51]. Temporal dynamics are more often associated with tree phenology and ecophysiology, accounting for the specific functional response to both seasonal dynamics of climate factors [51] and interannual climate variability. The dynamics in space are instead related to the complex interactions between the spatial distribution of the functional groups of trees on the urbanized territory and the seasonal dynamics in the amount of emitted pollutants [35]. Fusaro et al. [52] studied the PM_{10} removal by trees in the Municipality of Rome, Italy, and revealed that, with a total of 293.83 t/y (0.019 t/ha), evergreen trees were also found to be more effective. This removal quantity is similar to our data (301.1 t/y), but the effectiveness of urban trees in Plovdiv for PM_{10} removal was calculated to be 0.0296 t/ha. The higher values obtained in our study could be a consequence of different PM_{10} levels in the air, the biological properties of the plant species, local peculiarities, etc. The functional complementarity of the different functional tree groups (coniferous, evergreen and deciduous broad-leaved species) will support the functional stability of the phytocenosis under diverse climatic conditions and during the change of seasonal cycles in the vegetation. The observed synergism, as a consequence of the specific seasonal phenological and ecophysiological dynamics of the three tree groups, proves the need for biodiversity conservation, especially in urbanized ecosystems and in the context of global climate change.

It should be noted that there are some limitations to these estimates, but nevertheless, the results obtained are useful in terms of evaluating the magnitude of air pollutant removal by urban trees, as well as quantifying their added value to human health and well-being. Limitations of such studies include the following: (i) uncertainties related to tree cover estimation and/or leaf area volume; (ii) the formulas used to model particulate matter removal at a local and/or national level; (iii) the limited number of pollutant monitoring stations which provide pollutant concentration data; (iv) the negative effects (disservices) of urban forest not being accounted for (VOC emissions, pollens, reduced wind speed, etc.).

5. Conclusions

Trees are important for reducing urban pollution, and street trees play the leading role. Using detailed data provided a more accurate insight into the role of three widespread trees in air pollution mitigation. When comparing the plant species, we revealed a similar bioaccumulation of As, Cr, U and Zn. *Tilia tomentosa* was found to be a good accumulator of Fe, Mn, Ni, Pb and Sr; *Fraxinus excelsior* for Cu and Sr and *Pinus nigra* for Cd, Mn and Pb. Greater proximity was confirmed between the two broad-leaved species, and the coniferous species showed specific differences. The highest EF values were observed for U (3–18), followed by As (1.6–2.56) and Sr (0.87–2.46). The MAI values of the three studied species were found to be very similar, in the range of 22.35 to 23.08, which suggests that these species could be a good choice for planting in urban areas with worsened air quality.

The potential of the urban forest in countering air pollution was highlighted by three simulated scenarios for PM_{10}, $PM_{2.5}$ and NO_2 removal. The highest removal efficiency was calculated for evergreen species, followed by the mixed composition of deciduous (90%) and evergreen trees (10%), and the scenario with only deciduous trees had the lowest one. The contribution of nature-based solutions in meeting air quality standards and enhancing resilience in urban areas was clearly demonstrated. The functional complementarity of the different functional tree groups (coniferous, evergreen and deciduous broad-leaved species) was proven to be crucial for the support of both functional stabilities of the phytocenosis under diverse climatic conditions and during the change of seasonal cycles in the vegetation.

Funding: This research was funded by the Agricultural University, grant number 04-24.

Data Availability Statement: The dataset is available on request from the author.

Conflicts of Interest: The author declares no conflict of interest.

References

1. Millennium Ecosystem Assessment (MA). *Ecosystems and Human Well-Being: Synthesis*; Island Press: Washington, DC, USA, 2005.
2. Petrova, S. Algorithm for assessment and modeling of some ecosystem services in urban areas (Plovdiv, Bulgaria). *Ecol. Balk.* **2023**, *15*, 222–229.
3. Petrova, S.; Nikolov, B. Soil related ecosystem services in urban areas—A literature review. *Ecol. Balk.* **2023**, *15*, 203–231.
4. Maes, J.; Teller, A.; Erhard, M.; Liquete, C.; Braat, L.; Berry, P.; Egoh, B.; Puydarrieux, P.; Fiorina, C ; Santos, F.; et al. *Mapping and Assessment of Ecosystems and Their Services. An Analytical Framework for Ecosystem Assessments under Action 5 of the EU Biodiversity Strategy to 2020*; Publications Office of the European Union: Luxembourg, 2013.
5. Maes, J.; Teller, A.; Erhard, M.; Murphy, P.; Paracchini, M.L.; Barredo, J.I.; Grizzetti, B.; Cardoso, A.; Somma, F.; Petersen, J.E.; et al. *Mapping and Assessment of Ecosystems and Their Services. Indicators for Ecosystem Assessments under Action 5 of the EU Biodiversity Strategy to 2020, 2nd Report*; Publications office of the European Union: Luxembourg, 2014.
6. Maes, J.; Teller, A.; Erhard, M.; Conde, S.; Vallecillo Rodriguez, S.; Barredo Cano, J.I.; Paracchini, M.; Abdul Malak, D.; Trombetti, M.; Vigiak, O.; et al. *Mapping and Assessment of Ecosystems and Their Services: An EU Ecosystem Assessment*; EUR 30161 EN; Publications Office of the European Union: Luxembourg, 2020. [CrossRef]
7. Chen, B.W.; Wu, C.Y.; Huang, X.J.; Yang, X.F. Examining the relationship between urban land expansion and economic linkage using coupling analysis: A case study of the Yangtze River Economic Belt, China. *Sustainability* **2020**, *12*, 1227. [CrossRef]
8. Ji, X.; Wu, J.; Zhu, Q.; Sun, J. Using a hybrid heterogeneous DEA method to benchmark China's sustainable urbanization: An empirical study. *Ann. Oper. Res.* **2019**, *278*, 281–335. [CrossRef]
9. Makvandi, M.; Li, W.; Ou, X.; Chai, H.; Khodabakhshi, Z.; Fu, J.; Yuan, P.F.; Horimbere, E.D.L.J. Urban Heat Mitigation towards Climate Change Adaptation: An Eco-Sustainable Design Strategy to Improve Environmental Performance under Rapid Urbanization. *Atmosphere* **2023**, *14*, 638. [CrossRef]
10. Manes, F.; Silli, V.; Salvatori, E.; Incerti, G.; Galante, G.; Fusaro, L.; Perrino, C. Urban Ecosystem Services: Tree diversity and stability of PM10 removal in the Metropolitan Area of Rome. *Ann. Bot.* **2014**, *4*, 73–80.
11. Manes, F.; Salvatori, E. Ecosystem services of urban trees: The case of Rome. *Agrochimica* **2014**, *58*, 222–233.
12. Al-Dabbous, A.N.; Kumar, P. The influence of roadside vegetation barriers on airborne nanoparticles and pedestrian's exposure under varying wind condition. *Atmos. Environ.* **2014**, *90*, 113–124. [CrossRef]
13. Gallagher, J.; Baldauf, R.; Fuller, C.; Kumar, P.; Gill, L.W.; McNabola, A. Passive methods of air pollution control in the built environment: A review of porous and solid barriers. *Atmos. Environ.* **2015**, *120*, 61–70. [CrossRef]
14. Vaz Monteiro, M.; Handley, P.; Morison, J.I.L.; Doick, K.J. *The Role of Urban Trees and Greenspaces in Reducing Urban Air Temperatures*; Forestry Commission: Surrey, UK, 2019; p. 19. ISBN 978-0-85538-984-0.
15. Pearlmutter, D.; Calfapietra, C.; Samson, R.; O'Brien, L.; Krajter Ostoić, S.; Sanesi, G.; Alonso del Amo, R. *The Urban Forest, Cultivating Green Infrastructure for People and the Environment*; Springer: Berlin/Heidelberg, Germany, 2017.
16. Petrova, S.; Velcheva, I.; Nikolov, B.; Marinov-Serafimov, P.; Golubinova, I.; Popov, V.; Valcheva, E.; Todorova, K. Assessment of the adaptation of some tree species to the urban environment (Plovdiv, Bulgaria). *Comptes Rendus L'academie Bulg. Sci.* **2019**, *72*, 1676–1683.
17. Petrova, S. Efficiency of *Pinus nigra* J.F. Arnold in removing pollutants from urban environment (Plovdiv, Bulgaria). *Environ. Sci. Pollut. Res.* **2020**, *27*, 39490–39506. [CrossRef]

18. Valente, D.; Pasimeni, M.R.; Petrosillo, I. The role of green infrastructures in Italian cities by linking natural and social capital. *Ecol. Indic.* **2020**, *108*, 105694. [CrossRef]
19. Barwise, Y.; Kumar, P. Designing vegetation barriers for urban air pollution abatement: A practical review for appropriate plant species selection. *NPJ Clim. Atmos. Sci.* **2020**, *3*, 12. [CrossRef]
20. Tremper, A.H.; Green, D.C.; Eleftheriou-Vaus, K. *Impact of Green Screens on Concentrations of Particulate Matter and Oxides of Nitrogen in Near Road Environments*; King's College London: London, UK, 2015.
21. Tremper, A.H.; Green, D.C. *The Impact of a Green Screen on Concentrations of Nitrogen Dioxide at Bowes Primary School, Enfield*; King's College London: London, UK, 2018.
22. Abhijith, K.V.; Kumar, P. Field investigations for evaluating green infrastructure effects on air quality in open-road conditions. *Atmos. Environ.* **2019**, *201*, 132–147. [CrossRef]
23. Baro, F. Urban Green Infrastructure: Modelling and Mapping Ecosystem Services for Sustainable Planning and Management in and around Cities. Ph.D. Thesis, Universitat Autonoma de Barcelona, Barcelona, Spain, 2016; 227p. [CrossRef]
24. Petrova, S.; Velcheva, I.; Nikolov, B.; Vasileva, T.; Bivolarski, V. Antioxidant Responses and Adaptation Mechanisms of *Tilia tomentosa* Moench, *Fraxinus excelsior* L. and *Pinus nigra* J. F. Arnold towards Urban Air Pollution. *Forests* **2022**, *13*, 1689. [CrossRef]
25. ISTAT. Available online: http://www.en.istat.it (accessed on 11 September 2023).
26. European Environmental Agency. Report 2018. Available online: http://eea.government.bg/bg/soer/2017/air/emisii-na-vredni-veshtestva-vav-vazduha (accessed on 28 September 2023).
27. Petrova, S.; Petkova, M. Plant Traits of *Tilia tomentosa* Moench, *Fraxinus excelsior* L., and *Pinus nigra* J.F. Arnold as a Proxy of Urbanization. *Forests* **2023**, *14*, 800. [CrossRef]
28. Serbula, S.M.; Miljkovic, D.D.; Kovacevic, R.M.; Ilic, A.A. Assessment of Airborne Heavy Metal Pollution Using Plant Parts and Topsoil. *Ecotoxicol. Environ. Saf.* **2012**, *76*, 209–214. [CrossRef] [PubMed]
29. *NCS DC73348*; Bush Branches and Leaves—Trace Elements. LGC Standards Ltd.: London, UK, 2016.
30. Mingorance, M.D.; Valdés, B.; Rossini Oliva, S. Strategies of Heavy Metal Uptake by Plants Growing under Industrial Emissions. *Environ. Int.* **2007**, *33*, 514–520. [CrossRef]
31. Hu, R.; Yan, Y.; Zhou, X.; Wang, Y.; Fang, Y. Monitoring Heavy Metal Contents with *Sphagnum Junghuhnianum* Moss Bags in Relation to Traffic Volume in Wuxi, China. *Int. J. Environ. Res. Public Health* **2018**, *15*, 374. [CrossRef]
32. Liu, Y.-J.; Zhu, Y.-G.; Ding, H. Lead and cadmium in leaves of deciduous trees in Beijing, China: Development of a metal accumulation index (MAI). *Environ. Pollut.* **2007**, *145*, 387–390. [CrossRef]
33. Municipal Program of Air Quality Management in the City of Plovdiv 2018–2023. Available online: https://www.plovdiv.bg/wp-content/uploads/2018/09/programa-plovdiv-kav-version5-final.png (accessed on 24 April 2023). (In Bulgarian).
34. Nowak, D.J.; Hirabayashi, S.; Bodine, A.; Greenfield, E. Tree and forest effects on air quality and human health in the United States. *Environ. Pollut.* **2014**, *193*, 119–129. [CrossRef] [PubMed]
35. Manes, F.; Blasi, C.; Salvatori, E.; Capotorti, G.; Galante, G.; Feoli, E.; Incerti, G. Natural vegetation and ecosystem services related to air quality improvement: Tropospheric ozone removal by evergreen and deciduous forests in Latium (Italy). *Ann. Bot.* **2012**, *2*, 79–86.
36. Zinke, P.J. Forest interception studies in the United States. In *Forest Hydrology*; Sopper, W.E., Lull, H.W., Eds.; Pergamon Press: Oxford, UK, 1967; pp. 137–161.
37. Program for the Development, Maintenance and Protection of the Green System of the City of Plovdiv. Available online: https://www.plovdiv.bg/proekt-programa-zelena-sistema/ (accessed on 25 April 2023).
38. Fussell, J.C.; Franklin, M.; Green, D.C.; Gustafsson, M.; Harrison, R.M.; Hicks, W.; Kelly, F.J.; Kishta, F.; Miller, M.R.; Mudway, I.S.; et al. A Review of Road Traffic-Derived Non-Exhaust Particles: Emissions, Physicochemical Characteristics, Health Risks, and Mitigation Measures. *Environ. Sci. Technol.* **2022**, *56*, 6813–6835. [CrossRef]
39. Adamiec, E.; Jarosz-Krzemińska, E.; Wieszała, R. Heavy metals from non-exhaust vehicle emissions in urban and motorway road dusts. *Environ. Monit. Assess.* **2016**, *188*, 369. [CrossRef]
40. Hu, Y.; Wang, D.; Wei, L.; Zhang, X.; Song, B. Bioaccumulation of heavy metals in plant leaves from Yanan city of the Loess Plateau, China. *Ecotoxicol. Environ. Saf.* **2014**, *110*, 82–88. [CrossRef]
41. Alahabadi, A.; Ehrampoush, M.; Miri, M.; Aval, H.; Yousefzadeh, S.; Ghaffari, H.; Ahmadi, E.; Talebi, P.; Fathabadi, Z.; Babai, F.; et al. A comparative study on capability of different tree species in accumulating heavy metals from soil and ambient air. *Chemosphere* **2017**, *172*, 459–467. [CrossRef]
42. Serbula, S.M.; Kalinovic, T.S.; Ilic, A.A.; Kalinovic, J.V.; Steharnik, M.M. Assessment of Airborne Heavy Metal Pollution Using *Pinus* spp. and *Tilia* spp. *Aerosol Air Qual. Res.* **2013**, *13*, 563–573. [CrossRef]
43. Roy, A.; Bhattacharya, T.; Kumari, M. Air pollution tolerance, metal accumulation and dust capturing capacity of common tropical trees in commercial and industrial sites. *Sci. Tot. Environ.* **2020**, *722*, 37622. [CrossRef]
44. Hofman, J.; Stokkaer, I.; Snauwaert, L.; Samson, R. Spatial distribution assessment of particulate matter in an urban street canyon using biomagnetic leaf monitoring of tree crown deposited particles. *Environ. Pollut.* **2013**, *183*, 123–132. [CrossRef]
45. Yin, S.; Shen, Z.; Zhou, P.; Zou, X.; Che, S.; Wang, W. Quantifying air pollution attenuation within urban parks: An experimental approach in Shanghai, China. *Environ. Pollut.* **2011**, *159*, 2155–2163. [CrossRef]
46. Bell, M.L.; Peng, R.D.; Dominici, F. The exposure-response curve for ozone and risk of mortality and the adequacy of current ozone regulations. *Environ. Health Persp.* **2006**, *114*, 532–536. [CrossRef] [PubMed]

47. Escobedo, F.J.; Kroeger, T.; Wager, J.E. Urban forest and pollution mitigation. Analyzing ecosystem services and disservices. *Environ. Pollut.* **2011**, *159*, 2078–2087. [CrossRef] [PubMed]
48. Nowak, D.J.; Greenfield, E.J.; Hoehn, R.E.; Lapoint, E. Carbon storage and sequestration by trees in urban and community areas of the United States. *Environ. Pollut.* **2013**, *178*, 229–236. [CrossRef] [PubMed]
49. Yang, J.; McBride, J.; Zhou, J.; Sun, Z. The urban forest in Beijing and its role in air pollution reduction. *Urban For. Urban Green.* **2005**, *3*, 65–78. [CrossRef]
50. Escobedo, F.J.; Nowak, D.J. Spatial heterogeneity and air pollution removal by an urban forest. *Landsc. Urban Plan.* **2009**, *90*, 102–110. [CrossRef]
51. Fares, S.; Matteucci, G.; Scarascia-Mugnozza, G.; Morani, A.; Calfapietra, C.; Salvatori, E. Testing of models of stomatal ozone fluxes with field measurements in a mixed Mediterranean forest. *Atmos. Environ.* **2013**, *67*, 242–251 [CrossRef]
52. Fusaro, L.; Marando, F.; Sebastiani, A.; Capotorti, G.; Blasi, C.; Copiz, R.; Congedo, L.; Munafò, M.; Ciancarella, L.; Manes, F. Mapping and Assessment of PM_{10} and O_3 Removal by Woody Vegetation at Urban and Regional Level. *Remote Sens.* **2017**, *9*, 791. [CrossRef]

forests

MDPI

Article

Nature-Based Solutions to Reduce Air Pollution: A Case Study from Plovdiv, Bulgaria, Using Trees, Herbs, Mosses and Lichens

Slaveya Petrova [1,2,*], Iliana Velcheva [1] and Bogdan Nikolov [1]

[1] Department of Ecology and Environmental Conservation, Faculty of Biology, Plovdiv University "Paisii Hilendarski", 4000 Plovdiv, Bulgaria; anivel@uni-plovdiv.bg (I.V.); nikolov81bg@uni-plovdiv.bg (B.N.)

[2] Department of Microbiology and Ecological Biotechnologies, Faculty of Plant Protection and Agroecology, Agricultural University, 4000 Plovdiv, Bulgaria

* Correspondence: slaveya_petrova@uni-plovdiv.bg or sl_petrova@au-plovdiv.bg

Abstract: Nature-based solutions (NBSs) are becoming more and more recognized as useful instruments to address the challenges that urban areas are currently facing, i.e., climate change adaptation, flood mitigation, etc. In the present study, we aimed to: (1) compare the efficiency of mosses, lichens, herbs and trees in removing pollutants from the urban air using their biomonitoring potential; (2) assess their efficiency as nature-based solutions to mitigate urban air pollution; and (3) propose a framework for implementing such NBSs in urban areas. The first step involved analyses of the concentrations of 20 potentially toxic elements in eight selected biomonitors. After that, an assessment of their removal capacity was made on the basis of elements accumulation. This is the first complex study in an urban area involving the simultaneous application of organisms of eight different species and four different systematic groups (lichenized fungi, mosses, herbaceous plants, woody species) as well as such a large number of potentially hazardous elements. The present study sheds new light on some well-known biomonitors in the context of their application for air pollution mitigation. The great potential of the eight studied plant species for efficient removal of potentially toxic elements is highlighted and their implementation into NBS frameworks is recommended.

Keywords: biomonitoring; heavy metals; moss-bags; lichen-bags; ecosystem services; potentially toxic elements

1. Introduction

Urbanization is a process leading to an increased concentration of population into relatively small geographical areas, such as cities and megacities, as a result of which natural ecosystems are altered and replaced by anthropogenic ones. With the development of the city, natural habitats are fragmented, isolated and destroyed; the species composition becomes impoverished and homogenized; the hydrological regime is disturbed; and the flow of energy and the circulation of substances are modified [1,2]. Although urban areas occupy only 1–6% of the Earth's mainland, cities use a much larger share of the planet's capacity in terms of input of raw materials and release of waste. Because humanity is dependent on the biosphere for supplies of food, water, raw materials and other essential goods as well as ecosystem services, the anthropogenic-induced changes in urban areas sooner or later affect human well-being [3].

Air pollution is probably the leading environmental problem in human settlements, affecting both the population and economy, as well as the biodiversity and ecosystem processes [4]. Many chemicals, including greenhouse gases, organic compounds and fine particles, are emitted from natural and anthropogenic sources [5]. Moreover, after their release, these pollutants undergo various transformations in the atmosphere (physical, chemical, photochemical, etc.), determining their final concentrations and behavior [6]. Air pollutants do not stay close to emitters; they can diffuse long distances, depending on the topography and meteorological conditions, transported by winds as well as vertical and horizontal temperature gradients, which turns the problem of anthropogenic air pollution from local (regional) to global. Globally, more than 5 million deaths per year are associated with air pollution [7]. A report from the WHO [8] revealed that about 80% of the urban population worldwide inhabit areas with exceedance of the hygiene norms of air pollutant concentrations.

Urban vegetation is recognized as important for cities' resilience and many other ecosystem services, but there is a lack of detailed studies and recommendations on how city planners and practitioners can take the best measures for air quality improvement [9]. Nature-based solutions (NBSs) are becoming more and more recognized as a useful instrument to address the challenges that urban areas are currently facing, i.e., climate change adaptation, flood mitigation, air purification, etc. [10,11]. The benefits of NBSs are well recognized in terms of ecosystem services assessment and include improved air quality, life quality, and human health, amongst others [12]. NBSs are especially effective in tackling environmental issues such as climate change resilience and air pollution mitigation due to their potential co-benefits [11,12]. The European Commission has defined NBSs as "Solutions that are inspired and supported by nature, which are cost-effective, simultaneously provide environmental, social and economic benefits and help build resilience" [13]. NBSs are regarded as cheap and useful instruments to address the potential hazards of climate change and environmental pollution, as they could simultaneously contribute to human health and well-being, enhance biodiversity and promote urban ecosystems resilience in the most resource efficient way [14].

Nature-based approaches (involving urban green infrastructure) could be effectively applied in reducing exposure to fine particles, potentially toxic elements (PTEs) and gaseous pollutants in urban environments. Urban vegetation is well-known to contribute to air quality monitoring [15–19] as well as to air pollution mitigation [20–22]. Some plant species with high sensitivity to environmental pollution (lichens, algae, herbs and trees) have been applied as bio-indicators of air quality for more than 40 years [23–26]. Other plant species are found to be tolerant and thus they have been used to reduce air and/or soil pollution (phytoremediation). The effectiveness of a variety of plant species for bioaccumulation of heavy metals and other toxic elements has also been assessed [4,5,15–18,27]. These authors have evaluated some species with a significant capacity for metal accumulation and tolerance to environmental pollution, thus recommending them to be used not only for biomonitoring purposes, but also for the construction of green belts, buffer green patches, green walls and others, especially in so-called "hot spot" areas (heavily polluted, mainly industrial).

The biomonitoring of air pollution, based on the analysis of micro- and macroelements in plant tissues, has been increasingly applied in Europe and worldwide [4,28–31]. One of the main advantages of biomonitors is related to their potential to act as air samplers, so called "green filters," and their widespread distribution, providing a high density of parallel measurements. The relatively easy and inexpensive sampling as well as the accumulated and time-integrated behavior of biomonitor species are prerequisites for the application of air pollution biomonitoring, especially in large-scale studies of potentially hazardous pollutants [32–34]. Phytomonitoring, as a part of biomonitoring, is a system for assessing the environmental status using the plant objects in their natural environment (passive monitoring) or after their transfer from a control to the studied area where they will be exposed for a certain period (active monitoring). Biomonitors are very often applied for the monitoring of urban pollution with potentially toxic elements as they can provide a huge number of sampling points [15,19,30,35].

The aim of the present study was to: (1) compare the efficiency of mosses, lichens, herbs and trees in removing pollutants from the urban air using their biomonitoring potential; (2) assess their efficiency as nature-based solutions to mitigate urban air pollution; and (3) propose a framework for implementing such NBSs in urban areas. Thus, the content of 20 potentially toxic elements (Al, As, B, Bi, Cd, Co, Cr, Cu, Fe, Hg, Mn, Mo, Na, Ni, Pb, Se,

Sr, U, V, and Zn) in eight biomonitors have been analyzed and their removal effectiveness have been evaluated on the basis of elements accumulation. The passive biomonitors included three deciduous trees (*Acer platanoides* L., *Aesculus hippocastanum* L. and *Betula pendula* Roth) and three herbal species (*Taraxacum officinale* (L.) Weber ex F.H.Wigg., *Plantago lanceolata* L. and *Capsella bursa-pastoris* Medik.), while the active biomonitors were one moss (*Sphagnum girgensohnii* Russow) and one lichen (*Pseudevernia furfuracea* (L.) Zopf), applied as moss-bags or lichen-bags with dried material, respectively. This is the first complex study in an urban area involving the simultaneous application of organisms of eight different species and four different systematic groups (lichenized fungi, mosses, herbaceous plants, woody species) as well as such a large number of potentially hazardous elements.

2. Materials and Methods

2.1. Study Area

The object of the present study is the city of Plovdiv, Bulgaria, which occupies the second place by area and population number in the country after the city of Sofia (the state's capital). The city of Plovdiv (42°8′9.9492″ N; 24°44′31.8048″ E) is located along the Maritsa River, in the central south part of Bulgaria. It has a population of 346,893 inhabitants (as of 2018) but the real number of citizens, including students, workers, etc. passes half a million. The city of Plovdiv is well-known as a cultural hub in Bulgaria, and this was confirmed by its recognition as the European Capital of Culture in 2019. Due to its strategic geographic position, Plovdiv is also regarded as an important industrial, transport, tourist and educational center.

The climate in Plovdiv is classified as humid subtropical, where the significant humid continental influences are manifested. The gentle winds (0 to 5 m/s) and fog that are frequently present during the cooler months are some of the factors that contribute to the deterioration of air quality and it was rated as one of the European cities with worsened living conditions since 2008 [36]. The specific topography is also a factor related to the worsened air quality (elevated PM_{10} and $PM_{2.5}$ levels) in the city of Plovdiv (barrier effect of hills, lack of winds, canyon street effect, etc.). The main inputs of atmospheric pollutants are found to be derived from traffic emissions and residential heating, while the industrial sector has a minor impact [37].

2.2. Sampling Procedure

For the purposes of the present study, the city's area was divided into urban, suburban and industrial sectors, following the urban gradient hypothesis [38], and in every one of them, two representative sites were chosen. Thus, the present study involved six experimental sites, selected according to the type of anthropogenic pressure as well as the similarity of abiotic environment (light, temperature, and exposure) and the presence of selected biomonitor species (trees and herbs). One site from the city's outskirts was selected as a control (urban background level).

Tree leaves sampling: Three trees per species per site were sampled using an Expert Tree Pruner with a telescopic handle 1.5–2.5 m (Draper Tools, Hampshire, UK) to cut off some branches from the crown. We selected branches from the outer part of the canopies and from the four sides of the tree (N, S, E, W), as well as situated 2.5–3 m above the ground. In order to prepare a representative sample, about 80 mature leaves per species per site were collected [39].

Herbs sampling: At each experimental site, the aboveground biomass of 3–4 fully grown plants per species was sampled. One composite sample per species per site was prepared for the analyses [17,18].

Moss-bags preparation, exposition and sampling: Some specimens of *Sphagnum girgensohnii* Russ. were collected from the background area of the Vitosha Nature Park (42°37′ N, 23°19′ E, altitude 1710 m). About 3 g of unwashed plant material was placed in nylon mesh (10 × 10 cm) and the prepared moss-bags were exposed at the same six experimental sites for a period of 3 months [40].

Lichen-bags preparation, exposition and sampling: Lichen material (*Pseudevernia furfuracea* (L.) Zopf.) was also collected from the background area of the Vitosha Nature Park, Bulgaria (42°37′ N, 23°19′ E, altitude 1710 m) and transported to the laboratory. The specimens were carefully cleaned, removing all bark residues. About 3 g of unwashed plant material was placed in a nylon mesh (10 × 10 cm) and the prepared lichen-bags were exposed at the same six experimental sites for a period of 3 months [41].

2.3. Analyses of Chemical Content

About 1 g of plant tissues were placed in 5 mL of nitric acid and left for 24 h using a microwave digestion system (Microwave Digestion System, CEM MDS 81D). The first step of wet digestion consisted of 5 min at 600 W, followed by cooling for 1 h. After that, 2 mL HNO_3 and 3 mL 30% H_2O_2 were added to the plant samples and left for another 1 h. The second step included a 10 min microwave treatment at 600 W in order to achieve full decomposition of the plant sample. The filtrate was separated carefully and double distilled water was added up to a volume of 50 mL.

The elements' concentrations were measured using the ICP-MS apparatus (Agilent 7700, DF 1000, Agilent Technologies Inc., Waldbronn, Germany). Calibration was performed by Multy VI (MERCK, Darmstadt, Germany) calibration standards. Quality control was ensured with a plant standard (NCS DC73348) [42].

The data for each site was the arithmetic mean of three samples, and the data for each individual sample was the arithmetic mean of three analytical measurements.

2.4. Bioaccumulation Patterns Analysis

To assess the bioaccumulation potential of the studied plant species, three different parameters were applied—enrichment factor, relative accumulation factor and metal accumulation factor.

2.4.1. Enrichment Factor (EF) of Plants Species Used as Passive Biomonitors

The enrichment factor (EF) values were calculated by dividing the average elements content in each plant species from the urban area by the concentration of the elements in the same species from the control area, using the formula of Mingorance et al. [43]:

$$EF = \frac{C_{urban}}{C_{control}}$$

where: C_{urban} represents the concentration of an element in the plant biomass sampled from the urban area, while $C_{control}$ represents the concentration of the same element in the plant samples from the rural/control area.

2.4.2. Relative Accumulation Factor (RAF) of Plant Species Used as Active Biomonitors

Relative accumulation factors (RAF) were applied for moss-bag and lichen-bag samples collected after 90 days of exposure at the six experimental urban sites. They were calculated by dividing the average element concentration ($C_{exposed}$) from all urban sites by the content of the same element in the unexposed plant tissues ($C_{unexposed}$). using the formula of Culicov et al. [44]:

$$RAF = \frac{C_{exposed}}{C_{unexposed}}$$

These formulas (EF and RAF) are quite similar and they correspond to the formula proposed by Ma et al. [45] for Concentration factor (CF) calculations. So, they fit well for all studied plant species and allowed us to compare their efficiency in removing potentially toxic elements from the urban air.

We used the following scale for CF to assess the environmental pollution level [46]: (1) CF ≤ 1.2—no pollution; (2) 1.2 < CF ≤ 2.2—low pollution; (3) 2.2 < CF ≤ 3.3—medium pollution; (4) 3.3 < CF ≤ 4.3—heavy pollution; (5) CF > 4.3—very heavy pollution.

2.4.3. Metal Accumulation Index (MAI)

Metal accumulation index (MAI) values were calculated according to the formula of Liu et al. [47]:

$$MAI = \left(\frac{1}{N}\right)\sum_{j=1}^{N} I_j$$

where N—total number of analyzed elements, and I_j—sub-index for variable j.

The following equation was used for calculation of the sub-index:

$$I_j = \frac{x}{\delta x}$$

where x—average content of each element, and δx—standard deviation.

In the present study, we have used the MAI in order to compare the remediation ability of different plant species toward 20 potentially toxic elements from urban air.

2.5. Data Analysis

The raw data obtained have been processed using the SPSS 21 software package. The relationships between the analyzed variables have been assessed through Spearman's correlation coefficients. Student's t test was applied for discovering significant differences in PTEs accumulation by plant species as well as between the control and experimental sites ($p < 0.05$).

3. Results

3.1. Content of Potentially Toxic Elements (PTEs) in Plants

As a whole, all studied elements' contents varied among the plant species (Table 1), depending on their biological features. Significant differences in PTEs content were observed when comparing vascular plants (herbs) and cryptogams (moss and lichen) ($p < 0.05$). A wider range of variation was observed for Na (from 31.2 mg/kg in *B. pendula* leaves to 1539 mg/kg in *C. bursa-pastoris*), V (from 0.06 mg/kg in *A. hippocastanum* leaves to 1.58 in *S. girgensohnii*), Pb (from 2–3 mg/kg in higher plants to 28–29 mg/kg in moss and lichen), Hg (from 0.01 in *P. lanceolata* to 0.21 in *Ps. furfuracea*), B (from 5.0 mg/kg in *Ps. furfuracea* to 61 mg/kg in *A. platanoides* leaves), Zn (from 16.8 mg/kg in *A. hippocastanum* to 116.7 mg/kg in *B. pendula*), and As (from 0.25 mg/kg in higher plants to 2.18 in *S. girgensohnii*). The content of two elements was found to be less variable—Cu (from 4.53 mg/kg in *B. pendula* to 15.1 mg/kg in *Ps. furfuracea*) and Cd (from 0.23–0.24 mg/kg in tree leaves to 0.52 mg/kg in *T. officinale*).

Table 1. Average values of elements' contents in studied plant species (Plovdiv, Bulgaria).

Element	*Acer platanoides*		*Aesculus hippocastanum*		*Betula pendula*		*Capsella bursa-pastoris*		*Plantago lanceolata*		*Taraxacum officinale*		*Sphagnum girgensohnii*		*Pseudevernia furfuracea*	
	mg/kg	SD	mg/kg	SD	mg/kg	SD	mg/kg	SD	mg/kg	SD	mg/kg	SD	mg/kg	SD	mg/kg	SD
B	61.0	2.33	20.5	0.85	33.4	0.61	19.7	0.70	25.3	1.02	24.6	0.77	7.35	0.29	5.00	0.05
Na	34.9	0.96	71.2	1.25	31.2	0.41	1539	58.0	415	13.2	369.7	13.3	722	19.7	451	14.1
Al	55.3	1.21	49.0	1.23	62.7	5.69	171	6.82	224	6.28	342.3	13.5	396	11.4	405	10.6
V	0.08	0.01	0.06	0.002	0.17	0.01	0.46	0.02	0.77	0.05	0.896	0.04	1.58	0.05	1.70	0.07
Cr	0.49	0.05	0.24	0.02	0.59	0.05	1.49	0.09	1.25	0.12	2.23	0.19	1.15	0.05	1.43	0.04
Fe	128	3.71	105	3.37	130	2.69	227	11.8	340	11.8	428.9	16.8	475	8.31	529	9.12
Mn	64.0	1.72	48.7	1.56	70.6	0.68	33.0	1.21	26.6	1.04	53.3	1.69	293	5.05	100	2.06
Co	0.091	0.01	0.17	0.01	0.16	0.01	0.267	0.02	0.277	0.01	0.289	0.02	0.56	0.02	0.44	0.02
Ni	0.61	0.04	0.42	0.03	1.10	0.03	1.76	0.11	1.56	0.09	3.14	0.19	1.93	0.15	1.55	0.12
Cu	7.66	0.41	8.20	0.35	4.53	0.30	7.71	0.53	9.86	0.37	12.4	0.33	11.8	0.24	15.1	0.25
Zn	19.3	0.55	16.8	0.45	116.7	1.15	48.86	1.26	44.0	1.58	47.0	1.78	117	1.91	111	1.91
As	0.336	0.02	0.25	0.003	0.286	0.01	0.25	0.003	0.329	0.01	0.250	0.003	2.18	0.29	0.66	0.02
Se	0.486	0.03	0.44	0.03	0.300	0.003	0.30	0.003	0.600	0.04	0.357	0.01	1.58	0.15	0.30	0.003
Sr	50.4	1.46	58.6	1.59	53.43	0.38	52.29	1.43	53.0	2.14	39.286	1.36	27.9	0.54	11.5	0.24
Mo	0.229	0.01	0.28	0.006	0.33	0.01	1.46	0.06	0.786	0.1	1.029	0.04	0.88	0.21	0.20	0.002
Cd	0.231	0.01	0.24	0.012	0.243	0.01	0.434	0.02	0.306	0.03	0.516	0.03	0.49	0.02	0.37	0.02
Pb	2.71	0.07	2.75	0.075	2.32	0.02	2.214	0.09	3.529	0.15	3.143	0.12	29.3	0.48	28.0	0.53
Bi	0.31	0.01	0.29	0.003	0.82	0.03	0.216	0.01	0.271	0.01	0.200	0.002	0.89	0.09	0.20	0.002
U	0.016	0.002	0.02	0.002	0.014	0.001	0.024	0.002	0.037	0.003	0.034	0.003	0.09	0.003	0.06	0.003
Hg	0.047	0.002	0.04	0.003	0.032	0.002	0.020	0.001	0.010	0.001	0.041	0.004	0.1	0.014	0.21	0.03

3.2. Bioaccumulation Capacity of Potentially Toxic Elements by Plant

The Concentration factor (CF) values (bioaccumulation capacity) of the studied plant species in relation to the 20 PTEs are summarized in Table 2. As explained above, the bioaccumulation capacity could be assessed using different indices (EF, RAF, or CF) but all of them represent the ratio between the element's content in plants from an urban area and the content of the same element in plants from a control area.

Table 2. Concentration factor (CF) values of studied plant species (Plovdiv, Bulgaria).

Element	*Acer platanoides*	*Aesculus hippocastanum*	*Betula pendula*	*Capsella bursa-pastoris*	*Plantago lanceolata*	*Taraxacum officinale*	*Sphagnum girgensohnii*	*Pseudevernia furfuracea*
B	0.924	0.586	0.760	1.095	0.744	1.170	0.328	0.459
Na	2.681	2.967	0.985	9.216	4.029	9.992	1.645	6.098
Al	1.152	1.581	1.742	1.881	1.580	0.542	1.245	1.113
V	2.633	3.867	1.442	1.523	1.972	0.597	1.575	0.773
Cr	0.992	4.800	2.189	1.858	4.984	0.769	1.643	1.018
Fe	1.422	1.360	1.673	1.611	1.486	0.578	1.439	0.923
Mn	2.462	4.056	1.604	1.500	1.329	0.952	0.836	2.638
Co	1.517	0.246	0.413	0.834	1.539	0.615	1.191	1.926
Ni	0.761	1.273	0.649	1.952	1.730	1.572	1.750	1.192
Cu	0.815	1.708	1.332	1.205	0.986	0.886	2.034	1.861
Zn	0.946	1.099	0.957	1.629	1.000	0.855	1.655	1.493
As	1.344	1.000	1.144	1.000	0.411	1.000	0.702	0.332
Se	0.694	1.467	1.000	1.000	0.353	1.190	5.250	1.000
Sr	1.401	1.118	1.571	1.687	1.472	1.455	0.919	2.290
Mo	1.145	5.540	0.733	1.619	0.342	0.686	0.461	1.000
Cd	2.310	2.350	1.350	1.669	0.927	1.075	1.190	0.354
Pb	1.229	1.249	1.696	1.476	1.765	0.952	1.721	0.875
Bi	1.520	1.465	1.218	1.080	0.630	1.000	0.481	1.000
U	3.200	1.400	1.400	2.400	0.740	0.567	1.257	1.160
Hg	1.567	4.000	1.280	2.000	1.000	1.025	4.900	0.626

Obviously, there are some limitations to applying the assessment scale [46] when using different plant categories such as in our study. For example, the CF for Cd in *P. lanceolata, T. officinale, S. girgensohnii* and *Ps. furfuracea* has values <1.2 = no contamination; the CF for Cd in *C. bursa-pastoris* and *B. pendula* has values in the range 1.35–1.67 = slight contamination; and the CF for Cd in *A. platanoides* and *A. hippocastanum* has values between 2.2–3.2 = moderate contamination (Table 2). Nevertheless, the CF values are quite informative in assessing the potential of plant species to reduce air pollution based on their bioaccumulation capacity.

Acer platanoides was found to be the most efficient bioaccumulator of U (CF = 3.2), Bi (CF = 1.52) and As (CF = 1.34) among all studied plant species, demonstrating significant capacity (CF > 2) for Na, V, Mn and Cd, as well as moderate capacity (CF > 1.2) for Fe, Hg, Pb, Sr and Co.

Aesculus hippocastanum was proven to be an excellent bioaccumulator of Mo (CF = 5.54), Cr (CF = 4.8), Mn (CF = 4.06), Hg (CF = 4.0) and V (CF = 3.87). This species also showed a significant accumulation (CF > 2) of Na and Cd, while a moderate capacity (CF > 1.2) for Cu, Al, Fe, Ni, U, Pb, Se and Bi.

Betula pendula can be rated as the species with the lowest air pollutant removal capacity among the plants studied as no maximum accumulation was observed, as well as a CF > 2 was found for only one chemical element—Cr. However, CF values in the range 1.2–2 were obtained for 11 potentially toxic elements (Al, Fe, Mn, V, Cd, Pb, Sr, Hg, U, Bi and Cu), some of them being the highest for all trees studied—Al, Fe, Sr and Pb.

Notably, *Capsella bursa-pastoris* demonstrated CF values >1.0 for 19 of the studied 20 potentially toxic elements (except Co). The maximal value was its CF for Na (9.22), and significant (CF > 2) bioaccumulation was found for U and Hg, and it had a moderate capacity (CF > 1.2) to accumulate Al, Cr, Fe, Mn, V, Ni, Zn, Cu, Sr, Mo, Cd, and Pb.

Plantago lanceolata had a maximal capacity to accumulate Cr (CF = 4.98) and Pb (CF = 1.77) from all studied plant species, having significant potential for Na (CF = 4.03) and V (CF = 1.97), while only moderate (CF > 1.2) for Al, Fe, Co, Ni, Mn, and Sr.

In the present study, *Taraxacum officinale* was proven as the best accumulator of Na (CF = 9.99) and B (CF = 1.17) from all studied plant species, as well as a good accumulator of Ni and Sr (CF = 1.45–1.57).

The two cryptogam species also showed remarkable potential for removing PTEs from ambient air. Dried biomass of moss *Sphagnum girgensohnii* (exposed in moss-bags for a 90 day period) had the highest CF for Se (5.25), Hg (4.09), Cu (2.03) and Zn (1.66). A significant capacity (CF > 1.2) was also proven for Na, Al, V, Cr, Fe, Co, Ni, Cd, Pb and U. Dried biomass of *Pseudevernia furfuracea* (exposed in lichen-bags for a 90 day period) had the maximal CF for Mn (2.64), Sr (2.29) and Co (1.93), with excellent potential for Na (CF = 6.1). This species was found also to be a moderate (CF > 1.2) accumulator of Cu, Zn, Ni and U.

The MAI values of the eight studied plants are presented in Figure 1. The minimum was found in *P. lanceolata* (21.77), followed by *A. platanoides* (24.53); medium values were obtained for *A. hippocastanum* (33.76), *S. girgensohnii* (31.21), *T. officinale* (30.49) and *C. bursa-pastoris* (29.63); while the maximum was found for *B. pendula* (49.60) and *Ps. furfuracea* (48.47).

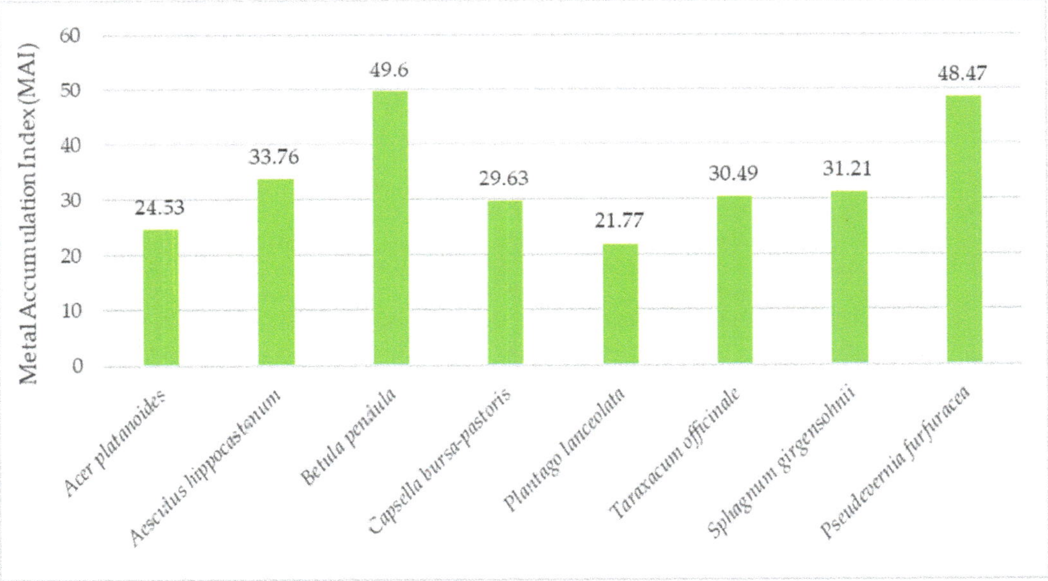

Figure 1. Metal accumulation index (MAI) for 20 potentially toxic elements in the studied plants.

4. Discussion

4.1. Assessment of Biomonitors' Efficiency as NBS for Air Pollution Mitigation

The MAI values obtained for Plovdiv's vegetation were quite higher when compared to some other studies [4,47,48]. Various factors could contribute to the observed differences, i.e., related to the local meteorology, local atmospheric chemistry, altitude, vegetation phase, plant performance, etc., as they affect the removal capacity of plants toward air pollutants [4,21,49]. The number of analyzed elements was also different—20 elements in our study and quite lower numbers (four to seven) of elements in the others. According to our MAI values, all studied plant species are quite tolerant of the contaminated urban environment and have significant accumulation efficiency, so they can be used as bioindicators or biomonitors for potentially toxic element contamination in urban areas. Thus, the present study confirms the other authors' recommendations [4] for the integration of such species as NBSs when constructing green buffers and barriers, for instance, between heavily impacted and vulnerable areas (hospitals, schools, kindergartens, parks, etc.).

Mosses and lichenized fungi are the most applied biomonitoring organisms worldwide. The main reason is their lack of roots, which should mean that they obtain nutrients only from the atmosphere and not from the substrate (unlike higher plants) [34,50]. However, the uptake of elements or other contributions from the substrate cannot be completely excluded [46]. Lichenized fungi and mosses are thought to accumulate elements mainly by passive processes [24,28], so the capture of relatively insoluble and metal-rich fine particles rather than the retention of soluble forms of chemical elements prevails [46,51], which is also supported by their reported high retention rates [34]. Our findings of moss-bag contents of Cu, Pb, V, and Zn correlate with the findings of Hu et al. [48] in Wuxi, China, although our Cr was significantly lower (three to four times). Our data from the lichen-bags were quite a bit higher (nine-fold for Zn, four-fold for Cu, two-fold for Pb, 1.5-fold for Cd, 1.5-fold for Mn) in comparison with the reported data from Italy [52].

Vegetation, especially herbs, are very effective biomonitors of environmental pollution (air, soil, etc.) due to their widespread distribution and availability [5,53,54]. *Taraxacum officinale* (dandelion) is one of the most commonly used species in such studies, especially in terms of pollution with potentially toxic elements, because of its large areal and ease of identification [18,55]. Concentrations of the various heavy metals in the leaves and roots of *T. officinale* collected from 132 areas in the territory of Poland were investigated by Kabata-Pendias and Dudka [56]. A high indicator efficiency of both aboveground and underground plant organs was established, different for individual elements—leaves showed higher values of Cd, Pb and Zn compared to roots. The content of Zn and Ni in *T. officinale* leaves from Plovdiv corresponds well to data from some urban localities in the USA, while the elements Cd and Cr were less accumulated [57]. When compared to some Italian cities, our results for Cr and Zn content were also lower [58].

Plantago lanceolata L. is also regarded as an excellent biomonitor due to its higher bioaccumulation potential [5,17]. Samples of roots, leaves, stalks and flowers of *P. lanceolata*, grown around a copper smelter, have been collected and the concentrations of some potentially toxic elements (Al, Cu, Fe, Mn, Ni, Pb and Zn) have been measured [5]. The data revealed that the leaves contained high concentrations of Cu, Pb and Zn (even at toxic levels), especially in plants collected from the periphery of the smelter or from the transects along the prevailing wind directions. *P. lanceolata* demonstrated a significant potential for bioaccumulation, especially in roots (Cu), aboveground plant parts (Fe) and stalks (Zn). Concentrations of some PTEs (Al, Fe, Mn, Cu and Zn) in the *Plantago lanceolata* leaves from Plovdiv were extremely higher than the results of Jordanovic et al. [5], while those of Ni and Pb were quite similar.

Aksoy et al. [59] tested the potential of *Capsella bursa-pastoris* for biomonitoring of environmental pollution. That study was carried out on the grounds of the city of Bradford, England, and the results obtained were compared with data for *Poa annua*, a species with a similar ecology, whose potential for biomonitoring has been explored in previous studies. The authors confirmed the qualities of both species based on the high linear regression for Pb, Cd, Zn and Cu contents in soil and leaves (after washing), as well as a consequence of the possibility to distinguish atmospheric deposition from root exchange.

Tree leaves are well-known as reliable biomonitors to track air pollution with trace elements [4,15,16,27,35,60], although there are difficulties that arise when comparing data from different studies. The problems are connected to both the use of various tree species and various experimental methodologies [19]. Leaf accumulation of PTEs in some broadleaf species (*Acer platanoides, Aesculus hippocastanum, Betula pendula,* and *Tilia cordata*) from high-traffic areas in Belgrade, Serbia, was studied by Tomašević et al. [19]. The Serbian leaf contents of Cr, Cu, Fe and Pb were significantly higher than values found in the same plant species from Plovdiv (except Zn in *B. pendula*). Our data for Cu and Cd leaf contents in *A. hipocastanum* and *B. pendula* correspond well to the results from Wroclaw, Poland [61], although the Pb content was two and three-fold higher (in *B. pendula* and *A. hipocastanum*, respectively). *A. hippocastanum* leaves, collected from urban, suburban and roadsides in the Thrace region of Turkey (Istanbul, Edirne, Tekirdag, Corlu), have been studied by

Yilmaz et al. [62] and the concentrations of Pb, Cd, Zn and Cu are similar to our findings. Another study from Turkey [63] also evaluated the leaf contents of Cd, Ni, Pb and Zn (*A. platanoides*, *A. hippocastanum* and *B. pendula*), but the samples had been collected from the city of Istanbul.

According to Kabata-Pendias and Pendias [64], the Pb content of tree leaves (both deciduous and coniferous) in Europe varied in the range of 1.5–2.1 mg/kg. The highest Pb value, found in our study, was 5.0 $\pm$ 0.1 mg/kg (urban zone), which means that the urban air is quite polluted with this element (based on foliar bioaccumulation) in comparison with other European cities (2.5-fold higher).

4.2. Framework for Implementing NBSs to Mitigate Air Pollution in Urban Areas

Here we provide some recommendations to help planners and practitioners to apply the best measures when implementing NBSs for air quality improvement as well as for enhancing the ecosystem services derived by urban green infrastructure.

The first step in the proposed framework for NBS implementation (Figure 2) requires the identification of the priority urban areas with more deteriorated air quality. The next step is to elaborate the major sources of air pollution, urban meteorology, microclimate and heterogeneity. The third step includes the selection of a NBS strategy, which mainly depends on the location characteristics (topography, microclimate, etc.), type of source (domestic heating, traffic or industrial pollution) and the desired outcome (air purification, noise attenuation, urban heat island mitigation, etc.). After that, in step 4, a careful selection of plant species based on their adaptation and tolerance to an urban environment, physiological characteristics, and air pollutants removal efficiency will be crucial in the mitigation of air pollution to a greater extent [16–18,22,28].

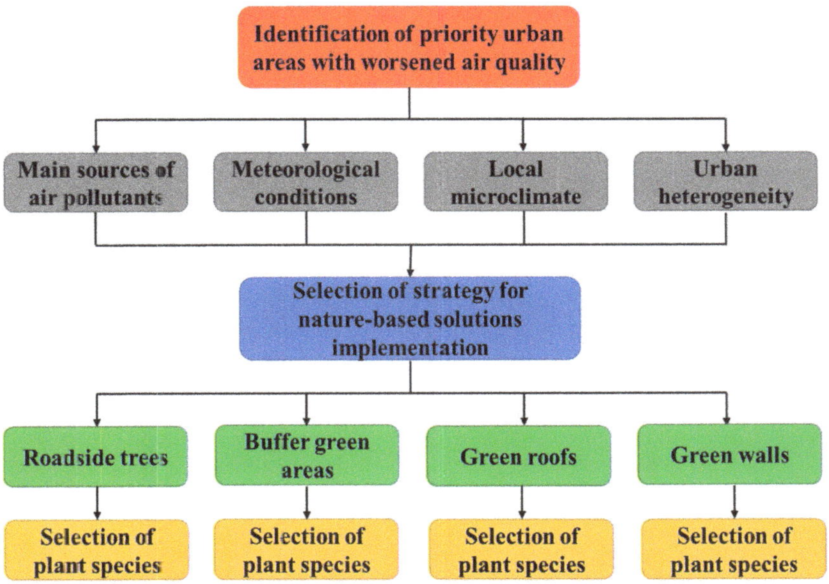

Figure 2. Framework for NBS implementation in urban areas for air pollution mitigation.

There is no doubt that plants and notably trees are particularly efficient solutions in removing pollutants from both air and soil [20,22,65]. They have large crowns, with a huge leaf area, and they act as pollutant capture devices for both gaseous and particulate substances from the ambient air. Hence, urban trees are found to be one of the most suitable NBSs for reducing roadside air pollution by both leaf surface retention, stoma absorption and wax accumulation [66]. Field measurements have shown that a wide vegetation protection belt along a road can markedly reduce the fine particle (PM_{10} and

$PM_{2.5}$) concentration. Avenue trees have lowered the PM_{10} levels by 39–89% (depending on the wind rose) and the $PM_{2.5}$ level by 17% [66]. Other benefits are related to the wind speed retention, noise attenuation, shading of sidewalks and bicycle lanes, lessened health risk, etc. Accordingly, the planting of roadside vegetation should be included in all NBS strategies for achieving urban sustainability. A medium spacing (approximately equal to the crown diameter) offers the highest impact in reducing air pollution on both bicycle lanes and roads [66]. Evergreen species have an advantage in filtering the air all year long [4,22], but deciduous trees have wider crowns and a larger leaf area, thus greater trapping efficiency [67]. Some previous studies also showed that urban vegetation is permanently exposed to the presence of plenty of toxic gases, aerosols and particulates, which even at chronic concentrations seriously affect plant physiology. For this reason, plants are involved in a constant process of adaptation toward the dynamics of an urban environment, aiming to achieve optimal development. There is a lack of knowledge about the proper adaptation mechanisms that take place, but some different signaling pathways should play role. In our previous study, we have proposed a reference scale of biochemical markers for the assessment of plant stress induced by urban air pollution [68]. Furthermore, they can be applied in landscaping activities and strategies for the sustainable management of urban forestry.

Various herb species are used for the construction of urban lawns, green walls and green roofs, especially when some limitations to tree development or maintenance problems persist [13]. Due to their biology, herbs are preferred over trees for soil phytoremediation purposes [10,69–71]. Urban lawn construction is a typical example of a NBS, aiming at sustainable management of urban soils, providing low-cost remediation, biodiversity support and other benefits. Herbal cover (i.e., urban lawns) could improve the soil structure by enhancing its porosity and aggregate stability, as well as by increasing the organic matter content and water holding capacity. It also influences many soil processes, soil functions, and soil microbiota as well, thus contributing to a more resilient soil ecosystem that has a better buffer capacity against external impacts and creates better livelihood conditions. Our previous work [72] describes an algorithm for urban lawns and buffer green patches construction. These buffer patches along roads led to a significant decrease in the traffic-derived pollutants that have been deposited on soil surface via ambient air. Furthermore, they exerted a positive effect on the soil microbiota's structure, enzymatic activities, etc. Furthermore, the sustainable management of urban soils is related to that of the urban air due to the carbon sequestration, greenhouse gases reduction, dust suspension and retention, and thus benefits air quality and human health. So, NBS integration into urban planning and management can lead to a significant enhancement of plenty of ecosystem services in urban areas.

In recent years, mosses are preferred and widely used for green walls construction in both indoor and outdoor conditions [71]. However, there are a lot of problems related to their indoor survival and their horticulture [73]. At the same time, their harvesting from nature is crucial to the environment and habitats [73]. In such cases, the usage of dried moss and/or lichen material can be implemented as an alternative to the living material without any need for periodic replacement. Furthermore, these non-vascular plants can accumulate plenty of PTEs from urban air in their dried tissues, thus reducing air pollution, especially in so-called hot spots. Possible removal of accumulated pollutants by washing can be applied to achieve recycling and reuse of dried wall material. That can also be an option for places with very poor environments where no plant will survive.

5. Conclusions

Urban vegetation is one of the most effective NBSs for reducing urban pollution and it provides many other ecosystem services to human well-being. A careful selection of plant species and design parameters, as well as proper planning of NBS strategies according to local conditions is needed for their successful and effective performance. The synergy between anthropogenic and natural processes when implementing NBSs reveals lower costs and lower maintenance, and they may even be more effective in achieving urban resilience.

(1) The present study sheds new light on some well-known biomonitors (trees, herbs, mosses and lichens) in the context of their application for air pollution mitigation.
(2) Ornamental trees demonstrated significant bioaccumulation of Al, Fe, and Pb (*Betula pendula*); Mo, V, and Cr (*Aesculus hippocastanum*); and As, Bi, and U (*Acer platanoides*), having Metal accumulation index (MAI) values in the range of 49.6, 33.76, and 24.53, respectively.
(3) Herbal species were found to be excellent accumulators of Na and B (*Taraxacum officinale*); Na, U and Hg (*Capsella bursa-pastoris*); and Cr and Pb (*Plantago lanceolata*). Their MAI values were in the range of 30.49, 29.63 and 21.77, respectively.
(4) Moss (*Sphagnum girgensohnii*) and lichen (*Pseudevernia furfuracea*) dried materials were the best accumulators of Se, Hg, Zn, and Cu (moss) and Mn, Sr, and Co (lichen), with MAI values of 31.21 and 48.47, respectively.

The great potential of the eight studied plant species for efficient removal of potentially toxic elements is highlighted and their implementation into NBS frameworks is recommended.

Author Contributions: All authors contributed to the study conception and design—S.P., B.N. and I.V. Material preparation, data collection and analysis were performed by S.P. and B.N. The first draft of the manuscript was written by S.P. and all authors (S.P., B.N. and I.V.) commented on previous versions of the manuscript. All authors have read and agreed to the published version of the manuscript.

Funding: This study was financed by the European Union-NextGenerationEU, through the National Recovery and Resilience Plan of the Republic of Bulgaria, project No. BG-RRP-2.004-0001-C01.

Data Availability Statement: The dataset is available on request from the authors.

Conflicts of Interest: The authors declare no conflicts of interest.

References

1. Alberti, M. The effects of urban patterns on ecosystem function. *Int. Reg. Sci. Rev.* **2005**, *28*, 168–192. [CrossRef]
2. Cyrys, J.; Stolzel, M.; Heinrich, J.; Kreyling, W.; Menzel, N.; Wittmaack, K.; Tuch, T.; Wichmann, H. Elemental composition and sources of fine and ultrafine ambient particles in Erfurt, Germany. *Sci. Total Environ.* **2003**, *305*, 143–156. [CrossRef] [PubMed]
3. Chen, M.; Zhou, Y.; Hu, M.; Zhou, Y. Influence of urban scale and urban expansion on the urban heat island effect in metropolitan areas: Case study of Beijing–Tianjin–Hebei Urban agglomeration. *Remote Sens.* **2020**, *12*, 3491. [CrossRef]
4. Alahabadi, A.; Ehrampoush, M.H.; Miri, M.; Aval, H.E.; Yousefzadeh, S.; Ghaffari, H.R.; Ahmadi, E.; Talebi, P.; Fathabadi, Z.A.; Babai, F.; et al. A comparative study on capability of different tree species in accumulating heavy metals from soil and ambient air. *Chemosphere* **2007**, *172*, 459–467. [CrossRef]
5. Jordanovic, J.S.; Serbula, S.M.; Markovic, M.M.; Radojevic, A.A.; Kalinovic, J.V.; Kalinovic, T.S. The influence of the environmental factors on the accumulation patterns of toxic elements in *Plantago lanceolata* sampled in the area under strong anthropopressure. *Process Saf. Environ. Prot.* **2024**, *183*, 1239–1248. [CrossRef]
6. Tang, X.; Wu, X.; Xia, P.; Lin, T.; Huang, X.; Zhang, Z.; Zhang, J. Health risk assessment of heavy metals in soils and screening of accumulating plants around the Wanshan mercury mine in Northeast Guizhou Province, China. *Environ. Sci. Pollut. Res.* **2021**, *28*, 48837–48850. [CrossRef] [PubMed]
7. Balakrishnan, K.; Dey, S.; Gupta, T.; Dhaliwal, R.S.; Brauer, M.; Cohen, A.J. The impact of air pollution on deaths, disease burden, and life expectancy across the states of India: The Global Burden of Disease Study 2017. *Lancet Planet. Health* **2019**, *3*, 26–39. [CrossRef] [PubMed]
8. World Health Organization (WHO). *WHO Global Urban Ambient Air Pollution Database*; World Health Organization: Geneva, Switzerland, 2016.
9. Davies, C.; Lafortezza, R. Urban green infrastructure in Europe: Is greenspace planning and policy compliant? *Land Use Policy* **2017**, *69*, 93–101. [CrossRef]
10. Bush, J.; Doyon, A. Building urban resilience with nature-based solutions: How can urban planning contribute? *Cities* **2019**, *95*, 102483. [CrossRef]
11. Menon, J.S.; Sharma, R. Nature-Based Solutions for Co-mitigation of Air Pollution and Urban Heat in Indian Cities. *Front. Sustain. Cities* **2021**, *3*, 705185. [CrossRef]

12. Crouse, D.L.; Pinault, L.; Balram, A.; Brauer, M.; Burnett, R.T.; Martin, R.V. Complex relationships between greenness, air pollution, and mortality in a population-based Canadian cohort. *Environ. Int.* **2019**, *128*, 292–300. [CrossRef]
13. European Commission. *Nature-based Solutions for Microclimate Regulation and Air Quality: Analysis of EU-funded Projects*; European Commission: Brussels, Brussels, 2020; ISBN 978-92-76-18205-4. [CrossRef]
14. Keesstra, S.; Nunes, J.; Novara, A.; Finger, D.; Avelar, D.; Kalantari, Z.; Cerdà, A. The superior effect of nature based solutions inland management for enhancing ecosystem services. *Sci. Total Environ.* **2018**, *610–611*, 997–1009. [CrossRef]
15. Kalinovic, S.; Serbula, S.M.; Kalinovic, J.V.; Radojevic, A.A.; Petrovic, J.A.; Steharnik, M.M.; Milosavljevic, J.S. Suitability of linden and elder in the assessment of environmental pollution of Brestovac spa and Bor lake (Serbia). *Environ. Earth Sci.* **2017**, *76*, 178. [CrossRef]
16. Petrova, S.; Yurukova, L.; Velcheva, I. Possibilities of using deciduous tree species in trace element biomonitoring in an urban area (Plovdiv, Bulgaria). *Atmos. Pollut. Res.* **2014**, *5*, 196–202. [CrossRef]
17. Petrova, S.; Velcheva, I.; Yurukova, L.; Berova, M. *Plantago lanceolata* L. as a Biomonitor of Trace Elements in an Urban Area. *Bulg. J. Agric. Sci.* **2014**, *20*, 325–329.
18. Petrova, S.; Yurukova, L.; Velcheva, I. *Taraxacum officinale* as a biomonitor of metals and toxic elements (Plovdiv, Bulgaria). *Bulg. J. Agric. Sci.* **2013**, *19*, 241–247.
19. Tomašević, M.; Aničić, M.; Jovanović, L.; Perić-Grujić, A.; Ristić, M. Deciduous tree leaves in trace elements biomonitoring: A contribution to methodology. *Ecol. Indic.* **2011**, *11*, 1689–1695. [CrossRef]
20. Yang, J.; McBride, J.; Zhou, J.; Sun, Z. The urban forest in Beijing and its role in air pollution reduction. *Urban For. Urban Green.* **2005**, *3*, 65–78. [CrossRef]
21. Yin, S.; Shen, Z.; Zhou, P.; Zou, X.; Che, S.; Wang, W. Quantifying air pollution attenuation within urban parks: An experimental approach in Shanghai, China. *Environ. Pollut.* **2011**, *159*, 2155–2163. [CrossRef]
22. Petrova, S. Efficiency of *Pinus nigra* J.F. Arnold in removing pollutants from urban environment (Plovdiv, Bulgaria). *Environ. Sci. Pollut. Res.* **2020**, *27*, 39490–39506. [CrossRef]
23. Rühling, A.; Tyler, G. Sorption and retention of heavy metals in the woodland moss *Hylocomium splendens* (Hedw.). *Oikos* **1970**, *21*, 92–97. [CrossRef]
24. Sloof, J.E. Environment Lichenology: Biomonitoring Trace-Element Air Pollution. Ph.D. Thesis, Delft University of Technology, Delft, The Netherlands, 1993.
25. Steinnes, E. A critical evaluation of the use of naturally growing moss to monitor the deposition of atmospheric metals. *Sci. Total Environ.* **1995**, *160/161*, 243–249. [CrossRef]
26. Wittig, R. General aspects of biomonitoring heavy metals by plants. In *Plants as Biomonitors: Indicators for heavy metals in terrestrial environment*; Markert, B., Ed.; VCH Publishers Ltd.: Cambridge, UK, 1993; pp. 3–28.
27. Kumar, A.; Kumar, P.; Singh, H.; Kumar, N. Adaptation and mitigation potential of roadside trees with bio-extraction of heavy metals under vehicular emissions and their impact on physiological traits during seasonal regimes. *Urban For. Urban Green.* **2021**, *58*, 126900. [CrossRef]
28. Sloof, J.E. Lichens as quantative biomonitors for atmospheric trace-element deposition, using transplants. *Atmos. Environ.* **1995**, *29*, 11–20. [CrossRef]
29. Aničić, M.; Tasić, M.; Frontasyeva, M.V.; Tomašević, M.; Rajšić, S.; Mijić, Z.; Popović, A. Active moss biomonitoring of trace elements with *Sphagnum girgensohnii* moss bags in relation to atmospheric bulk deposition in Belgrade, Serbia. *Environ. Pollut.* **2009**, *157*, 673–679. [CrossRef] [PubMed]
30. Aničić, M.; Tomašević, M.; Tasić, M.; Rajšić, S.; Popović, A.; Frontasyeva, M.V.; Lierhagen, S.; Steinnes, E. Monitoring of trace element atmospheric deposition using dry and wet moss bags: Accumulation capacity versus exposure time. *J. Hazard. Mater.* **2009**, *171*, 182–188. [CrossRef]
31. Steinnes, E.; Hanssen, J.E.; Ramb, J.P.; Vogt, N.B. Atmospheric deposition of trace elements in Norway: Temporal and spatial trends studied by moss analysis. *Water Air Soil Pollut.* **1994**, *74*, 121–140. [CrossRef]
32. Thöni, L.; Yurukova, L.; Bergamini, A.; Ilyin, I.; Matthaei, D. Temporal trends and spatial patterns of heavy metal concentrations in mosses in Bulgaria and Switzerland: 1990–2005. *Atmos. Environ.* **2011**, *45*, 1899–1912. [CrossRef]
33. Rühling, A. Atmospheric heavy metal deposition in Europe—estimations based on moss analysis. NORD 9, Nordic Council of Ministers. AKA-Print, A/S, Arhus, 1994. Available online: https://books.google.com.hk/books/about/Atmospheric_Heavy_Metal_Deposition_in_Eu.html?id=Wsv1suCjzAIC&redir_esc=y (accessed on 14 March 2024).
34. Rühling, A.; Steinnes, E. *Atmospheric heavy metal deposition in Europe 1995–1996*. NORD 15, Nordic Council of Ministers, Copenhagen, 1998, Rounborgs Grafiske hus, Holstebo. Available online: https://core.ac.uk/download/pdf/62557.png (accessed on 14 March 2024).
35. Matic, M.; Pavlovic, D.; Perovic, V.; Cakmak, D.; Kostic, O.; Mitrovic, M.; Pavlovic, P. Assessing the Potential of Urban Trees to Accumulate Potentially Toxic Elements: A Network Approach. *Forests* **2023**, *14*, 2116. [CrossRef]
36. Instituto Nazionale di Statistica (ISTAT). Available online: http://www.en.istat.it (accessed on 19 September 2023).
37. Atanassov, D.; Spassova, S.; Grancharova, D.; Krastev, S.; Yankova, T.; Nikolov, L.; Chakarova, M.; Krasteva, P.; Genov, N.; Stamenov, J.; et al. Air pollution monitoring and modeling system of the town of Plovdiv (Phase I). *J. Environ. Prot. Ecol.* **2006**, *7*, 260–268.

38. McDonnell, M.J.; Hahs, A.K. The use of gradient analysis studies in advancing our understanding of the ecology of urbanizing landscapes: Current status and future directions. *Landsc. Ecol.* **2008**, *23*, 1143–1155. [CrossRef]
39. De Nicola, F.; Maisto, G.; Prati, M.V.; Alfani, A. Leaf accumulation of trace elements and polycyclic aromatic hydrocarbons (PAHs) in *Quercus ilex* L. *Environ. Pollut.* **2008**, *153*, 376–383. [CrossRef] [PubMed]
40. Yurukova, L.; Petrova, S.; Velcheva, I.; Aleksieva, I. Preliminary data of moss-bags technique in an urban area (Plovdiv, Bulgaria). *Comptes Rendus De L'academie Bulg. Des Sci.* **2013**, *66*, 1135–1138.
41. Petrova, S.; Yurukova, L.; Velcheva, I. Lichen-bags as a biomonitoring technique in an urban area. *Appl. Ecol. Environ. Res.* **2015**, *13*, 915–923. [CrossRef]
42. *NCS DC73348*; Bush branches and leaves—Trace elements. LGC Standards: Wesel, Germany, 2008.
43. Mingorance, M.D.; Valdés, B.; Rossini, O.S. Strategies of heavy metal uptake by plants growing under industrial emissions. *Environ. Int.* **2007**, *33*, 514–520. [CrossRef] [PubMed]
44. Culicov, O.A.; Mocanu, R.; Frontasyeva, M.V.; Yurukova, L.; Steinnes, E. Active moss biomonitoring applied to an industrial site in Romania: Relative accumulation of 36 elements in moss-bags. *Environ. Monit. Assess.* **2005**, *108*, 229–240. [CrossRef]
45. Ma, Y.L.; Jia, G.M.; Wang, Y.P.; Liu, H.P. Contents of heavy metal in leaves of plants and air pollution evaluation in Guangzhou region. *Urban Environ. Urban Ecol* **2001**, *14*, 23–30.
46. Prussia, C.M.; Killingbeck, K.T. Concentrations of ten elements in two common foliose lichens: Leachability, seasonality and the influence of the rock and tree bark substrates. *Bryologist* **1991**, *94*, 135–142. [CrossRef]
47. Liu, Y.-J.; Zhu, Y.-G.; Ding, H. Lead and cadmium in leaves of deciduous trees in Beijing, China: Development of a metal accumulation index (MAI). *Environ. Pollut* **2007**, *145*, 387–390. [CrossRef]
48. Hu, R.; Yan, Y.; Zhou, X.; Wang, Y.; Fang, Y. Monitoring Heavy Metal Contents with *Sphagnum Junghuhnianum* Moss Bags in Relation to Traffic Volume in Wuxi, China. *Int. J. Environ. Res. Public Health* **2018**, *15*, 374. [CrossRef]
49. Hofman, J.; Stokkaer, I.; Snauwaert, L.; Samson, R. Spatial distribution assessment of particulate matter in an urban street canyon using biomagnetic leaf monitoring of tree crown deposited particles. *Environ. Pollut.* **2013**, *183*, 123–132. [CrossRef] [PubMed]
50. Puckett, K.J.; Burton, M.A.S. The effect of trace elements on lower plants. In *Effect of Heavy Metal Pollution on Plants, vol.2. Metals in the Environment*; Lepp, N.W., Ed.; Applied Science Publishers: London, UK, 1981; pp. 213–238.
51. Brown, D.H.; Brown, R.M. Reproducibility of sampling for element analysis using bryophytes. In *Element Concentration Cadasters in Ecosystems*; Lieth, H., Market, B., Eds.; VCH Publishers: Weinheim, Germany, 1990; pp. 55–62.
52. Basile, A.; Sorbo, S.; Aprile, G.; Conte, B.; Castaldo Cobianchi, R. Comparison of the heavy metal bioaccumulation capacity of an epiphytic moss and an epiphytic lichen. *Environ. Pollut.* **2008**, *151*, 401–407. [CrossRef] [PubMed]
53. Hubai, K.; Kováts, N.; Sainnokhoi, T.-A.; Teke, G. Accumulation pattern of polycyclic aromatic hydrocarbons using *Plantago lanceolata* L. as passive biomonitor. *Environ. Sci. Pollut. Res.* **2022**, *29*, 7300–7311. [CrossRef] [PubMed]
54. Kardel, F.; Wuyts, K.; Babanezhad, M.; Vitaharana, U.W.A.; Wuytack, T.; Potters, G.; Samson, R. Assessing urban habitat quality based on specific leaf area and stomatal characteristics of *Plantago lanceolata* L. *Environ. Pollut.* **2010**, *158*, 788–794. [CrossRef] [PubMed]
55. Ianovici, N.; Vereş, M.; Catrina, R.G.; Pîrvulescu, A.-M.; Tănase, R.M.; Datcu, D.A. Methods of biomonitoring in urban environment: Leaf area and fractal dimension. *Ann. West Univ. Timişoara Ser. Biol.* **2015**, *XVIII*, 169–178.
56. Kabata-Pendias, A.; Dudka, S. Trace metal contents of *Taraxacum officinale* (dandelion) as a convenient environmental indicator. *Environ. Geochem. Health* **1991**, *13*, 108–113. [CrossRef] [PubMed]
57. Keane, B.; Collier, M.H.; Shann, J.R.; Rogstad, S.H. Metal content of dandelion (*Taraxacum officinale*) leaves in relation to soil contamination and airborne particulate matter. *Sci. Total Environ.* **2001**, *281*, 63–78. [CrossRef]
58. Vanni, G.; Cardelli, K.; Marchini, F.; Saviozzi, A.; Guidi, L. Are the physiological and biochemical characteristics in dandelion plants growing in an urban area (Pisa, Italy) indicative of soil pollution? *Water Air Soil Pollut.* **2015**, *226*, 124. [CrossRef]
59. Aksoy, A.; Hale, W.H.G.; Dixon, J.M. *Capsella bursa-pastoris* (L.) Medic. as a biomonitor of heavy metals. *Sci. Total Environ.* **1999**, *226*, 177–186. [CrossRef]
60. Markert, B. *Plants as Biomonitors/Indicators for Heavy Metals in the Terrestrial Environment*; VCH Press: Weinheim, Germany, 1993.
61. Piczak, K.; Leśniewicz, A.; Zyrnicki, W. Metal concentrations in deciduous tree leaves from urban areas in Poland. *Environ. Monit. Assess.* **2003**, *86*, 273–287. [CrossRef]
62. Yilmaz, R.; Sakcali, S.; Yarci, C.; Aksoy, A.; Ozturk, M. Use of *Aesculus hippocastanum* L. as a biomonitor of heavy metal pollution. *Pak. J. Bot.* **2006**, *38*, 1519–1527.
63. Baycu, G.; Tolunay, D.; Ozden, H.; Gunebakan, S. Ecophysiological and seasonal variations in Cd, Pb, Zn, and Ni concentrations in the leaves of urban deciduous trees in Istanbul. *Environ. Pollut.* **2006**, *143*, 545–554. [CrossRef]
64. Kabata–Pendias, A.; Pendias, H. *Trace Elements in Soils and Plants*, 3rd ed.; CRC Press: Boca Raton, FL, USA, 2001.
65. Kroeger, T. Reforestation as a novel abatement and compliance measure for ground level ozone. *PNAS* **2014**, *111*, 4204–4213. [CrossRef] [PubMed]
66. Ren, F.; Qiu, Z.; Liu, Z.; Bai, H.; Gao, H.O. Trees help reduce street-side air pollution: A focus on cyclist and pedestrian exposure risk. *Build. Environ.* **2023**, *229*, 109923. [CrossRef]
67. Tomasevic, M.; Rajsic, S.; Dordevic, D.; Tasic, M.; Krstic, J.; Novakovic, V. Heavy metals accumulation in tree leaves from urban areas. *Environ. Chem. Lett.* **2004**, *2*, 151–154. [CrossRef]

68. Petrova, S.; Petkova, M. Plant Traits of *Tilia tomentosa* Moench, *Fraxinus excelsior* L., and *Pinus nigra* J.F. Arnold as a Proxy of Urbanization. *Forests* **2023**, *4*, 800. [CrossRef]
69. Arienzo, M.; Adamo, P.; Cozzolino, V. The potential of Lolium perenne for revegetation of contaminated soil from a metallurgical site. *Sci. Total Environ.* **2004**, *319*, 13–25. [CrossRef] [PubMed]
70. Burges, A.; Alkorta, I.; Epelde, L.; Garbisu, C. From phytoremediation of soil contaminants to phytomanagement of ecosystem services in metal contaminated sites. *Int. J. Phytopharm.* **2018**, *20*, 384–397. [CrossRef]
71. Suchkova, N.; Darakas, E.; Ganoulis, J. Phytoremediation as a prospective method for rehabilitation of areas contaminated by long term sewage sludge storage: Ukrainian–Greek case study. *Ecol. Eng.* **2010**, *36*, 373–378. [CrossRef]
72. Petrova, S.; Nikolov, B.; Velcheva, I.; Angelov, N.; Valcheva, E.; Katova, A.; Golubinova, I.; Marinov-Serafimov, P. Buffer Green Patches around Urban Road Network as a Tool for Sustainable Soil Management. *Land* **2022**, *11*, 343. [CrossRef]
73. Zechmeister, H.G.; Möslinger, L.; Korjenic, A.; Streit, E.; Sulejmanovski, A.; Frank, P.N.; Hummel, E. Viability of Living Moss for Indoor GreenWalls: A Study on Temperature, Humidity, and Irrigation. *Sustainability* **2023**, *15*, 15625. [CrossRef]

forests

MDPI

Article

Impacts of Plant Configuration on the Outdoor Wind Comfort of Subtropical Coastal Campuses: Evidence from a Study of Quanzhou

Jing Chen [1], Jiushan Zeng [2], Tiantian Huang [3,*], Yaolong Wang [3], Haosen Yang [3], Xiaofang Yu [3] and Zefa Wang [2,3,*]

1 School of Geographical Sciences, Fujian Normal University, Fuzhou 350007, China
2 School of Resources and Environmental Sciences, Quanzhou Normal University, Quanzhou 362000, China
3 School of Architecture, Tianjin University, Tianjin 300072, China
* Correspondence: h_06042@tju.edu.cn (T.H.); zefawang@qztc.edu.cn (Z.W.)

Abstract

Even though the interaction between plants and the outdoor wind environment has been a focus of interest for scholars from various disciplines in recent years, the relationship between campus outdoor wind comfort and plant configuration in subtropical coastal areas remains poorly understood. Using the outdoor space of a typical subtropical coastal campus (the Donghai Campus of Quanzhou Normal University) as a case study, we explore the connection between plant configuration and outdoor wind comfort. The campus outdoor area is segmented into roads, squares, and courtyards to investigate this relationship. To achieve this goal, a 9-h fixed-point measurement method and the PHOENICS software (2016) were utilized. The following are the findings of the research: (1) Within the realm of trees, the banyan, *Bischofia javanica*, and kapok species exhibit a notable impact on wind speed reduction, with respective wind reduction ratios of 1.22, 1.31, and 1.29. Notably, among shrubs, waringin stands out with a wind reduction ratio of 1.83. (2) The tree + shrub + grass combination is the most effective method for reducing wind among the three plant facade configurations. Specifically, the combination of *Bischofia javanica*, waringin, and carpet grass has the best wind reduction effect, with a wind reduction ratio of 2.55. (3) Adding *Bischofia javanica*, waringin, and grass plants in areas with high wind speeds can effectively improve wind comfort. This provides directions for creating a comfortable wind environment on university campuses situated in subtropical coastal areas.

Keywords: plant configuration; wind comfort; subtropical coastal campuses; outdoor space; Quanzhou

1. Introduction

The campus outdoor space, being the largest public area within the school, plays a pivotal role in hosting educational activities, as well as serving as a hub for the students' daily outdoor pursuits and interpersonal engagements. When evaluating the quality of the campus outdoor environment, the wind environment is a key factor to consider [1], as it often interferes with the safety and comfort of outdoor activities for teachers and students. The adverse wind environment directly affects the utilization rate of campus outdoor space activities. In subtropical coastal areas, such as Quanzhou in the summer and winter, this problem is particularly prominent.

At present, research on the campus wind environment in Chinese universities mostly focuses on cold regions, in North China, and humid and hot regions in South China. The research mainly involves three aspects: campus architectural form and layout, landscape planning, optimization strategies, and suggestions. In terms of architectural form and layout, the study by Xiaoyu Ying et al. revealed that the comfort of the outdoor wind environment on campus is associated with specific building characteristics, such as the transparency of the windward side, the size of the ventilation gap, the presence of a ventilation corridor, and the positioning of the vertical connection volume [2]. Through the design of external clusters, Shuo Chen et al. optimized the configuration of campus clusters, resulting in an increase in the proportion of spatial wind shadow areas from 20% to 53% [3]. Cheng Sun et al. discovered that achieving a D/H ratio of 2.50 enhances the outdoor thermal comfort and spatial experience of the teaching building cluster. The study suggests that L-shaped and three-sided enclosures are the optimal grouping forms, striking a balance between outdoor thermal comfort, energy efficiency, and spatial needs [4]. Regarding landscape planning, Binyi Liu et al. investigated how landscape layout influences the near-ground wind conditions in residential settings [5]. Qiang Zeng et al. investigated the influence of landscape planning and design on environmental ventilation conditions and examined the underlying causes of unfavorable working conditions [6]. Jia Guo et al. maintained the overall concepts of land use indicators and early campus planning and design unchanged. Under this premise, they proposed optimization strategies from the perspectives of campus clusters and individual buildings [7]. With regard to optimization strategies and recommendations, Mingya Liu et al. suggested optimization approaches to enhance the campus wind environment, encompassing architectural layout and green configuration [8]. Neng Zhang et al. discovered that enhancing campus greenery and establishing pocket parks can enhance ventilation efficiency in regions where wind conditions fall below prescribed standards [9]. Wang et al. discovered that the average wind speed on the floor plan at 6 m and 10 m above the roof was generally above 3 m/s, making it a favorable condition for harnessing wind energy [10]. The research conducted thus far has highlighted a lack of studies focusing on campus wind conditions in areas with hot summers and mild winters. Thus, additional research is imperative.

Despite the lack of research on the effects of vegetation greening on the wind environment, Yuehua Han et al. discovered that, during the summer season, implementing a double-row planting arrangement with a tree crown spacing of 2 m effectively enhances wind speed on the windward side of the gymnasium. Additionally, in the winter, the optimal planting configuration involves a double-row setup with a spacing of 2 m [11]. Ruijie Liu et al. discovered that the combination of small trees and large shrubs, as well as medium trees along with small trees and large shrubs, exhibits optimal wind speed reduction efficiency for pedestrian comfort in urban green spaces [12]. The findings of Zhang Li et al. suggest that the influence of vegetation on thermal conditions and airflow is contingent upon factors such as tree distribution, leaf area index (LAI), crown width, and tree height [13]. Ahmad Hami et al. found that the cooling efficiency of plants depends on factors such as plant species, tree and shrub arrangement, the distribution and connectivity of green patches in the landscape, and orientation of the planting rows [14]. The research findings of Murtaza Mohammadi et al. show that, although hedges and trees can impede airflow, trees have a greater capacity to generate cooling effects [15]. Liyan Rui et al. discovered that when grass and shrub coverage decreases under equivalent green space coverage, trees take their place. This substitution, while having a minor effect on thermal comfort, wind speed, and air pollution, enhances the recreational areas available to residents [16]. The study by Hao Li et al. revealed a 43% reduction in the wind speed amplification factor when protective forests were present, resulting in a notable decrease in amplification factors for the key hazardous wind directions [17]. Woei-Leong Chan et al. used porous tetrahedral elements to model tree canopies and calculate aerodynamic wind forces on trees, with the goal of reducing the risk of tree failure [18]. There is still a need for more research on how plant configuration affects the wind environment on university campuses located in hot summer and warm winter areas.

In summary, this study utilizes both CFD simulation and on-site measurement methods, focusing on a study of the impact of different plant configurations on the wind environment of the campus outdoor spaces in subtropical coastal universities (specifically the Donghai Campus of Quanzhou Normal University). It was discovered through initial experiments that the comfort of outdoor spaces on campus plays a significant role in student outdoor activities. Therefore, by combining CFD simulation and on-site measurements in both the summer and winter, this study explores the impact of plant configuration on the wind environment in order to provide a reference for creating a comfortable wind environment for outdoor campus spaces in subtropical coastal areas.

2. Research Object and Methods

2.1. Research Object: Quanzhou Normal University Donghai Campus

2.1.1. Spatial Overview of Quanzhou Normal University, Quanzhou City, Fujian Province

Located on the southeast coast of China, Quanzhou City ranges from 117°25′ to 119°04′ east longitude and from 24°30′ to 25°56′ north latitude. It is classified as a Cfa type according to the Köppen–Geiger climate classification system. The city is recognized by the United Nations as the beginning of the Maritime Silk Road, and it is also acknowledged as a World Heritage city. It belongs to a subtropical marine monsoon climate (Figure 1a) and is deeply affected by sea winds, especially the East China Sea campus of Quanzhou Normal University, which is more affected by strong wind erosion (Figure 1b,c). The Donghai Campus of Quanzhou Normal University, located just 1 km from the coast, is prone to frequent summer typhoons and winter gales. The average annual temperature is approximately 19.8 °C, with an average annual precipitation of around 1030 mm.

Figure 1. (**a**) Map of China. (**b**) Map of Fujian Province. (**c**) Map showing the location of Donghai Campus of Quanzhou Normal University.

2.1.2. Outdoor Space Types of Donghai Campus of Quanzhou Normal University

Through an analysis of the current overall plan of the Donghai Campus of Quanzhou Normal University (Figure 2a), it is found that the outdoor space can be divided into three categories based on the different plant enclosure methods used: square space, road space, and courtyard space. The square space is an open outdoor area mainly surrounded by plants, such as the main entrance square in Zone A and the sports square in Zone B. The road space is a linear outdoor area mainly enclosed by plants on both sides, such as the east–west roads in Zone C and the north–south roads in Zone D. The courtyard space is an enclosed outdoor area surrounded by buildings and plants, such as individual buildings in Zone E and group buildings in Zone F (Figure 2b). These categories will be used as measurement points to compare and analyze the wind environment of different types of outdoor spaces on campus.

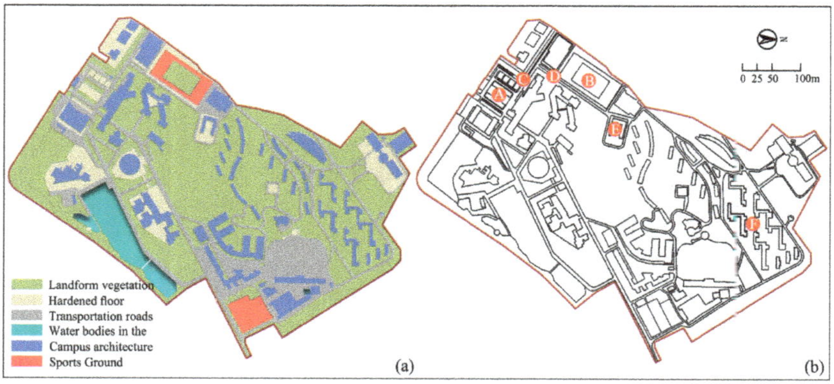

Figure 2. (**a**) Overall plan of Donghai Campus of Quanzhou Normal University. (**b**) Map showing measurement points for the wind environment in campus outdoor spaces (A–F are wind environment detection points).

2.2. Contrastive Research Method

This study focuses on three types of enclosed planting spaces (square, road, and courtyard) in the Donghai Campus of Quanzhou Normal University. By conducting on-site measurements, the wind environment comfort in these spaces can be assessed to identify both their advantages and disadvantages. Based on this, the impact of plants on the spatial wind comfort is explored, as well as whether plant configuration can improve the wind environment comfort in uncomfortable spaces.

To evaluate how different plant configurations affect the wind environment, the flat configurations were modeled using SketchUp v.2021 software (solitary planting, opposite planting, and row planting) and vertical configurations (tree + shrub + grass, tree + grass, and shrub + grass). The plants were simplified into rectangular forms, with a focus on the relationship between different plant configurations and summer and winter monsoon comfort. Afterwards, the 3D models were brought into PHOENICS software to carry out CFD simulation calculations.

2.3. Research Framework

This study analyzed the simulation results of three spatial models and two plant configuration models for the Donghai Campus of Quanzhou Normal University using a combination of mathematical analysis, on-site measurements, and simulations, as well as comparative research methods. The CFD simulation results demonstrated the instantaneous wind speed on site, and a comparison was conducted to validate the reliability of the simulation by comparing the measured and simulated wind speed values. Then, combined with practical cases, it is demonstrated that the plant configuration can improve the universality of the wind environment comfort. Figure 3 illustrates the research framework of this study.

2.4. Evaluation Methods and CFD Simulation

2.4.1. Wind Environment Evaluation Methods in the Summer and Winter

This study primarily examines how various plant configurations affect the outdoor wind comfort of subtropical coastal campuses. Based on the Chinese green building evaluation system and existing research standards, a standard for assessing the summer and winter wind environment was developed. In accordance with the "Green Building Evaluation Standards" (GB/T 50378-2019), it is recommended that, during typical summer wind conditions, the human activity area of a site should be free from vortices or windless areas. Similarly, during standard winter wind conditions, the wind speed should not exceed 5 m/s at a height of 1.5 m above ground level [19]. Studies by Tan and Liu have suggested that low wind speeds, termed quiet wind zones (below 1 m/s), can result in poor air circulation, prolonged accumulation of pollutants, and adverse effects on human comfort and health [20,21]. According to Soligo et al., wind speeds of 1.6 m/s to 3.4 m/s are ideal for prolonged standing or sitting, while speeds between 3.4 m/s and 5.4 m/s are suitable for shorter periods [22]. Consequently, the criteria for evaluating the summer and winter monsoon environments are proposed as follows: wind speeds exceeding 1 m/s in the summer and between 1 m/s and 5 m/s in the winter. Specifically, wind speeds of 1.6 m/s to 3.4 m/s are deemed suitable for extended periods of standing or sitting, whereas speeds of 3.4 m/s to 5.0 m/s are suitable for shorter durations. Furthermore, in alignment with the "Green Building Evaluation Standards" (GB/T 50378-2019) and established research practices, the standard for evaluating the wind environment is the wind speed data collected at a height of 1.5 m above ground level.

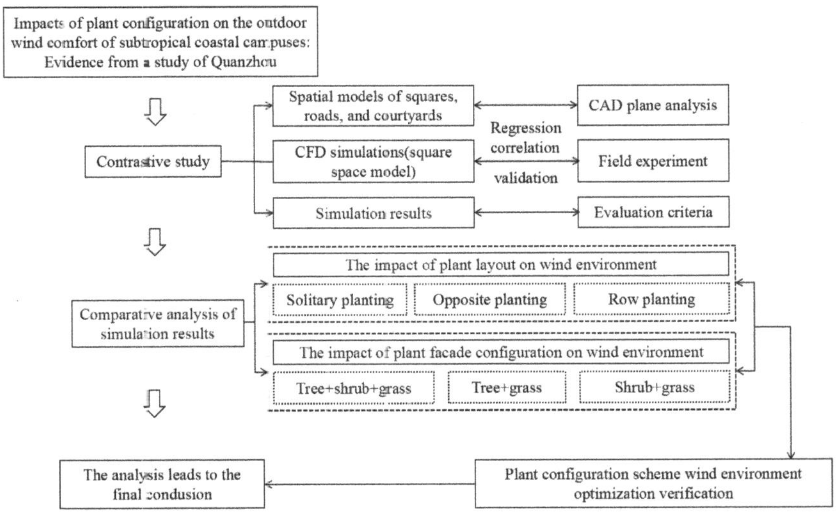

Figure 3. Research framework.

2.4.2. Details of Wind Environment Experiment and Simulations

(1) Measured Points Selection

CFD was used to simulate the wind environment at the main gate square of the Donghai Campus of Quanzhou Normal University in order to clarify the horizontal flow field division in the outdoor area and determine the measurement points for the summer and winter wind conditions. Representative wind speed detection points were chosen according to the distinct flow patterns generated by natural winds during the summer and winter as they interact with the primary entrance square, individual building courtyards, and varied plant configurations: ① In terms of outdoor space, the main entrance square space in Zone A and the courtyard space of individual buildings in Zone E were selected, and 19 points most commonly involved in crowd activities in the square space (Figure 4a) and 13 points most commonly involved in crowd activities in the courtyard space (Figure 4b) were chosen; ② In terms of plant configuration, wind speed detection points were selected as the wind shadow area of plant configuration according to seasonal changes (Figure 4c–f) to explore the wind reduction efficiency of plants.

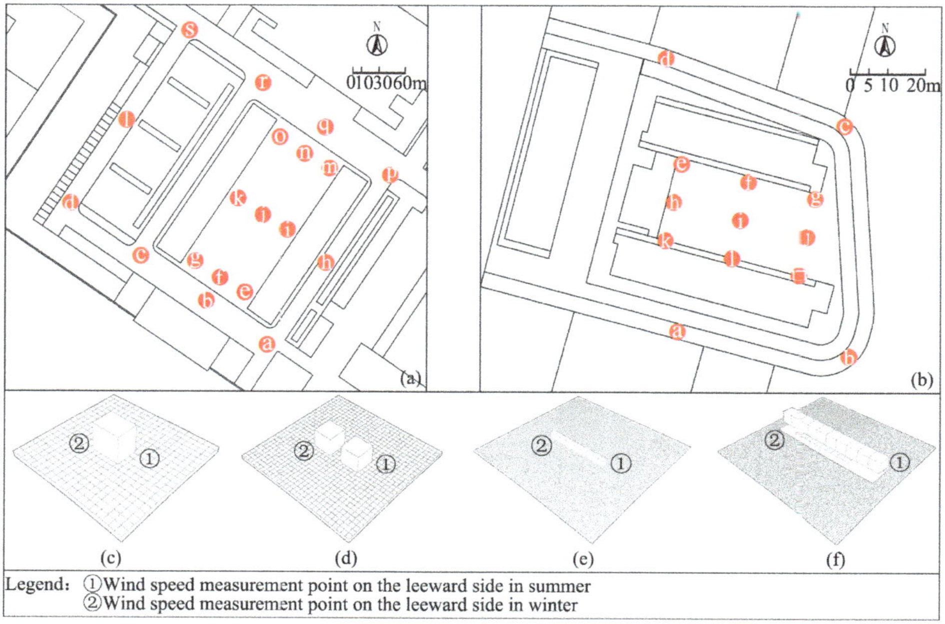

Figure 4. Wind environment measurement points: (**a**) main entrance square space (a–s are wind environment detection points); (**b**) courtyard space of an individual building (a–m are wind environment detection points); (**c**) solitary planting; (**d**) opposite planting; (**e**) row planting; (**f**) combination of trees, shrubs, and grasses.

(2) Onsite Experiment Method

To validate the correlation between observed and modeled wind speeds, we conducted measurements of the outdoor wind conditions at the Donghai Campus of Quanzhou Normal University during sunny and partly cloudy summer days. From 9:00 a.m. to 5:00 p.m. on 30 September 2022, the NK 5500 LINK wind speed meteorological instrument was used, and multiple people were simultaneously positioned and observed to conduct fixed-point measurements in six measurement areas, identified as A, B, C, D, E, and F, on the campus. Firstly, a tripod-fixed anemometer was used for testing in each of the six measurement areas, positioning the inlet of the anemometer at a height of 1.5 m above the ground. Secondly, data was collected hourly, and the average wind speed within a one-minute timeframe of each measurement point was recorded for every data point (wind speed values were measured at minimum, average, and maximum levels every 10 s over the course of 1 min). During the hours of 9:00 a.m. to 5:00 p.m., 6 groups and 54 valid wind speed data points were obtained. Then, an evaluation was performed to compare and analyze the authentic wind conditions in different outdoor locations on campus, utilizing the measured data. The NK500LNK anemometer has a wind speed measurement standard range from 0.6 m/s to 40.0 m/s, featuring a resolution of 0.1 and an accuracy of ±3%.

On 6 December 2024, leaf area index (LAI) measurements were conducted on typical plants in the main entrance square of the campus (Figure 5) using the LAI-2200C Plant Canopy Analyzer (LI-COR Biosciences, Lincoln, Nebraska, USA). Measurements were taken in four directions beneath the canopy of each plant to determine the corresponding leaf area index. Subsequently, the measured data from the analyzer was transferred to a computer for analysis using the FV2200 V 2.1.1 software to derive the LAI values for each plant.

(a) (b) (c)

Figure 5. (**a**) LAI-2200C Plant Canopy Analyzer. (**b**) On-site measurement. (**c**) FV2200 V 2.1.1 software data analysis.

(3) 3D Model Construction

Establishment of Plant Configuration Model: Through on-site investigation of the plants in the main entrance square, it was found that the plants are mainly divided into three categories: herbs, shrubs, and trees. Among them, the herbs include carpet grass, the shrubs include waringin and *Aglaia odorata* lour, and the trees include *Erythrina indica*, banyan, *Bischofia javanica*, mango trees, *Chorisia speciosa*, and kapok. Aerial surveys were conducted on the canopy height, width, and trunk height of various plants, and the leaf area index of different plants was measured using the LAI-2200C Plant Canopy Analyzer (Table 1). The research by Li et al. indicates that the rectangular model has advantages such as simple modeling, fast calculation, and good convergence [23]. Therefore, the simulation employed the technique of simplifying the rectangular model and utilized SketchUp v.2021 software to create a three-dimensional model for calculating the outdoor wind field for 34 plant configurations, categorized into planar and vertical configurations (refer to Table 2 and Figure 6a–d). In terms of the planar configuration, under opposite planting conditions, the crown spacing between two plants is 3 m. Under the row planting condition, it is set to single-row dense planting, with a crown spacing of 0 m, for a total of 10 trees. In terms of the facade configuration, select plants with better wind regulation efficiency when planting alone to further explore the facade configuration scheme of plants with better wind reduction efficiency. Through solitary planting experiments, it was found that the wind regulation efficiency of banyan, *Bischofia javanica*, and kapok was relatively good, while that of waringin in shrubs was relatively good. Therefore, the facade configuration scheme of trees + shrubs + grass includes three types: banyan + waringin + carpet grass, *Bischofia javanica* + waringin + carpet grass, and kapok + waringin + carpet grass. The facade configuration scheme of trees + grass includes three types: banyan + carpet grass, *Bischofia javanica* + carpet grass, and kapok + carpet grass. The facade configuration scheme of shrubs + grass only includes one type: waringin + carpet grass.

Establishment of an outdoor space model: Using the measured CAD v.2021 topographic map of the campus, create a three-dimensional model with SketchUp v.2021 software to simulate outdoor wind flow in the main square and courtyard spaces of a single building. Make necessary simplifications and rounding of the model details without impacting the calculation results (Table 2, Figure 6e,f).

Change the outdoor space and plant configuration models to STL format, and then import them into PHOENICS software for simulating the wind environment. Retrieve the simulation results for the wind environment at a pedestrian height of 1.5 m (Z = 1.5 m) above ground level.

Table 1. Typical plant parameters.

Typical Plants	Crown Height (m)	Crown Diameter (m)	Tree Trunk Height (m)	LAI (Summer)	LAI (Winter)
Carpet grass	0.2	1.0	0	-	-
Aglaia odorata lour	0.8	1.0	0	4.25	4.85
Waringin	4.5	4.5	0	5.98	3.35
Erythrina indica	4.0	3.0	2.5	4.10	1.45
Banyan	4.5	3.5	2.5	5.60	2.56
Bischofia javanica	5.0	4.0	3.5	1.22	3.14
Mango trees	6.5	7.5	4.0	4.10	4.50
Chorisia speciosa	7.0	8.0	5.0	2.22	1.28
Kapok	8.0	8.0	6.0	0.70	1.56

Table 2. Simplified integer value table for two outdoor space and plant scene models on campus.

Project Name	Dimension (m)	Project Name	Dimension (m)	Project Name	Dimension (m)
A main entrance square space	410 × 460 × 108	*Aglaia odorata* lour (opposite planting)	30 × 30 × 1.8	Banyan (row planting)	100 × 100 × 8.0
E single building courtyard space	120 × 104 × 48	Waringin (opposite planting)	30 × 30 × 5.5	*Bischofia javanica* (row planting)	100 × 100 × 9.5
Carpet grass (solitary planting)	20 × 20 × 1.2	*Erythrina indica* (opposite planting)	30 × 30 × 7.5	Mango trees (row planting)	100 × 100 × 11.5
Aglaia odorata lour (solitary planting)	20 × 20 × 1.8	Banyan (opposite planting)	30 × 30 × 8.0	*Chorisia speciosa* (row planting)	100 × 100 × 13.0
Waringin (solitary planting)	20 × 20 × 5.5	*Bischofia javanica* (opposite planting)	30 × 30 × 9.5	Kapok (row planting)	100 × 100 × 15.0
Erythrina indica (solitary planting)	20 × 20 × 7.5	Mango trees (opposite planting)	30 × 30 × 11.5	Banyan + Waringin + Carpet grass	100 × 100 × 8.0
Banyan (solitary planting)	20 × 20 × 8.0	*Chorisia speciosa* (opposite planting)	30 × 30 × 13.0	*Bischofia javanica* + Waringin + Carpet grass	100 × 100 × 9.5
Bischofia javanica (solitary planting)	20 × 20 × 9.5	Kapok (opposite planting)	30 × 30 × 15.0	Kapok + Waringin + Carpet grass	100 × 100 × 15.0
Mango trees (solitary planting)	20 × 20 × 11.5	Carpet grass (row planting)	100 × 100 × 1.2	Banyan + Carpet grass	100 × 100 × 8.0
Chorisia speciosa (solitary planting)	20 × 20 × 13.0	*Aglaia odorata* lour (row planting)	100 × 100 × 1.8	*Bischofia javanica* + Carpet grass	100 × 100 × 9.5
Kapok (solitary planting)	20 × 20 × 15.0	Waringin (row planting)	100 × 100 × 5.5	Kapok + Carpet grass	100 × 100 × 15.0
Carpet grass (opposite planting)	30 × 30 × 1.2	*Erythrina indica* (row planting)	100 × 100 × 7.5	Waringin + Carpet grass	100 × 100 × 5.5

(4) CFD Simulation Settings

There are several software choices for carrying out CFD simulations, such as PHOENICS v.2016, ANSYS FLUENT v.2019, ENVI-met v.2021, Airpak v.2016, and Butterfly v.2018. PHOENICS, considered the original commercial software for fluid and heat transfer calculations, is increasingly favored by researchers for simulating wind conditions in different settings like residential areas, school campuses, and landscape gardens. Through the use of PHOENICS for CFD simulation, researchers have discovered that different height distribution patterns can influence alterations in both wind speed and pressure [24]. Making changes to the architectural layout can be a useful strategy for decreasing community energy consumption and carbon emissions [25]. The arrangement of courtyard plants has a significant effect on the outdoor microclimate, residential wind, and thermal comfort [26].

Changes in the arrangement and proportion of architectural spaces can create efficient ecological buffer zones [27]. ENVI-Met is primarily used for thermal comfort and has extensive experience and data backing in simulating urban microclimates. It offers precise simulation results for specific urban environments and climate conditions [28]. PHOENICS software, in contrast to ENVI-met software 4.0, is capable of modeling intricate three-dimensional fluid dynamics, encompassing turbulence and eddies, essential for the precise evaluation of wind conditions. By employing meticulous grid segmentation and sophisticated numerical techniques, PHOENICS can deliver detailed wind parameters, such as speed and direction, thus augmenting the assessment of wind comfort. Hence, the PHOENICS software is ideal for modeling wind conditions within the square area under investigation. The setup of the CFD simulation and the creation of the model will be outlined in the following section.

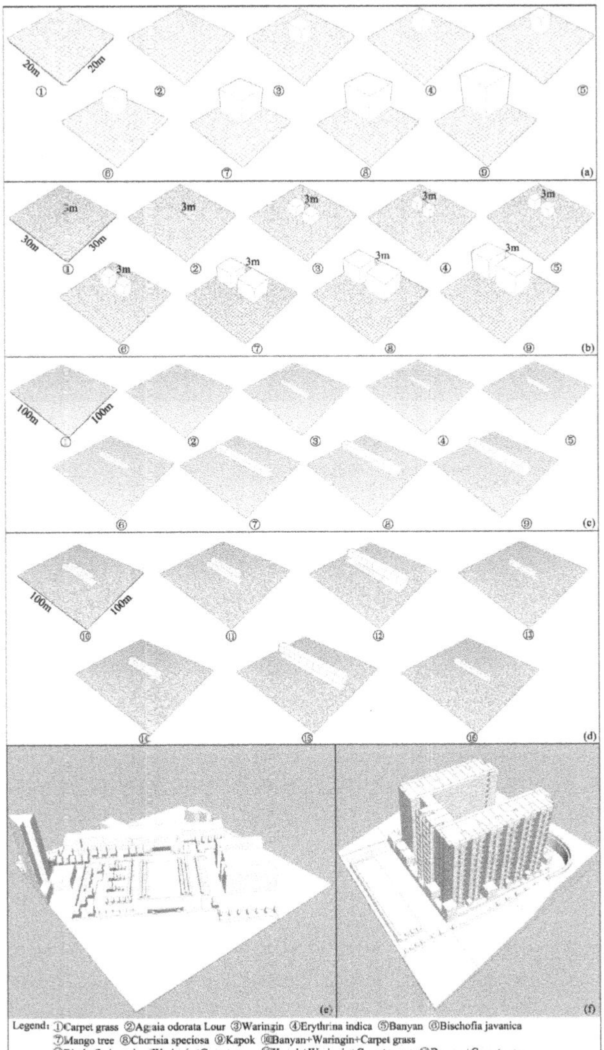

Figure 6. Simplified 3D model: (**a**) A main entrance square space; (**b**) E single building courtyard space; (**c**) solitary plant; (**d**) opposite plant; (**e**) row plant; (**f**) tree + shrub + grass, tree + grass, and shrub + grass.

Model selection: In PHOENICS, the RNG k-ε turbulence model is employed, utilizing Equations (1) and (2) alongside the PRESTO discrete equation and the integrated PARSOL function settings for simulating velocity–pressure coupling. Implement small-scale network configurations in the core area to enhance calculation accuracy. The automatic convergence detection feature in PHOENICS ensures that simulation results converge reasonably with an accuracy level of 10^{-5} [29].

$$\frac{\partial(\rho k)}{\partial t} + \frac{\partial(\rho k u_i)}{\partial x_i} = \frac{\partial}{x_j}\left(\alpha_k \eta_{eff} \frac{\partial k}{\partial x_j}\right) + G_k + \rho\varepsilon \quad (1)$$

$$\frac{\partial(\rho\varepsilon)}{\partial t} + \frac{\partial(\rho\varepsilon v_i)}{\partial x_i} = \frac{\partial}{\partial x_j}\left(\alpha_\varepsilon \eta_{\Delta f} \frac{\partial\varepsilon}{\partial x_j}\right) + \frac{C^*_{1s}\varepsilon}{k} G_k - C_{2s}\rho\frac{\varepsilon^2}{k} \quad (2)$$

The equations utilize the symbols k for turbulent kinetic energy and ε for turbulent dissipation rate, and are solved using the software PHOENICS v.2016.

Grid setting: The length and width of the calculation domain are five times larger than those of the corresponding scene model, and the height is three times greater than the model's height (Figure 7) [30]. The calculation domain measures 5 W × 5 L × 3 H (Table 3 shows a comparative analysis of simulation data using various computational domain heights, indicating that the height of the computational domain has no impact on simulation outcomes). The computational domain grid is divided into two regions: the center and the edge, with specific grid densities assigned to each. The grid densities are as follows: coarse meshes, $X_{min} = Y_{min} = Z_{min}$ = 0.27 H; fine meshes, $X_{min} = Y_{min} = Z_{min}$ = 0.13 H; and finest meshes, $X_{min} = Y_{min} = Z_{min}$ = 0.06 H (Table 4 presents a comparative analysis of simulated data using various computational domain grid densities, showing that higher grid resolutions result in improved accuracy at the cost of longer simulation durations) [31]. Simulation accuracy is enhanced and the number of grid segments is decreased by the grid setting, which improves time efficiency. Using a fixed pressure and zero gradients, an outlet boundary condition was defined. The ground roughness parameter α, was set to 0.2.

The inflow profile at the top boundary had a constant horizontal velocity and turbulent kinetic energy, while the left and right symmetric boundaries were considered slip walls without gradients.

By utilizing Equation (3), it is possible to calculate the gradient of the oncoming wind at the inlet:

$$u(z) = u_0(z/z_0)^\alpha \quad (3)$$

where $u(z)$ is the horizontal velocity at height z, and u_0 is the horizontal velocity at height z_0. In this model, u_0 = 3.4/3.5 m/s (summer/winter), z_0 = 108.4 m, and α = 0.25 [32].

The turbulent kinetic energy, k (m^2/s^2), and its dissipation rate, ε (m^2/s^3), are set as follows:

$$k = \frac{{u_*}^2}{\sqrt{C_\mu}}\left(1 - \frac{z}{\delta}\right) \quad (4)$$

$$\varepsilon = \frac{{u_*}^3}{kz}\left(1 - \frac{z}{\delta}\right) \quad (5)$$

where u_* is the friction velocity, δ is the depth of the boundary layer, and K is the von Karman's constant. In this model, u_* = 2.7/2.9 m/s (summer/winter), K = 0.4, and C_μ = 0.09 [30].

Trees in the model were parameterized as a one-dimensional column, where the normalized LAI was scaled to tree height [33]. Observing the non-uniform vertical distribution of LAI allows for easy identification of different types of vegetation, as LAI changes depending on height, influenced by crown shape, height, and canopy edges [34]. When it comes to turbulence models, tree crowns are seen as porous media, with individual tree branches being compared to the crown [35]. The tree canopy causes a reduction in the kinematic energy of airflow due to drag and pressure, leading to utilization of resistance in the momentum equation to model the impact of vegetation on turbulent flow patterns. The momentum equation includes a sink term to factor in turbulence resistance caused by the canopy layer as follows:

$$S_{d,i} = -C_d \times LAD \times |U| \times U_i \tag{6}$$

where C_d is the drag coefficient, $|U|$ is the vector speed on foliage surface (m/s), and U_i is the Cartesian velocity in i direction (m/s).

Additional source terms in the momentum equation can demonstrate the correlation between airflow and tree canopy turbulence as follows:

$$S_k = C_d \times LAD \times \left(\beta_P|U|^3 - \beta_d|U|k\right) \tag{7}$$

$$S_k = C_d \times LAD \times \left(C_{4\varepsilon}\beta_P|U|^3\frac{\varepsilon}{k} - C_{5\varepsilon}\beta_d|U|\varepsilon\right) \tag{8}$$

In this investigation, β_p, β_d, $C_{4\varepsilon}$, and $C_{5\varepsilon}$ are empirical constants with values of 1.0, 3.0, 1.5, and 1.5, respectively. β_p denotes the average fluid kinetic energy of the wake flow, k, produced by the drag force of the canopy; whereas β_d indicates the kinetic energy, k, dissipated by the short circuits of Kolmogorov energy gradients [36–39].

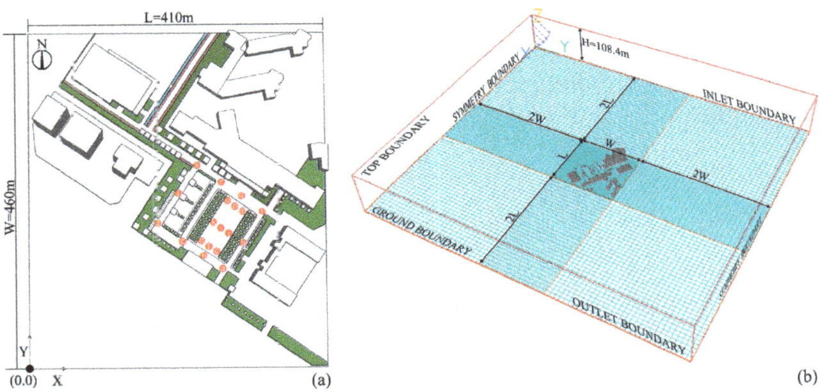

Figure 7. (**a**) Original coordinate (a–s are wind environment detection points). (**b**) Computational domain.

Table 3. Comparison of simulation results for different calculation domain heights.

Detection Points	The Calculation Domain Measures 5 W × 5 L × 3 H (Simulation Results m/s)	The Calculation Domain Measures 5 W × 5 L × 5 H (Simulation Results m/s)
b	3.3/3.5	3.3/3.5
e	3.7/4.4	3.7/4.5
h	1.6/1.2	1.8/1.2
m	2.4/1.7	2.4/1.7
p	3.0/2.0	3.0/2.2

Wind conditions setting: As per the "Design Code for Heating, Ventilation and Air Conditioning of Civil Buildings" (GB507360-2012, China), "China Weather Network", "China Meteorological Network", and "China Meteorological Data Network", Quanzhou experiences an average wind speed of 4.7 m/s and an average air temperature of 27.8 °C during the summer, with a relative humidity of 85% and a mean radiant temperature of 31 °C. The prevailing wind direction is southwest. In the winter, the average wind speed increases to 6.13 m/s, while the average air temperature drops to 14.6 °C. The average relative humidity is 75% and the mean radiant temperature is 17.8 °C, with the prevailing wind direction shifting to the northeast. The seasonal dominant wind direction refers to the range of wind direction angles with the highest wind frequency within a specific season. The dominant wind direction in Quanzhou is southwest in the summer and northeast in the winter. These values are used as inflow boundary conditions for 2000 iterations.

Table 4. Comparison of simulation results for different grids.

Detection Points	Coarse Meshes (Simulation Results m/s)	Fine Meshes (Simulation Results m/s)	Finest Meshes (Simulation Results m/s)
b	4.3/4.6	2.4/1.8	1.9/1.6
e	4.0/3.9	1.8/0.8	1.7/0.6
h	4.2/3.6	1.5/0.9	1.5/0.9
m	5.0/4.3	0.8/1.7	0.8/1.7
p	4.3/4.5	0.8/1.7	0.8/1.7

3. Results and Discussion

3.1. Comparative Analysis of Measured and Simulated Values

On 30 September 2022, between 9:00 a.m. and 5:00 p.m., we employed an anemometer to conduct on-site measurements of the main square at the Donghai Campus of Quanzhou Normal University, followed by numerical simulations using CFD. In the CFD simulation procedure, the inflow velocity was determined by the average wind speeds per hour (0.9 m/s, 1.0 m/s, 0.8 m/s, 3.2 m/s, 2.9 m/s, 2.4 m/s, 2.6 m/s, 2.7 m/s, and 2.7 m/s) observed from 9:00 am to 5:00 pm. After completing nine simulation experiments, detection points (b, e, h, m, p) were positioned at the main entrance square of the wind farm to gauge the average wind speed at each designated point. Comparing the simulated wind field data of each detection point at the main entrance plaza for nine time periods with the corresponding measurement data, Figure 8 displays high R2 values for each time period, all surpassing 0.80. This suggests that more than 80% of the observed wind speed changes can be elucidated by the simulated wind speed. The linear regression equations derived are as follows: $y = 1.09x - 0.53$, $y = 0.58x + 0.02$, $y = 0.78x - 0.43$, $y = 2.20x + 0.24$, $y = 3.27x - 0.65$, $y = 0.76x + 1.15$, $y = 0.80x + 0.96$, $y = 2.35x - 0.31$, and $y = 1.20x + 0.54$. The analysis in Table 5 reveals that all p-values for each time period were below 0.05, indicating significant linear regression features. The score deviation range is [−0.9, 0.7] and the mean square error ranges from 0.61 to 0.31, suggesting a small deviation between simulated and observed values. The prediction error of CFD simulation values is relatively low, signifying the suitability of PHOENICS software for simulating outdoor wind environments. Furthermore, this aids in validating the effectiveness of CFD simulation data in assessing outdoor campus areas.

3.2. The Simulation Results of Whether There Are Plants in the Main Entrance Square and the Wind Environment Simulation Results of the Young Teacher's Apartment

Import the models of the main entrance square, whether with or without trees, and the individual building (such as the Youth Teacher Apartment) into the PHOENICS software to conduct wind environment simulations. Acquire data on wind speeds during the summer and winter for each model at a height of 1.5 m (Z = 1.5 m) above the ground for pedestrians (Figures 9 and 10). Through data analysis, the following results have been obtained.

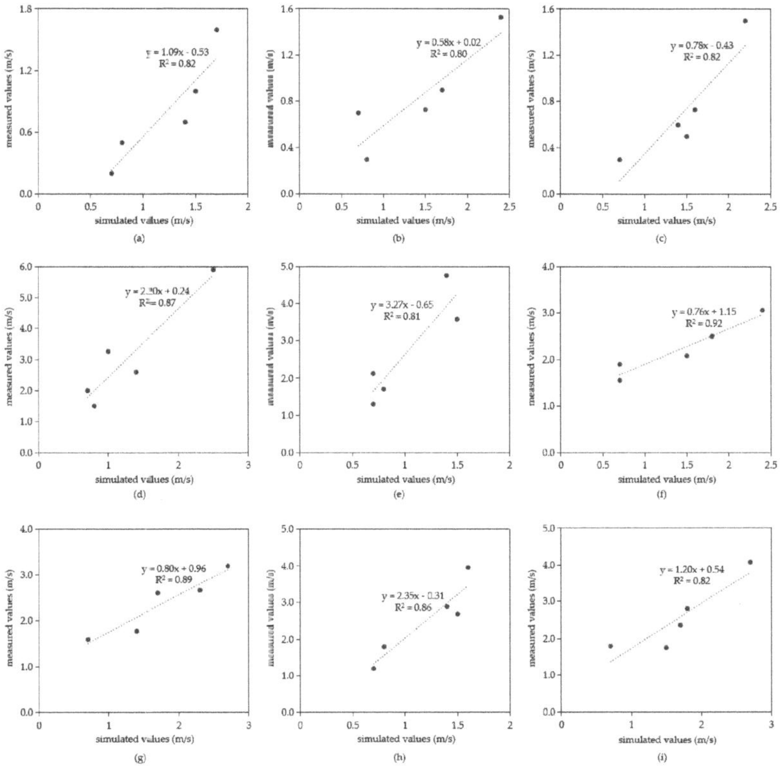

Figure 8. Linear regression plot of measured and CFD-simulated values ((**a**–**i**) correspond to the time periods of 9:00 a.m. and 5:00 p.m., respectively).

Table 5. Statistical analysis of winter measured and simulated average wind speed in the main entrance square space [a].

Model		Sum of Squares	df	Mean Squared Error	Fractional Bias	F	Sig.
09:00	Regression	0.94	1	0.94	0.42	13.98	0.03 [b]
10:00		0.61	1	0.61	0.52	11.37	0.04 [b]
11:00		0.69	1	0.69	0.68	13.42	0.04 [b]
12:00		10.31	1	10.31	−0.82	19.56	0.02 [b]
13:00		6.96	1	6.96	−0.90	13.17	0.04 [b]
14:00		1.25	1	1.25	−0.44	35.28	0.01 [b]
15:00		1.59	1	1.59	−0.30	26.68	0.01 [b]
16:00		3.98	1	3.98	−0.70	18.56	0.02 [b]
17:00		3.00	1	3.00	−0.42	14.75	0.03 [b]

[a] Dependent Variable: Measured wind speed; [b] Predictors. (Constant), Simulated wind speed.

In Figure 9, the wind speeds in the main square during the summer without trees range from 2.9 m/s to 4.6 m/s. The average wind speed is 4.4 m/s, with a difference of 1.7 m/s. With trees present, the wind speeds in the summer range from 2.7 m/s to 3.4 m/s. The average wind speed is 2.9 m/s, showing a difference of 0.7 m/s. In the winter, the wind speeds without trees range from 3.8 m/s to 5.8 m/s. The average wind speed is 5.6 m/s, with a difference of 2.0 m/s. With trees in the winter, the wind speeds range from 2.9 m/s to 3.5 m/s. The average wind speed is 3.0 m/s, with a difference of 0.6 m/s. The presence of trees in the main square leads to better wind comfort in both the summer and winter, as opposed to when there are no trees. This suggests that vegetation has the ability to block strong winds, lower wind speeds, and enhance wind speeds in localized areas.

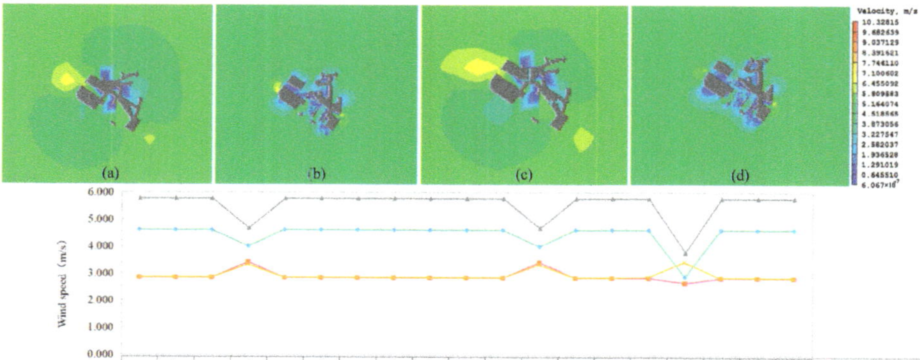

	a	b	c	d	e	f	g	h	i	j	k	l	m	n	o	p	q	r	s	average wind speed	wind difference
no trees in summer	4.6	4.6	4.6	4.0	4.6	4.6	4.6	4.6	4.6	4.6	4.6	4.0	4.6	4.6	4.6	2.9	4.6	4.6	4.6	4.4	1.7
trees in summer	2.9	2.9	2.9	3.4	2.9	2.9	2.9	2.9	2.9	2.9	2.9	3.4	2.9	2.9	2.9	2.7	2.9	2.9	2.9	2.9	0.7
no trees in winter	5.8	5.8	5.8	4.7	5.8	5.8	5.8	5.8	5.8	5.8	5.8	4.7	5.8	5.8	5.8	3.8	5.8	5.8	5.8	5.6	2.0
trees in winter	2.9	2.9	2.9	3.4	2.9	2.9	2.9	2.9	2.9	2.9	2.9	3.3	2.9	2.9	2.9	3.5	2.9	2.9	2.9	3.0	0.6

There is a plant wind speed detection point in the main entrance square

(e)

Figure 9. (**a**) Map showing summer wind speeds without plants. (**b**) Map showing summer wind speeds with plants. (**c**) Map showing winter wind speeds without plants. (**d**) Map showing winter wind speeds with plants. (**e**) Simulation results of having or not having a plant in the wind environment during the summer and winter at the main gate square (red represents the highest wind difference and average wind speed, while green represents the lowest wind difference and average wind speed).

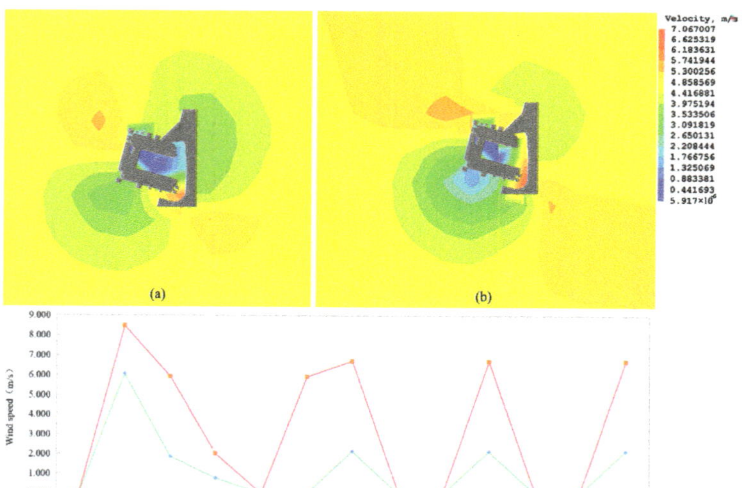

	a	b	c	d	e	f	g	h	i	j	k	l	m	average wind speed	wind difference
Summer	0.0	6.1	1.9	0.8	0.0	0.0	2.2	0.0	0.0	2.2	0.0	0.0	2.2	1.2	6.1
Winter	0.0	8.5	5.9	2.0	0.0	5.9	6.7	0.1	0.1	6.7	0.1	0.1	6.7	3.3	8.5

Wind speed at monitoring points before and after optimization in summer and winter for teacher apartments

(c)

Figure 10. (**a**) Map showing summer wind speeds. (**b**) Map showing winter wind speeds. (**c**) Simulation results of having or not having a plant in the wind environment during the summer and winter at the Youth Teacher Apartment (red represents the highest wind difference and average wind speed, while green represents the lowest wind difference and average wind speed).

Based on Figure 10, at the Youth Teacher Apartment, the wind speeds vary between 0–6.1 m/s in the summer and 0–8.5 m/s in the winter. The average wind speed in the summer is 1.2 m/s, and in the winter is 3.3 m/s, showing a difference of 6.1 m/s in the summer and 8.5 m/s in the winter. In the summer, there are 8 instances of uncomfortable wind speeds, compared to 12 in the winter. Consequently, it can be concluded that the winter wind speed performance at the Youth Teacher Apartment is the most unfavorable. With minimal large tree coverage, the Youth Teacher Apartment's outdoor space is characterized by consistently high wind speeds in the summer and winter, interspersed with some tranquil spots. As a result, the area is only conducive to brief periods of use or simple exercises.

3.3. The Influence of Plant Configuration on the Wind Environment

The impact of different plant configurations on the wind environment in the square was investigated using PHOENICS software for CFD simulations on flat configurations (solitary planting, opposite planting, row planting) and facade configurations (tree + shrub + grass, tree + grass, shrub + grass), and to obtain wind speed data at a height of 1.5 m (Z = 1.5 m) above the ground for each pedestrian model. Through data analysis, the following results have been obtained.

3.3.1. The Influence of Plant Plane Configuration on Wind Environment

(1) Solitary Planting

As depicted in Figure 11, with regards to wind speed, among tree species, and under the same conditions where only the height of the trunk remains the same, the overall wind speed in the summer is around 4.0 m/s, and in the winter wind speeds range from 5.1 m/s to 5.2 m/s. Compared to *Erythrina indica*, the banyan performs better. Under the same conditions, where only the height and width of the trunk change, the overall wind speeds with *Bischofia javanica* and mango trees in the summer range from 3.7 m/s to 4.5 m/s, and in the winter wind speeds range from 4.8 m/s to 4.9 m/s. *Bischofia javanica* performs better. Under the same conditions and maintaining a consistent tree trunk width, the overall wind speed in the summer is around 4.6 m/s, and in the winter it is around 5.0 m/s. Compared with kapok, *Chorisia speciosa* performs better. Among shrubs, the overall wind speeds in the summer range from 3.8 m/s to 4.0 m/s, and in the winter wind speeds range from 5.0 m/s to 5.2 m/s. The impact of solitary *Aglaia odorata* lour and waringin on wind speeds is not significantly different. The wind speed with carpet grass is 3.8 m/s in the summer and 4.9 m/s in the winter. Overall, the wind speed of *Bischofia javanica* is the best, followed by carpet grass, and the performance of *Erythrina indica* is the worst. Considering the reality of strong winds in Quanzhou, when the trees are too high, there may be a phenomenon of lodging or drying due to being heavy on the head and light on the feet. Therefore, it is more appropriate to limit the height of the trees. Among them, trees with a height of 6.5–8.5 m are more suitable, and *Bischofia javanica* is the most suitable.

When it comes to wind regulation amplitude, Table 6 displays the wind regulation amplitudes of various plant varieties under specific conditions. In the summer, the wind regulation amplitudes for plant species such as *Erythrina indica*, banyan, *Bischofia javanica*, mango trees, *Chorisia speciosa*, kapok, *Aglaia odorata* lour, waringin, and carpet grass are 14.89%, 14.89%, 21.28%, 4.26%, 2.13%, 2.13%, 19.15%, 14.89%, and 19.15%, respectively. With a wind regulation amplitude 1.43 times higher than that of *Erythrina indica*, *Bischofia javanica* demonstrates the greatest ability to withstand wind, while *Chorisia speciosa* and kapok shows the lowest (0.14 times that of *Erythrina indica*). In the winter, the wind regulation amplitudes for the same plant varieties are 15.17%, 16.80%, 21.70%, 20.07%, 18.43%, 18.43%, 18.43%, 15.17%, and 20.07%, respectively. *Bischofia javanica* demonstrates the highest wind regulation amplitude (1.43 times that of *Erythrina indica*), while *Erythrina indica* and waringin has the lowest wind regulation amplitude (1.0 times).

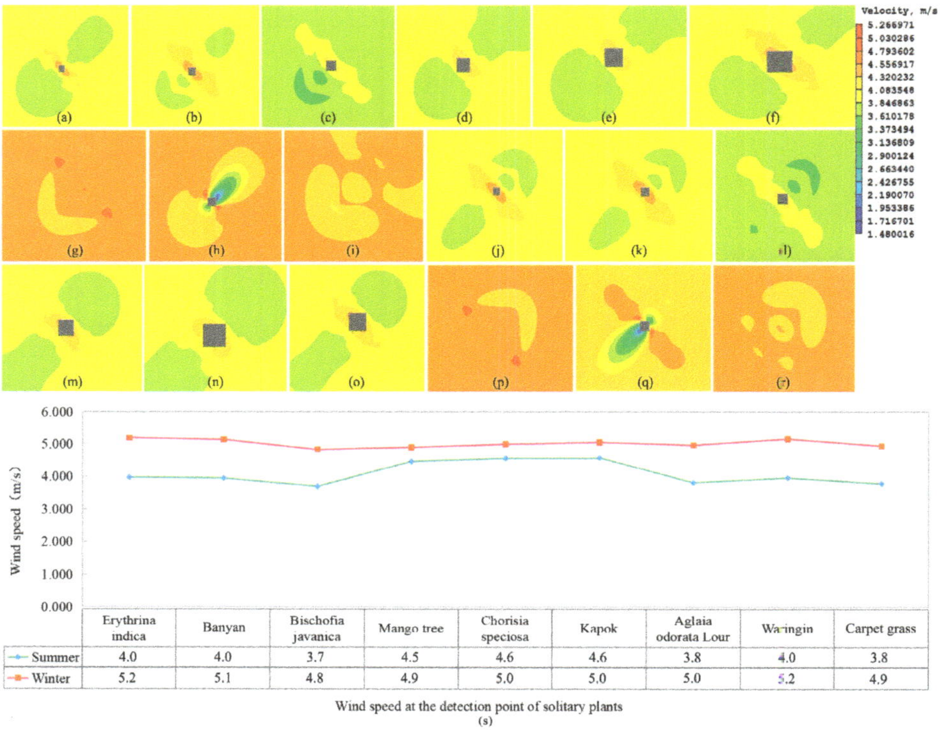

Figure 11. (**a–i**) Maps showing summer wind speeds. (**j–r**) Maps showing winter wind speeds. (**s**) Simulated data graph of the wind reduction for individual trees and shrubs (wind speeds at a line of sight height of 1.5 m for each tree and shrub).

Table 6. Statistical table for the summer/winter wind regulation amplitude of solitary plant species configurations.

Figure No.	Plant Varieties	Summer/Winter Input Wind Speeds	Summer/Winter Average Wind Speeds	Summer/Winter Wind Regulation Amplitude	Summer/Winter Wind Regulation Rate
1	*Erythrina indica*	4.70/6.13	4.0/5.2	14.89%/15.17%	1.00/1.00
2	Banyan	4.70/6.13	4.0/5.1	14.89%/16.80%	1.00/1.11
3	*Bischofia javanica*	4.70/6.13	3.7/4.8	21.28%/21.70%	1.43/1.43
4	Mango trees	4.70/6.13	4.5/4.9	4.26%/20.07%	0.29/1.32
5	*Chorisia speciosa*	4.70/6.13	4.6/5.0	2.13%/18.43%	0.14/1.22
6	Kapok	4.70/6.13	4.6/5.0	2.13%/18.43%	0.14/1.22
7	*Aglaia odorata* lour	4.70/6.13	3.8/5.0	19.15%/18.43%	1.29/1.22
8	Waringin	4.70/6.13	4.0/5.2	14.89%/15.17%	1.00/1.00
9	Carpet grass	4.70/6.13	3.8/4.9	19.15%/20.07%	1.29/1.32

Note: The red color signifies the highest wind regulation rate, while the green color signifies the lowest wind regulation rate among the nine plant varieties in the isolated plant configuration (the wind regulation amplitude equals 1 minus the average wind speed compared to the input wind speed. Positive numbers indicate a downwind amplitude, while negative numbers indicate a lift amplitude opposite to the downwind amplitude. The wind regulation rate is the wind regulation amplitude n divided by the wind regulation amplitude 1. Afterwards, data were obtained using the above method).

(2) Opposite Planting

As depicted in Figure 12, when it comes to wind speed, under the same conditions, with only the same height of the trunk, the overall wind speeds in the summer are between 3.8 m/s and 3.9 m/s, and in the winter wind speeds are between 4.9 m/s and 5.0 m/s. The wind speeds with *Erythrina indica* and banyan are similar, with no significant changes. Under the same conditions, where only the height and width of the trunk change, the overall wind speeds in the summer range from 3.9 m/s to 4.2 m/s, and in the winter, wind speeds vary between 4.5 m/s and 5.0 m/s. The performance of *Bischofia javanica* and mango trees is similar. Under the same conditions, with only the same trunk width, the overall wind speeds in the summer range from 4.1 m/s to 4.2 m/s, and in the winter wind speeds range from 5.4 m/s to 5.5 m/s. Compared to the kapok, the *Chorisia speciosa* performs better. Among shrubs, the overall wind speeds in the summer range from 2.7 m/s to 3.8 m/s, and in the winter wind speeds range from 3.4 m/s to 5.0 m/s. Compared to *Aglaia odorata* lour, waringin performs better. Among herbs, the wind speed of carpet grass in the summer is 3.8 m/s, while in the winter it is 4.9 m/s. Overall, the wind speed performance of the waringin is the best, followed by *Bischofia javanica*, and the kapok performs the worst.

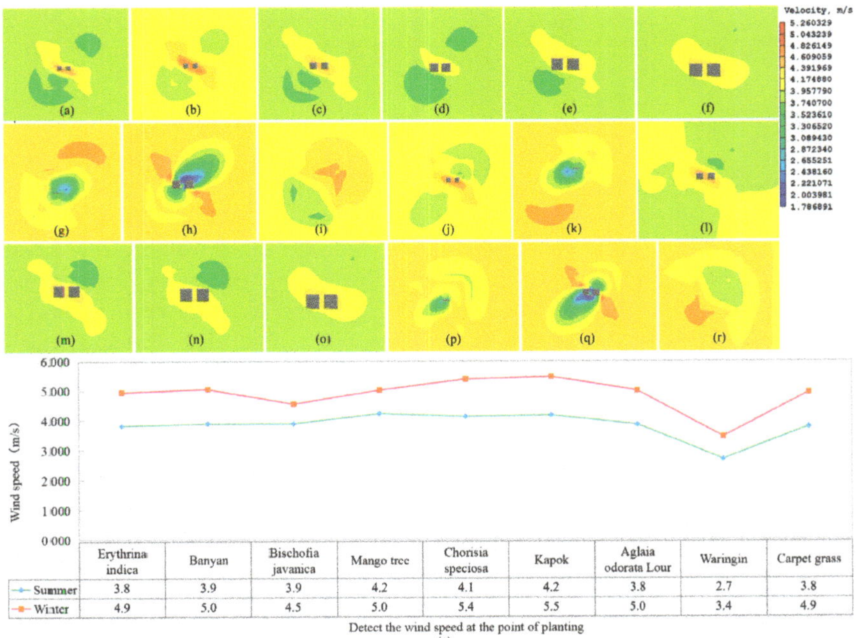

Figure 12. (**a–i**) Maps showing summer wind speeds. (**j–r**) Maps showing winter wind speeds. (**s**) Simulated data graph of wind reduction for opposite planting of trees and shrubs (wind speeds at a line of sight height of 1.5 m for opposite planting of trees and shrubs).

When it comes to wind regulation amplitude, according to Table 7, when considering only changes in plant varieties under consistent conditions, the wind regulation amplitudes in the summer are as follows: *Erythrina indica* (19.15%), banyan (17.02%), *Bischofia javanica* (17.02%), mango trees (10.64%), *Chorisia speciosa* (12.77%), kapok (10.64%), *Aglaia odorata* lour (19.15%), waringin (42.55%), and carpet grass (19.15%). Among these, waringin exhibits the highest wind regulation amplitude (2.22 times that of *Erythrina indica*), while mango trees and kapok show the lowest (0.56 times that of *Erythrina indica*). In the winter, the wind regulation amplitudes for the same plant varieties are 20.07%, 18.43%, 26.59%,

18.43%, 11.91%, 10.28%, 18.43%, 44.54%, and 20.07%, respectively. Waringin demonstrates the highest wind regulation amplitude (2.22 times that of *Erythrina indica*), while kapok has the lowest wind regulation amplitude (0.51 times that of *Erythrina indica*).

Table 7. Statistical table for the summer/winter wind regulation amplitude of opposite plant species configurations.

Figure No.	Plant Varieties	Summer/Winter Input Wind Speeds	Summer/Winter Average Wind Speeds	Summer/Winter Wind Regulation Amplitude	Summer/Winter Wind Regulation Rate
1	*Erythrina indica*	4.70/6.13	3.8/4.9	19.15%/20.07%	1.00/1.00
2	Banyan	4.70/6.13	3.9/5.0	17.02%/18.43%	0.89/0.92
3	*Bischofia javanica*	4.70/6.13	3.9/4.5	17.02%/26.59%	0.89/1.33
4	Mango trees	4.70/6.13	4.2/5.0	10.64%/18.43%	0.56/0.92
5	*Chorisia speciosa*	4.70/6.13	4.1/5.4	12.77%/11.91%	0.67/0.59
6	Kapok	4.70/6.13	4.2/5.5	10.64%/10.28%	0.56/0.51
7	*Aglaia odorata* lour	4.70/6.13	3.8/5.0	19.15%/18.43%	1.00/0.92
8	Waringin	4.70/6.13	2.7/3.4	42.55%/44.54%	2.22/2.22
9	Carpet grass	4.70/6.13	3.8/4.9	19.15%/20.07%	1.00/1.00

Note: The red color signifies the highest wind regulation rate among the nine plant varieties in the plant configuration, whereas the green color signifies the lowest wind regulation rate among the nine plant varieties in the plant configuration.

(3) Row Planting

As depicted in Figure 13, in relation to wind speeds, under the same conditions, with only the same height of the tree trunk, the overall wind speeds in the summer range from 3.2 m/s to 3.5 m/s, while in the winter wind speeds range from 4.1 m/s to 4.5 m/s. Compared to *Erythrina indica*, the banyan performs better. Under the same conditions, where only the height and width of the trunk change, the overall wind speeds in the summer range from 3.8 m/s to 4.0 m/s, while in the winter wind speeds vary from 4.9 m/s to 5.2 m/s. Compared to the mango trees, *Bischofia javanica* performs better. Under the same conditions, with only the same trunk width, the overall wind speeds in the summer range from 3.9 m/s to 4.1 m/s, and in the winter wind speeds range from 5.2 m/s to 5.4 m/s. Compared to *Chorisia speciosa*, kapok performs better. Among shrubs, the overall wind speeds in the summer range from 3.3 m/s to 3.9 m/s, and in the winter wind speeds range from 4.3 m/s to 5.1 m/s. Compared to *Aglaia odorata* lour, waringin performs better. While the overall wind speed with carpet grass is 3.8 m/s in the summer, wind speed increases to 4.9 m/s in the winter. Overall, the banyan performs the best in terms of wind speeds, followed by waringin, and *Chorisia speciosa* performs the worst.

When it comes to wind regulation amplitude, Table 8 displays the wind regulation amplitudes of various plant varieties under specified conditions, focusing solely on changes. In the summer, plant varieties such as *Erythrina indica*, banyan, *Bischofia javanica*, mango trees, *Chorisia speciosa*, kapok, *Aglaia odorata* lour, waringin, and carpet grass exhibit amplitudes of 25.53%, 31.91%, 19.15%, 14.89%, 12.77%, 17.02%, 17.02%, 29.79%, and 19.15%, respectively. Banyan demonstrates the highest wind regulation amplitude (1.25 times that of *Erythrina indica*), while *Chorisia speciosa* shows the lowest (0.50 times the amplitude of *Erythrina indica*). In the winter, the wind regulation amplitudes for the same plant varieties are 26.59%, 33.12%, 20.07%, 15.17%, 11.19%, 15.17%, 16.80%, 29.85%, and 20.07%, respectively. Banyan maintains the highest wind regulation amplitude (1.25 times that of *Erythrina indica*), whereas *Chorisia speciosa* exhibits the lowest (0.45 times the amplitude of *Erythrina indica*).

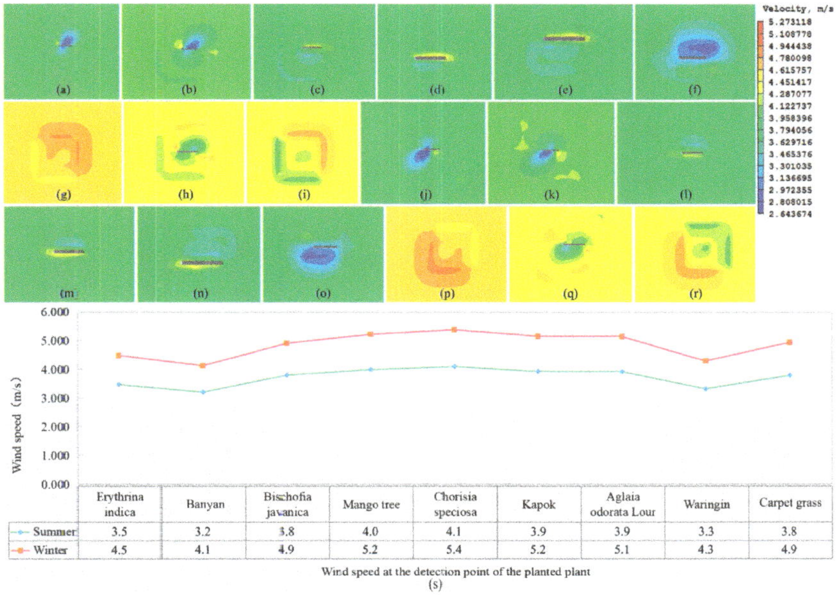

Figure 13. (**a**–**i**) Maps showing summer wind speeds. (**j**–**r**) Maps showing winter wind speeds. (**s**) Simulated data graph of wind reduction for row planting of trees and shrubs (wind speed at a line of sight height of 1.5 m for row planting of trees and shrubs).

Table 8. Statistical table for the summer/winter wind regulation amplitude of row plant species configurations.

Figure No.	Plant Varieties	Summer/Winter Input Wind Speeds	Summer/Winter Average Wind Speeds	Summer/Winter Wind Regulation Amplitude	Summer/Winter Wind Regulation Rate
1	*Erythrina indica*	4.70/6.13	3.5/4.5	25.53%/26.59%	1.00/1.00
2	Banyan	4.70/6.13	3.2/4.1	31.91%/33.12%	1.25/1.25
3	*Bischofia javanica*	4.70/6.13	3.8/4.9	19.15%/20.07%	0.75/0.75
4	Mango tree	4.70/6.13	4.0/5.2	14.89%/15.17%	0.58/0.57
5	*Chorisia speciosa*	4.70/6.13	4.1/5.4	12.77%/11.19%	0.50/0.45
6	Kapok	4.70/6.13	3.9/5.2	17.02%/15.17%	0.67/0.57
7	*Aglaia odorata* lour	4.70/6.13	3.9/5.1	17.02%/16.80%	0.67/0.63
8	Waringin	4.70/6.13	3.3/4.3	29.79%/29.85%	1.17/1.12
9	Carpet grass	4.70/6.13	3.8/4.9	19.15%/20.07%	0.75/0.75

Note: The red color signifies the highest wind regulation rate among the nine plant varieties in the column planting configuration, whereas the green color signifies the lowest wind regulation rate among the nine plant varieties in the column planting configuration.

In both the summer and winter seasons, under the same conditions and having the same trunk height, banyan performs better in regulating wind speeds. When both height and width change, compared to mango trees, *Bischofia javanica* has a better wind regulating effect. Under the same conditions and having the same width, it can be seen that kapok has a better wind regulating effect compared to *Chorisia speciosa*. Among the two shrubs, *Aglaia odorata* lour and waringin have a higher range of wind regulation compared to waringin. Among herbs, carpet grass has a relatively stable range of wind regulation.

3.3.2. The Impact of Plant Facade Configuration on Wind Environment

The plant facade configuration includes three types: tree + shrub + grass, tree + grass, and shrub + grass. As can be seen from the previous text, the banyan, *Bischofia javanica*, and kapok, in the tree category, and waringin, in the shrub category, are the species with better wind speed reduction effects, while the herb species is carpet grass. This is combined into seven plant facade configuration model groups (Figure 6d), namely tree + shrub + grass: banyan + waringin + grass, *Bischofia javanica* + waringin + grass, kapok + waringin + grass; tree + grass: banyan + grass, *Bischofia javanica* + grass, kapok + grass; shrub + grass: waringin + grass. Then, they are imported separately into PHOENICS software for wind environment simulation, the models are analyzed separately to obtain wind speed data for each pedestrian at a height of 1.5 m (Z = 1.5 m) above the ground, illustrated in Figure 14.

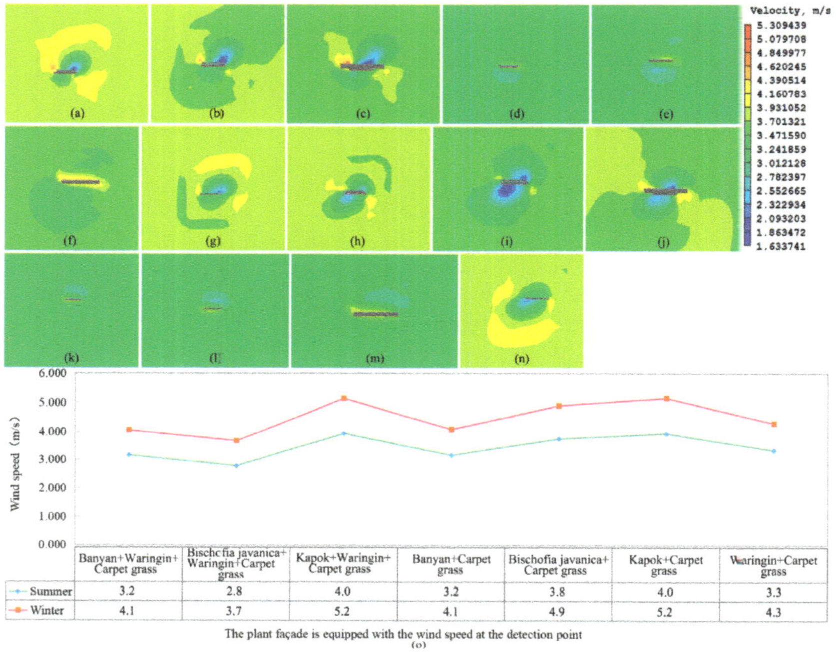

Figure 14. (**a–g**) Maps showing summer wind speeds. (**h–n**) Maps showing winter wind speeds. (**o**) Simulated data graph of wind reduction for seven plant facade configuration models (wind speed at a line of sight height of 1.5 m).

(1) Tree + Shrub + Grass

The simulation results depicted in Figure 13 indicate that the combined wind speeds for tree + shrub + grass during the summer vary between 2.8 m/s and 4.0 m/s, while in the winter, wind speeds range from 3.7 m/s to 5.2 m/s. The combination of kapok, waringin, and grass performs poorly in the winter, with wind speeds exceeding 5 m/s, which is considered an unsuitable wind speed. The configuration of banyan + waringin + grass and *Bischofia javanica* + waringin + grass has wind speeds not exceeding 5 m/s in both the summer and winter seasons, and there are no strong wind situations. Among them, the combination of *Bischofia javanica*, waringin, and grass performs the best. During the summer, a wind speed of 2.8 m/s allows individuals to comfortably remain outdoors for extended durations; whereas, in the winter, a wind speed of 3.7 m/s permits only brief outdoor standing or sitting sessions. This indicates that the vertical configuration of tree + shrub + grass can all play a role in reducing wind speeds. Regardless of summer or winter, the configuration mode of *Bischofia javanica* + waringin + carpet grass has the best effect on reducing wind speeds, significantly better than banyan + waringin + carpet grass and kapok + waringin + carpet grass. Therefore, the plant configuration of *Bischofia javanica*, waringin, and carpet grass is adopted in the tree + shrub + grass mode.

(2) Tree + Grass

The simulation results depicted in Figure 14 illustrate that the tree + grass configuration effectively reduces wind speeds. During the summer, the wind speeds vary from 3.2 m/s to 4.0 m/s, while in the winter, wind speeds range from 4.1 m/s to 5.2 m/s. Notably, the kapok and grass combination exhibits poor performance in the winter, with wind speeds exceeding 5 m/s, Which is considered unsuitable. Conversely, the banyan and grass combination performs optimally. A wind speed of 3.2 m/s in the summer is conducive to prolonged outdoor activities, whereas 4.1 m/s in the winter is suitable for brief outdoor exposure. Based on this, it can be seen that the tree + grass configuration mode has a higher or lower wind reduction effect compared to the same variety of the tree + shrub + grass plants. When comparing banyan and carpet grass, the wind reduction effect of banyan and waringin and carpet grass is similar. However, the wind reduction effect of *Bischofia javanica* + carpet grass is significantly lower than that of *Bischofia javanica* + waringin + carpet grass. Additionally, the wind reduction effect of kapok + carpet grass is slightly higher than that of kapok + waringin + carpet grass. Among the three configuration modes of tree + grass, banyan and carpet grass have the best wind reduction effect.

(3) Shrub + Grass

The simulation results depicted in Figure 14 indicate that the combination of shrub + grass has the potential to mitigate wind speeds. By adopting a configuration consisting of waringin and carpet grass, wind speeds during the summer and winter remain below 5 m/s, eliminating instances of strong winds. Specifically, the average wind speed in the summer measures 3.3 m/s, providing a conducive environment for prolonged outdoor activities. During the winter season, the average wind speed is 4.3 m/s, which is ideal for short periods of time outdoors. Therefore, it indicates that, compared to the same tree + shrub + grass configuration mode, the shrub + grass configuration mode is even better than some tree + shrub + grass configuration modes. For example, the wind reduction effect of waringin + carpet grass is higher than that of kapok + waringin + carpet grass, slightly lower than that of banyan + waringin + carpet grass, and significantly lower than that of *Bischofia javanica* + waringin + carpet grass.

(4) Wind Regulation Amplitude

When it comes to wind regulation amplitude, Table 9 illustrates the impact of different facade configurations on wind regulation amplitudes of various plant varieties during the summer and winter. In the summer, the wind regulation amplitudes for plant combinations, such as banyan + waringin + grass, *Bischofia javanica* + waringin + grass, kapok + waringin + grass, banyan + grass, *Bischofia javanica* + grass, kapok + grass, and waringin + grass are 31.91%, 40.43%, 14.89%, 31.91%, 19.15%, 14.89%, and 29.79%, respectively. It is worth noting that the wind regulation amplitude is highest in the combination of *Bischofia javanica* + waringin + grass, at 1.27 times that of the plant combination banyan + waringin + grass, while kapok + waringin + grass and *Bischofia javanica* + grass show the lowest performance (0.47 times that of the plant combination banyan + waringin + grass). During the winter, the wind regulation amplitudes for the same plant combinations are 33.12%, 39.64%, 15.17%, 33.12%, 20.07%, 15.17%, and 29.85%, respectively. Among these, *Bischofia javanica* + waringin + grass shows the highest wind regulation amplitude, which is 1.20 times that of the plant combination banyan + waringin + grass, while kapok + waringin + grass and *Bischofia javanica* + grass perform the poorest (0.46 times that of the plant combination banyan + waringin + grass).

Table 9. Statistical table for summer/winter wind regulation amplitude of plant facade configurations.

Figure No.	Plant Varieties	Summer/Winter Input Wind Speeds	Summer/Winter Average Wind Speeds	Summer/Winter Wind Regulation Amplitude	Summer/Winter Wind Regulation Rate
1	Banyan + waringin + grass	4.70/6.13	3.2/4.1	31.91%/33.12%	1.00/1.00
2	*Bischofia javanica* + waringin + grass	4.70/6.13	2.8/3.7	40.43%/39.64%	1.27/1.20
3	Kapok + waringin + grass	4.70/6.13	4.0/5.2	14.89%/15.17%	0.47/0.46
4	Banyan + grass	4.70/6.13	3.2/4.1	31.91%/33.12%	1.00/1.00
5	*Bischofia javanica* + grass	4.70/6.13	3.8/4.9	19.15%/20.07%	0.60/0.61
6	Kapok + grass	4.70/6.13	4.0/5.2	14.89%/15.17%	0.47/0.46
7	Waringin + grass	4.70/6.13	3.3/4.3	29.79%/29.85%	0.93/0.90

Note: The red color shows the highest wind regulation rate among the seven plant varieties in the facade configuration, whereas the green color represents the lowest wind regulation rate of the seven plant varieties in the facade configuration.

In summary, the configuration mode of tree + shrub + grass performs the best at a Z = 1.5 m wind speed, while the plant variety combination of *Bischofia javanica* + waringin + carpet grass performs the best and has the most effective wind regulation. This is due to the ventilation provided by the *Bischofia javanica* forest, and the waringin plays a significant role in guiding the wind and preventing direct exposure to it for people (Figure 15).

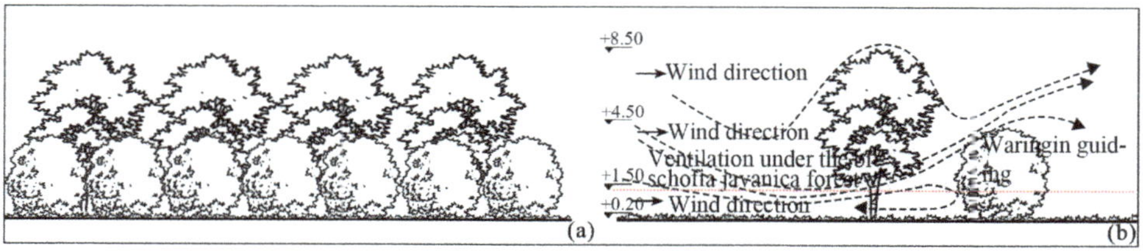

Figure 15. Vertical and side schematic diagrams of ventilation, wind blocking, and wind guidance for *Bischofia javanica* + waringin + grass plant configuration: (**a**) front elevation view; (**b**) side elevation view.

3.4. Plant Configuration Scheme Wind Environment Optimization Verification

It is evident from Figure 10 that the current single-building courtyard space (Youth Teacher Apartment) experiences wind speeds exceeding 5 m/s in both the summer and winter. During the summer, winds are situated in the southeast region (point b), while in the winter, winds extend from the southeast to northeast regions (points b and c), as well as encompassing the southeast to northeast corner of the courtyard (points f, g, j, and m). A targeted plant configuration optimization plan (Figure 16a) is proposed to address the current wind environment issues in the courtyard space of a single building (Youth Teacher Apartment): *Bischofia javanica*, waringin, and grass plants will be added to two areas where wind speeds exceed 5 m/s during both the summer and winter seasons. Then, the plant configuration optimization plan model was inputted into PHOENICS software to simulate the wind environment. Subsequently, the simulation values at Z = 1.5 m following optimization were contrasted with the initial simulation values (Figure 16b).

Upon optimizing the plant configuration scheme, improvements in wind speed are evident in the courtyard space of the single building (Youth Teacher Apartment) during both the summer and winter, with no detection points recording wind speeds exceeding 5 m/s, and the average wind speed has significantly decreased, as depicted in Figure 16b. Therefore, increasing the plant configuration of *Bischofia javanica*, waringin, and grass plants in areas with high wind speeds can effectively reduce their wind speed and improve wind comfort. This indicates that the plant configuration can not only beautify the campus environment, but it can be a cost-effective way to improve the wind comfort for outdoor campus spaces.

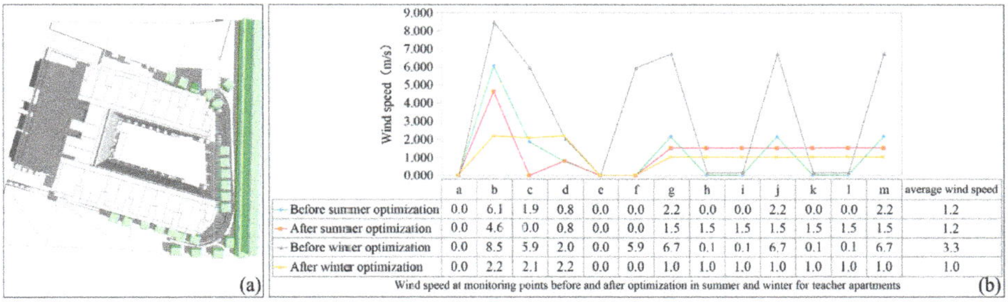

Figure 16. (**a**) Optimization plan for the plant configuration for the Youth Teacher Apartment. (**b**) Comparison of wind speed at measurement points before and after optimization in the summer and winter for the Youth Teacher Apartment.

4. Conclusions

This study delves into the outdoor environment of the Donghai Campus of Quanzhou Normal University, specifically examining spatial models for roads, squares, and courtyards, along with two types of plant configurations. The primary objective is to analyze how different plant arrangements affect the campus's outdoor wind conditions. A model incorporating both flat and vertical plant configurations is established for this purpose, and CFD simulations are conducted using PHOENICS software to quantify wind speeds. The research yields the following conclusions:

(1) Among trees, under the same conditions and with a consistent trunk height, the wind speeds with banyan are more suitable than that of *Erythrina indica*. When both the height and width change, *Bischofia javanica* has a better wind speed compared to the mango tree. Under the same conditions and with a consistent width, the wind speeds with kapok are more suitable compared to that of the *Chorisia speciosa*. Among shrubs, waringin has a more comfortable wind speed compared to *Aglaia odorata* lour.

(2) Among the three configuration modes of tree + shrub + grass, *Bischofia javanica* + waringin + carpet grass has the best wind reduction effect. Among the three configuration modes of tree + grass, the wind reduction effect of banyan + carpet grass is significant. When using three types of plants—*Bischofia javanica*, waringin, and carpet grass—for the configuration of tree + shrub + grass, tree + grass, and shrub + grass, the wind reduction effect of tree + shrub + grass is the best.

(3) In specific plant variety combinations, adding *Bischofia javanica*, waringin, and grass plant configurations in areas with high wind speeds can effectively reduce wind speeds and improve the comfort of the wind environment.

The above conclusion indicates that the selection of different plant varieties and their configuration modes can achieve different wind regulation effects. In this study, we only analyzed the impact of different plant species and configuration modes on the comfort of the outdoor wind environment in campus spaces. In the future, we will compare the effects of other factors, such as plant layout, plant orientation, among other factors, on the internal wind comfort. This study's limitation lies in the investigation of pedestrians' perception of outdoor space comfort. Subsequently, a comprehensive analysis from both natural and cultural viewpoints will be conducted on this matter. The above results have verified the advantages of plant configuration in optimizing the wind comfort of outdoor spaces on campus and offer insights for designing outdoor environments in subtropical coastal regions.

Author Contributions: Conceptualization, J.C., T.H. and Z.W.; methodology, J.Z., Y.W. and T.H.; software, T.H. and J.C.; validation, H.Y., Z.W., J.Z. and T.H.; formal analysis, X.Y., J.C., T.H. and Y.W.; investigation, Z.W., H.Y. and Y.W.; resources, Z.W.; data curation, J.C., X.Y., Z.W. and T.H.; writing—original draft preparation, H.Y., Z.W. and J.Z.; writing—review and editing, Z.W., J.Z. and J.C.; visualization, T.H.; supervision, X.Y., T.H. and J.C.; project administration, Z.W.; funding acquisition, Z.W. All authors have read and agreed to the published version of the manuscript.

Funding: This study received funding from the Ministry of Education through the Humanities and Social Science Research Youth Fund projects [20YJCZH009], the National Natural Science Foundation of China Youth Science Fund Project [42401236], and the General Project for Fujian Province Natural Science Foundation in China [2023J01894].

Data Availability Statement: Data is contained within the article.

Conflicts of Interest: No conflicts of interest have been declared by the authors.

References

1. Liu, X.; Bao, Y. Research on the overall design practice of green university campuses in humid and hot regions: A case study of China Capital Market Institute and The Chinese University of Hong Kong (Shenzhen). *South Archit.* **2019**, *5*, 60–67, 125.
2. Ying, X.; Shen, L.; Han, X.; Gao, J.; Qin, X.; Chen, J.; Huangpu, F.; Wang, H. Comparative analysis and strategic research on high-density primary school campus layout in humid and hot southern regions from the perspective of wind environment. *Huazhong Archit.* **2023**, *41*, 51–57.
3. Chen, S.; Mei, H. Research on the Design of Campus Cluster Form in Cold Regions Universities Based on Microclimate Environment Optimization. *Contemp. Archit.* **2020**, *7*, 19–23.
4. Sun, C.; Zhi, D.; Han, Y. Design Strategy for Teaching Group Form in Cold Regions Based on Outdoor Thermal Comfort. *South Archit.* **2020**, *4*, 80–85.
5. Liu, B.; Si, R. Adaptive Strategies for Wind Environment Landscape in Residential Areas Based on Data Measurement and CFD Simulation: A Case Study of Zhangwu Road Dormitory Area at Tongji University. *China Landsc. Archit.* **2018**, *34*, 24–28.
6. Zeng, Q.; Xu, J.; Li, C.; Chen, J. The impact of campus landscape design on spatial ventilation environment in humid and hot regions of South China. *Hubei Agric. Sci.* **2021**, *60* (Suppl. S1), 270–274.
7. Guo, J.; Li, X. CFD based simulation analysis and design optimization of outdoor wind environment in university campuses in hot summer and warm winter areas: Taking the planning and design scheme of Shantou University East Campus Project (Phase III) as an example. *Archit. Technol.* **2021**, *27*, 115–117.
8. Liu, M.; Cui, L. Analysis of Factors Influencing Campus Landscape Space Wind Environment and Improvement Strategies. *Value Eng.* **2023**, *42*, 128–130.
9. Zhang, N.; Wang, H.; Yu, R. Research on Campus Wind Environment Renovation and Optimization Based on CFD Numerical Simulation: A Case Study of China University of Petroleum. *Urban Archit.* **2021**, *18*, 126–129.
10. Wang, B.; Sun, S.; Li, Y.; Xie, Y.N.; Hou, Y.X.; Jin, Q.T.; Ren, Y.Y. Campus wind environment evaluation in north China—A case study of NCUT. *Energy Rep.* **2020**, *6* (Suppl. S1), 787–793. [CrossRef]
11. Han, Y.; Chen, Z.; Zhang, G. The impact of plant planting patterns on the outdoor wind environment of university sports halls. *Mod. Hortic.* **2024**, *47*, 49–51.
12. Liu, R.; Yan, X.; Wang, X. Study on the Impact of Urban Greening on Wind Environment Based on CFD Simulation Technology. *J. Shijiazhuang Railw. Univ. Soc. Sci. Ed.* **2022**, *16*, 104–108.
13. Zhang, L.; Zhan, Q.; Lan, Y. Effects of the tree distribution and species on outdoor environment conditions in a hot summer and cold winter zone: A case study in Wuhan residential quarters. *Build. Environ.* **2018**, *130*, 27–39. [CrossRef]
14. Hami, A.; Abdi, B.; Zarehaghi, D.; Bin Maulan, S. Assessing the thermal comfort effects of green spaces: A systematic review of methods, parameters, and plants' attributes. *Sustain. Cities Soc.* **2019**, *49*, 10163. [CrossRef]
15. Murtaza, M.; Wenbin, P.T.; Kaiser, J.C. Numerical evaluation of the use of vegetation as a shelterbelt for enhancing the wind and thermal comfort in peripheral and lateral-type skygardens in highrise buildings. *Build. Simul.* **2022**, *16*, 11–19.
16. Rui, L.; Buccolieri, R.; Gao, Z.; Ding, W.; Shen, J. The Impact of Green Space Layouts on Microclimate and Air Quality in Residential Districts of Nanjing, China. *Forests* **2018**, *9*, 224. [CrossRef]
17. Li, H.; Yan, Z.; Zhang, Z.; Lang, J.; Wang, X. A Numerical Study of the Effect of Vegetative Windbreak on Wind Erosion Over Complex Terrain. *Forests* **2022**, *13*, 1072. [CrossRef]

18. Chan, W.L.; Eng, Y.; Ge, Z.; Lim, C.W.C.; Gobeawan, L.; Poh, H.J.; Wise, D.J.; Burcham, D.C.; Lee, D.; Cui, Y.; et al. Wind Loading on Scaled Down Fractal Tree Models of Major Urban Tree Species in Singapore. *Forests* **2020**, *11*, 803. [CrossRef]
19. *GB/T 50378-2019*; Green Building Evaluation Standard. China Construction Industry Press: Beijing, China, 2019; pp. 126–127.
20. Tan, L. Simulation Study on Outdoor Wind Environment in Residential Areas Based on Reasonable Static Wind Rate Determination. Master's Thesis, Chongqing University, Chongqing, China, 2017.
21. Liu, B. Study on Outdoor Wind Environment in Winter in CBD of Cold Regions—Taking Xi'an High Tech Software Park CBD as an Example. Master's Thesis, Chongqing University, Chongqing, China, 2017.
22. Soligo, M.J.; Irwin, P.A.; Williams, C.J.; Schuyler, G.D. A comprehensive assessment of pedestrian comfort including thermal effects. *J. Wind Eng. Ind. Aerodyn.* **1998**, *77*, 753–766. [CrossRef]
23. Li, L.; Li, X.; Lin, B. Simulation of canopy flow using a two equation turbulence model with source term k-ε. *J. Tsinghua Univ. Nat. Sci. Ed.* **2006**, *46*, 753–756.
24. Lu, M.; Song, D.; Shi, J. Effect of high-rise residential building layout on the spatial vertical wind environment in Harbin, China. *Buildings* **2022**, *12*, 705. [CrossRef]
25. Wei, X. Research on reducing carbon consumption in residential community spaces as influenced by microclimate environments. *J. Urban Plan. Dev.* **2021**, *147*, 147. [CrossRef]
26. Li, J.; Liu, J.; Srebric, J.; Hu, Y.; Liu, M.; Su, L.; Wang, S. The effect of tree-planting patterns on the microclimate within a courtyard. *Sustainability* **2019**, *11*, 1665. [CrossRef]
27. Xu, X.; Luo, F.; Wang, W.; Hong, T.; Fu, X. Performance-based evaluation of courtyard design in China's cold-winter hot-summer climate regions. *Sustainability* **2018**, *10*, 3950. [CrossRef]
28. Norouziasas, A.; Ha, P.P.; Ahmadi, M.; Rijal, H.B. Evaluation of urban form influence on pedestrians' wind comfort. *Build. Environ.* **2022**, *224*, 109522. [CrossRef]
29. Chu, C.; Su, Z. Natural ventilation design for underground parking garages. *Build. Environ.* **2023**, *227 Pt 1*, 109784. [CrossRef]
30. Franke, J.; Hellsten, A.; Schlunzen, K.H.; Carissimo, B. The COST 732 Best Practice Guideline for CFD simulation of flows in the urban environment: A summary. *Int. J. Environ. Pollut.* **2011**, *44*, 419–427. [CrossRef]
31. Qin, H.; Hong, B.; Jiang, R.; Yan, S.; Zhou, Y. The Effect of Vegetation Enhancement on Particulate Pollution Reduction: CFD Simulations in an Urban Park. *Forests* **2019**, *10*, 373. [CrossRef]
32. Xue, F.; Li, X. The impact of roadside trees on traffic released PM10 in urban street canyon: Aerodynamic and deposition effects. *Sustain. Cities Soc.* **2017**, *30*, 195–204. [CrossRef]
33. Bruse, M.; Fleer, H. Simulating surface–plant–air interactions inside urban environments with a three dimensional numerical model. *Environ. Model. Softw.* **1998**, *13*, 373–384. [CrossRef]
34. Srivanit, M.; Hokao, K. Evaluating the cooling effects of greening for improving the outdoor thermal environment at an institutional campus in the summer. *Build. Environ.* **2013**, *66*, 158–172. [CrossRef]
35. Lin, B.; Li, X.; Zhu, Y.; Qin, Y. Numerical simulation studies of the different vegetation patterns' effect on outdoor pedestrian thermal comfort. *J. Wind Eng. Ind. Aerod.* **2008**, *96*, 1707–1718. [CrossRef]
36. Sanz, C. A note on k-ε modelling of vegetation canopy air-flows. *Bound. Layer Meteorol.* **2003**, *108*, 191–197. [CrossRef]
37. Katul, G.G.; Mahrt, L.; Poggi, D.; Sanz, C. One- and two-equation models for canopy turbulence. *Bound. Layer Meteorol.* **2004**, *113*, 81–109. [CrossRef]
38. Endalew, A.M.; Hertog, M.; Delele, M.; Baetens, K.; Persoons, T.; Baelmans, M.; Ramon, H.; Nicolaï, B.; Verboven, P. CFD modelling and wind tunnel validation of airflow through plant canopies using 3D canopy architecture. *Int. J. Heat Fluid Flow* **2009**, *30*, 356–368. [CrossRef]
39. Jin, W.; Zhou, C. Impact of administrative level and land marketization level on urban construction land utilization benefits. *Geogr. Sci.* **2023**, *43*, 2080–2090.

forests

MDPI

Article

A Study on the Effect of Green Plot Ratio (GPR) on Urban Heat Island Intensity and Outdoor Thermal Comfort in Residential Areas

Jian Zheng [1], Zilong Li [2,*] and Bohong Zheng [2,*]

[1] School of Architecture, Changsha University of Science and Technology, Changsha 410076, China; zhengjian@csust.edu.cn

[2] School of Architecture and Art, Central South University, Changsha 410075, China

* Correspondence: 221312024@csu.edu.cn (Z.L.); 203126@csu.edu.cn (B.Z.); Tel.: +86-178-7237-4349 (Z.L.); +86-138-7598-2310 (B.Z.)

Abstract: Greenery impacts the urban thermal environment, but the benefits of the three-dimensional green volume of space have not been effectively evaluated. In this paper, we analyzed the impact of 3D greenery on urban heat island intensity and thermal comfort in residential areas from the perspective of the green plot ratio (GPR). We selected a typical residential area, set up simulation models, and then analyzed the effect of different GPR values on the outdoor thermal environment using the validated ENVI-MET simulation. The results showed that increasing GPR in residential areas can effectively reduce the intensity of urban heat island and improve thermal comfort. When the GPR reaches 0.5 and 1.5, the thermal comfort level of the building overhead space and the north–south street space decreases from "very strong thermal stress" to "strong thermal stress". When the GPR reaches 2.5, the outdoor thermal comfort of the east–west street space and courtyard space is reduced to "hot". When the GPR is higher than 0.5, the urban heat island intensity in the north–south street space decreases by one level, from "very strong" to "strong". When the GPR reaches 3.5, all four types of spaces have "moderate" urban heat island intensity. Increased GPR exacerbates urban heat island intensity to some extent and worsens outdoor thermal comfort due to the nocturnal insulating effect of plants. Based on the results, the study proposes the bottom-line control of the GPR index from the perspective of urban heat island mitigation and thermal comfort improvement. This paper points out the benefits of GPR in residential areas in improving the human environment, which is of great practical value for developing urban residential environment from "increasing quantity" to "improving quality".

Keywords: green plot ratio; residential areas; urban heat island; thermal comfort; ENVI-MET

1. Introduction

1.1. The Necessity of Greenery in Residential Areas

As urbanization continues to increase, the construction of a healthy and livable urban environment is an essential element of future urban development [1–3]. However, rapid urbanization makes cities hotter than surrounding suburbs, affecting the urban thermal environment and threatening the health of urban residents [4]. In urban space, residential areas are the most relevant to the life and leisure of urban residents, and at the same time, they are the places where residents are most exposed to the outdoors [5]. As an important part of the urban ecosystem, residential greenery plays a vital role in regulating the urban climate, improving urban air quality, mitigating urban noise, improving urban landscapes, and increasing the thermal comfort of urban residents [6–8]. The urban residential area is a vital carrier with one of the four primary functions of the city (residence, work, recreation, and transport) [9]. Therefore, it is essential to clarify the relationship between greenery in residential areas and urban heat island intensity and outdoor thermal comfort. This is vital for improving the urban thermal environment and promoting the physical and mental health of urban residents.

1.2. Current Research Perspectives on Greenery in Residential Areas

Plants, as the main component of greenery in residential areas, have a cooling and humidifying effect on the surrounding environment through transpiration [10]. Trees improve outdoor thermal comfort by lowering temperatures through canopy shading and transpiration [11]. Current studies, mainly qualitative and quantitative, have examined the impact of residential greenery on the outdoor thermal environment. From the qualitative perspective, this includes plant species and greenery layout. Firstly, different types of plants affect the thermal environment differently. A study in Shanghai, China, suggests that deciduous plants can effectively improve year-round thermal comfort in hot-summer and cold-winter climate by reducing excessive outdoor shading in winter compared with evergreen plants [12]. Trees have a more apparent cooling effect than grasses because they provide more shade and effectively intercept more solar radiation [13]. Of the three plant types, trees, shrubs, and grasses, trees are more effective than shrubs in improving thermal comfort, and shrubs are better than grasses [14]. The differences in the characteristics of the different types of trees resulted in apparent differences in their impact on the microclimate of the settlements [15]. Secondly, greenery layout and green space structure have a specific regulatory effect on the thermal environment. In terms of greenery layout, a green space surrounded by trees provides optimal thermal comfort for activities; the thermal comfort under both greenery layout structures of trees surrounded by shrubs and shrubs surrounded by trees is better than that of a single shrub [8]. A study comparing 17 tree layouts in a hot-summer and cold-winter climate zone noted that evenly dispersed trees provided a limited increase in shade from 11% to 14% but did not improve thermal comfort [16]. Tree species and planting spacing have an effect on the outdoor thermal environment. It was found that reducing the planting spacing of trees can improve thermal comfort to a certain extent and that trees with high leaf area density have a more apparent effect on the outdoor thermal environment [17]. Compared to bare ground or asphalt, grass can effectively reduce ground temperatures and the surface temperatures of neighboring buildings [18]. There were distinct differences in the cooling of different greenery layouts, and among the shrub–grass, arbor–grass, and arbor–shrub–grass greenery layouts, arbor–shrub–grass had the most apparent cooling effect on the region as a whole [19].

From a quantitative perspective, green space ratio, green coverage, and tree canopy coverage are commonly used to measure the quality of urban green spaces. The green space ratio is the percentage of green plants and does not take into account the vertical projection of trees, whereas green coverage and tree canopy coverage take it into account. Many studies have explored the effects of these indicators on the thermal environment. Urban green space can reduce surface temperature by reducing incident radiation through shading and dissipating heat through evaporation, so increasing the green space ratio obviously mitigates the urban heat island effect in urban residential areas [20]. By constructing scenarios with different green space ratio, it was shown that increasing the green space ratio can effectively improve outdoor thermal comfort in semi-arid climates [21]. In the tropical city of Singapore, increasing green coverage reduces the UTCI by about 3.0 °C at midday, but continued increase in green coverage leads to an increase in relative humidity, resulting in no sustained reduction in thermal comfort [22]. Increased tree canopy coverage effectively reduced air temperature and mean radiant temperature in urban residence, with air temperature and mean radiant temperature with high tree canopy coverage reduced by 3.3 °C and 13.9 °C, respectively, compared to urban residence with no tree canopy coverage [23]. A study in Nigeria showed that outdoor air temperature could be reduced by up to 3.38 °C and mean radiant temperature by up to 24.24 °C when 45 percent of the canopy was covered with trees [24]. Based on the consideration of thermal dynamics and human thermal comfort, the study quantified the maximum effective canopy cover thresholds of 45 percent, 30 percent, and 25 percent for residential areas of 33 m (11 floors),

54 m (18 floors), and 100 m (33 floors) in height, respectively [25]. Additionally, increased plant canopy density can intercept more solar radiation, reducing outdoor air temperature in residential areas and improving thermal comfort [26]. Increasing tree canopy coverage can provide a more comfortable thermal environment for the area compared to increasing the green space ratio [22]. Studies on the thermal environment of subtropical settlements have found that even in neighborhoods with a high green space ratio, the maximum outdoor temperature can be up to 34 °C, and the total horizontal solar radiation intensity can be up to 1000 W/m^2 [27]. Actually, the green space ratio, green coverage, and tree canopy coverage quantify the rate of plant coverage per unit area. These indicators are only considered from the perspective of the horizontal surface but cannot reflect the actual three-dimensional green volume of plants. The three-dimensional green volume reveals the spatial volume of plants through the calculation of stem and leaf volume [28]. As a greenery index, it breaks through the limitations of the original two-dimensional greenery index and can more accurately reflect the reasonableness of the plant composition of the residential area. A synthesis of existing research results reveals fewer articles analyzing and evaluating the impact of greenery on the thermal environment of residential areas from the perspective of three-dimensional greenery, and there is a noticeable gap in research in this area.

1.3. A New Perspective on Measuring Plants Eco-Efficiency Using Green Plot Ratio (GPR)

To fill the gaps in the existing research referred to, related scholars have defined the green plot ratio (GPR) to measure the three-dimensional green volume of plants by combining the concept of building plot ratio (BPR, total building area divided by the building site area) in the field of urban planning. The GPR is calculated by drawing on the BPR method and introducing the leaf area index (LAI) of plants, that is, the total plant leaf area per unit of land [29]. The GPR can be obtained by calculating the overall amount of plant greenery in the area, thus enabling the assessment and control of green space quality. China's current design standards related to residential areas contain provisions for greenery in residential areas. The Urban Residential Planning and Design Standards states that the green space ratio in new urban areas should be no less than 30% and that the ratio for the reconstruction of old cities should be no less than 25% [30]. However, the outdoor space of many residential areas only meets the lower limit of the green space ratio indicator in the construction process in practice [31], which leads to a reduction in the quality of the outdoor space and a deterioration of the thermal environment [32]. Design standard for green building in Hunan province points out that the combination of trees, shrubs, and grasses should be used in a compound greenery way, and the green space should not be planted with less than 3 trees/100 m^2 to reduce the intensity of urban heat island [33]. In Jiangsu Province, China, the ratio of the canopy projection area of trees and shrubs to the lawn area should not be less than 4:1 [34]. Design standard for residential buildings in Sichuan Province states that the green plot ratio is the ratio of the total leaf area of various types of vegetation in the site (minus the building base area) to the site area, and stipulates that the GPR of the site in the residential area should not be less than 3.0 [35]. In Shenzhen, Shanghai and Fujian [36–38], the site GPR shall not be lower than 3.0. The above standards indicate that the evaluation standard of urban green space quality is developed from the traditional two-dimensional indexes, such as green space ratio and green coverage, to the three-dimensional indexes represented by GPR, and the benefits of urban green space will be analyzed more comprehensively. However, the current research has not been able to analyze the impact of GPR on the thermal environment of the residential area, and there are fewer studies on the measurement and simulation of the GPR as an evaluation index. Therefore, it is necessary to analyze the influence of GPR on the outdoor thermal environment.

1.4. Scientific Hypothesis and the Sub-Goals

From the above analysis, we realize a scientific question. GPR, as an indicator that is gradually being emphasized, we have paid little attention to its impact on the thermal environment of residential areas. Although GPR is required in some areas, we still do not know the effect of different values of GPR on the outdoor thermal environment, whether raising GPR all the time always has a positive effect, and if not, whether there is a lower or upper limit. These indicate that there is still a gap. Therefore, we conducted this study and proposed the following objectives.

Firstly, a research framework was established (shown in Figure 1), which briefly describes the main steps of this study. It is worth noting that this is applicable to specific climate zones and can be extended to more climate zones.

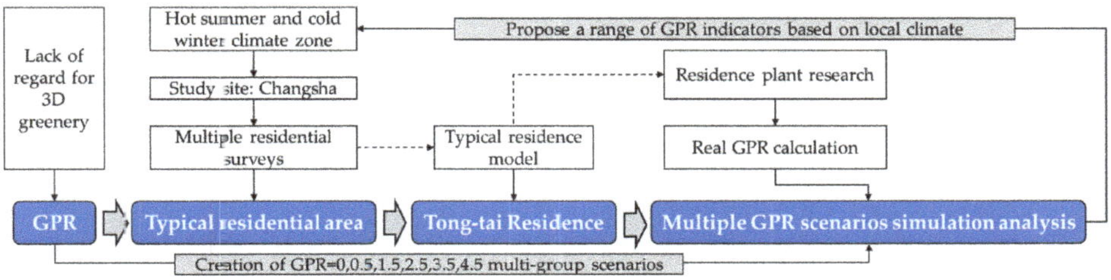

Figure 1. Research framework.

Secondly, we study the effects of different GPR values on the urban heat island effect and thermal comfort in residential areas and explore the relationship between GPR values and urban heat island intensity and thermal comfort. Based on the results, a reasonable range of GPR values is proposed.

Finally, we call on scholars in more countries or regions to emphasize the impact and benefits of GPR to help new urban development planning to cope with climate warming and thermal environmental degradation.

In the following sections, we took a residential area in a hot-summer and cold-winter climate zone as the research object and adopted the combination of field observation and numerical simulation. We took the GPR as the evaluation index, established the regional GPR scenario, and constructed the ENVI-MET model from the effect of different GPR values on the urban heat island intensity and thermal comfort in the residential area. This research aims to provide scientific guidance and suggestions for the greenery planning of residential areas to alleviate the urban heat island effect and improve the thermal comfort of residents.

2. Materials and Methods

2.1. Study Site and Typical Residential Area

The study site is located in Changsha, Hunan Province, China. Changsha is a typical subtropical monsoon climate zone [39], and meanwhile belongs to the hot-summer and cold-winter zone in China's building climate zones [40]. The average daily air temperature from June to September can reach over 25 °C, and there are nearly 84 days with an average daily temperature greater than 30 °C. Winter is cold, and the severe cold period below 0 °C is relatively short [41,42]. Changsha has high humidity, with the average annual precipitation reaching 1472.9 mm. Spring is cloudy and rainy, with very heavy precipitation; summer precipitation is uneven, with stormy weather mainly concentrated from May to June; autumn has little precipitation and mostly sunny weather with suitable temperatures; and winter precipitation accounts for 16% of the year [43].

Based on the introduction of the study site, we further analyzed the residential area. In this study, we introduced the concept of a typical residential area. It can be understood as a prototype similar to or close to most residences within a region. This is mainly in terms of spatial morphology and relevant settlement indicators. In order to understand the characteristics and indicators of Changsha's residential spatial morphology, we selected a number of residential areas for analysis based on field research, as shown in Figure 2. Firstly, we analyzed from the point of view of the residence plot area. Among the selected research samples, the number of residence plot area distribution of 4~4.5 hm^2 is the largest, which is 32%. Secondly, the shape of residential plots is affected by natural conditions, social and historical development, economic development mechanisms, and other factors, resulting in different plot shapes on the plane [44]. By analyzing the aspect ratios of the plots in the sample, it can be seen that the total proportion of residential areas with aspect ratios of 0.75~1.0 and 1.0~1.5 is 72%, meaning that residential plots are mostly rectangular. Analyzed from the perspective of the orientation of residential plots, the proportion of residential plots with a declination angle of ±5.4° in the samples is 60%, indicating that the orientation of Changsha's residential areas is primarily due south and north. The architectural layout of the residential areas in the sample mainly includes row-type, enclosed-type, and mixed-type (point-type + row-type). Among them, the proportion of row layout is 65.8%, and the proportion of enclosed layout is 7.9%. It can be seen that Changsha's residential buildings are dominated mainly by row and column layouts. The number of floors of residential areas in the statistical samples belongs to the multi-story and high-rise category. Multi-story category I accounts for 23.7%, multi-story category II accounts for 10.5%, high-rise category I accounts for 47.4%, and high-rise category II accounts for 13.2%, of which high-rise category I (10~18 floors) accounts for the largest proportion. Additionally, the proportion of residential areas with a building density of 25% or less in the sample was 60.5%.

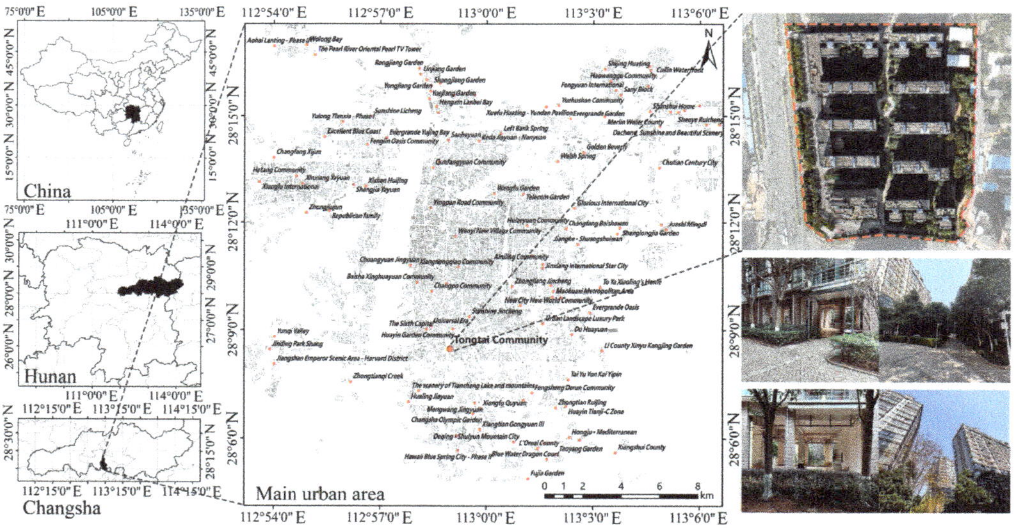

Figure 2. Residential area research and selection of research subjects, red dotted box represent residential areas.

Based on the above analysis, we selected the Tong-tai residence in the Tian-Xin District of Changsha City, China, as the typical residential area for the measurements and simulations. The typical residential area is located in the city center of Changsha (112°58′ E, 28°11′ N, 23.5 m above sea level), Hunan, China (Figure 2). From the aspects of spatial morphology and relevant indicators of the settlement, the land area of the residential area is about 4.43 hectares. The overall shape of the plot is regular, consisting of ten 14 floors (high-rise category I) residential buildings, with a building density of about 23.9%, and the buildings are arranged in rows and columns, facing south and north; the buildings on the north and south sides jointly enclose to form the space of the street, and the height-to-width ratio of the street is about 1.75. Concrete pavements, broadleaf trees, shrubs, and grasses are uniformly distributed in the street space, forming the lower bedding surface of the residential area. Based on the area of the residence, building layout form, building height, building density and other indicators, we believe that the Tong-tai residence is typical.

2.2. Calculation of GPR and Modeling of Six Scenarios

Green plot ratio (GPR) is a widely used indicator to quantify the three-dimensional green volume of green space. In 2003, a Singaporean scholar proposed the planning index of green plot ratio (GPR) based on the biological parameter leaf area index (LAI), defined as the average green area of a plot of land, which is the average of the green area of a site [29]. The leaf area index (LAI) can be used to measure the structural type of green space. The higher the value of LAI, the larger the total leaf area of plants per unit area and the richer the type of vegetation structure. The concept of building plot ratio is borrowed in the evaluation of urban green space, and GPR is defined as the green plot ratio, which is the total plant leaf area per unit of land. The formulas are shown in (1) and (2). We conducted field counts of five species of trees, one shrub, and one grass that dominate the residential area, and the counts are shown in Table 1. According to the formulas, we calculated the GPR for Tong-tai residence to be 3.43, which is close to 3.5.

$$GPR = LA/S \tag{1}$$

$$GPR = \frac{\sum_a N_a \cdot \pi R_a^2 \cdot LAI_a + \sum_b N_b \cdot \pi R_b^2 \cdot LAI_b + \sum_c S_c \cdot LAI_c}{TA} \tag{2}$$

where LAI represents the sum of leaf area in a given region, and S represents the total land area of the region. N_a is the number of trees (a), R_a is the canopy radius of trees (a), and LAI_a is the leaf area index of trees (a); N_b is the number of shrubs (b), R_b is the radius of shrubs (b), and LAI_b is the leaf area index of shrubs (b); S_c is the area covered by grass (c), and LAI_c is the leaf area index of grass (c). TA represents the total land area of the study area.

Table 1. Number or area of plants counted in the field.

Plants Name	Total Number or Total Area
Michelia maudiae Dunn.	121 trees
Koelreuteria bipinnata Franch.	67 trees
Cinnamomum camphora (L.) Presl.	76 trees
Prunus subg. *Cerasus* sp.	127 trees
Osmanthus sp.	140 trees
Camellia japonica L.	63 trees
Reineckea carnea (Andrews) Kunth	12,800 m^2

According to the GPR formulas, we can adjust the number of plants to construct different GPR scenarios. In order to facilitate the subsequent ENVI-MET modeling, we selected the above plants as the representative plants of the plot's greenery. Table 2 shows the total number of trees in different GPR scenarios. In addition to trees and shrubs, we used grass as a fixed quantity. When GPR = 0, there is no vegetation on the site. The size and amount of grass are the same in the other five scenarios.

Figure 3 illustrates the six GPR scenarios established by the study. At GPR = 0, there is no vegetation in the residence. This scenario is mainly used as a comparison to analyze the cooling and thermal comfort improvement effects of greenery. The number of plants on the site gradually increased as the GPR increased to 0.5, 1.5, 2.5, 3.5, and 4.5, and the increase in plants followed a randomized layout. We chose GPR = 4.5 as the upper limit because in the actual greenery layout of residential areas, few residential areas can be planted with such a large number of plants due to economic considerations. This study focuses on the effect of varying the value of GPR between 0 and 4.5 on the thermal environment of a residential area.

Table 2. Total number of trees for different GPR scenarios.

Different GPR Scenarios	Total Number of Trees
GPR = 0	0
GPR = 0.5	113
GPR = 1.5	229
GPR = 2.5	435
GPR = 3.5	610
GPR = 4.5	860

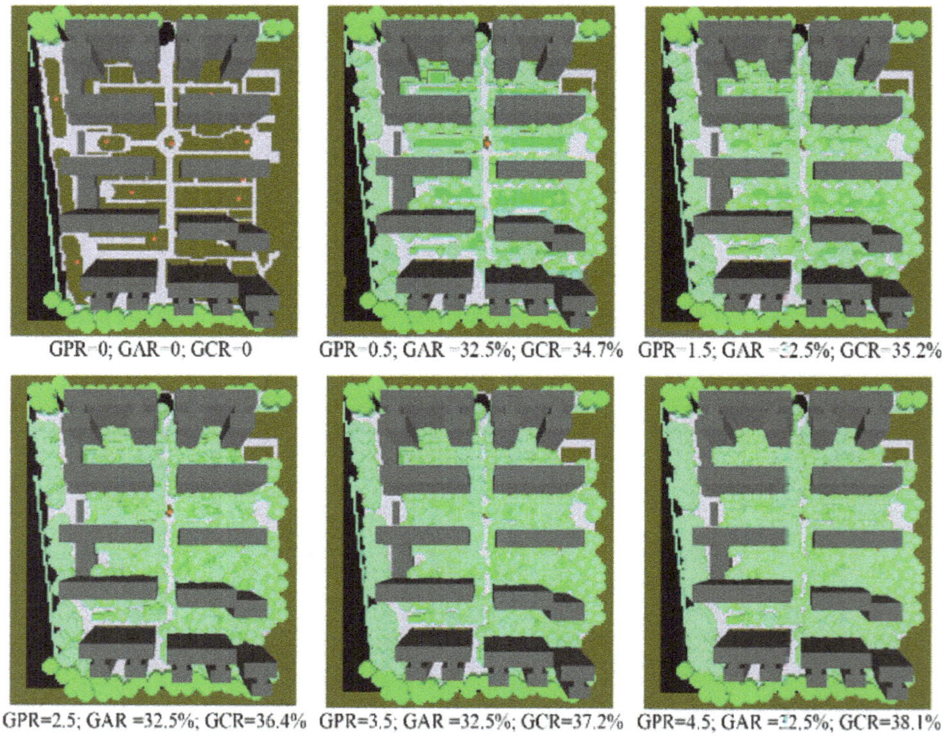

Figure 3. Aerial view of the ideal model and parameter indicators, GPR represents green plot ratio, GAR represents green area ratio, and GCR represents green coverage ratio.

2.3. Measurement and Accuracy Verification

On-site measurement is one of the essential ways to evaluate the characteristics of the thermal environment objectively, and at the same time, it is an essential basis for verifying the accuracy of numerical simulation. In the study, a typical weather day in summer was selected for the field measurement, and the air temperature (Ta), relative humidity (RH), black globe temperature (Tg), and wind speed (Va) at 1.5 m near the ground were measured by different testing instruments. Finally, the open-source software Rayman 1.2 was used to calculate the thermal comfort indicators mean radiant temperature (Tmrt) and physiological equivalent temperature (PET). The study focused on the microclimate distribution at pedestrian heights, so the experimental instruments were arranged at 1.5 m above the ground. The study area was surveyed, and the green points in Figure 4 are

DS1923 (i-Button) temperature and humidity data recorders, with a total of 10 groups; the blue points are hand-held temperature and humidity recorders, with a total of 12 groups; and the red points were Kestrel 5500 meteorological instrument, with a total of 3 groups. DS1923 (i-Button) were mainly arranged in the green group of the residential area. The Kestrel 5500 meteorological instruments were located at the entrance square on the west side of the residential area, the central square, and the first elevated floor of the residential area. Table 3 describes the metrics associated with the measurement instruments.

Figure 4. Layout of survey points and related instrumentation in the study area.

Table 3. Introduction to the relevant indicators of measuring instruments.

Measuring Instruments	Measurement Indicators	Measurement Range	Instrument Precision	Range of Error
JANTYTHCH 2022	air temperature	−20~100 °C	0.1 °C	±0.5 °C
	wind speed	0~10 m/s	0.1 m/s	±0.5 m/s
	black globe temperature	−20~100 °C	0.1 °C	±0.5 °C
Kestrel 5500	wind direction	0~360°	1°	±3°
	wind speed	0~70 m/s	0.1 m/s	±0.5 m/s
DS1923 (i-Button)	Ta/RH	−40~80 °C/0%~100%	0.1 °C/0.1%	±0.3 °C/±0.5%

The parameter settings for the study validation simulation are shown in Table 4, including geographic location, simulation time, and meteorological parameters.

Table 4. ENVI-MET simulation parameter settings.

Simulation Parameters	Settings
Geographic location	Changsha (28.11° N, 112.58° E)
Simulation time	Date: 7 August 2021 Start time: 2:00 Duration of simulation: 32 h
Meteorological parameters	Minimum air temperature 28.7 °C Maximum air temperature 34.45 °C
	Maximum humidity: 87.66% Minimum humidity 59.75%
	Wind speed: 2.0 m/s
	Wind direction: 180°

Comparing the measured and simulated values, the correlation coefficients of air temperature, mean radiant temperature, and physiological equivalent temperature were 0.833, 0.825 and 0.809, respectively (Figure 5). Since the R^2 is closer to one, the measured data fits the simulated data better, and the measured and simulated values in the present study have an obvious correlation and a high degree of fit. It could be seen that the simulation of ENVI-MET on thermal environment parameters is effective, and its simulated values are highly fitted to the measured values.

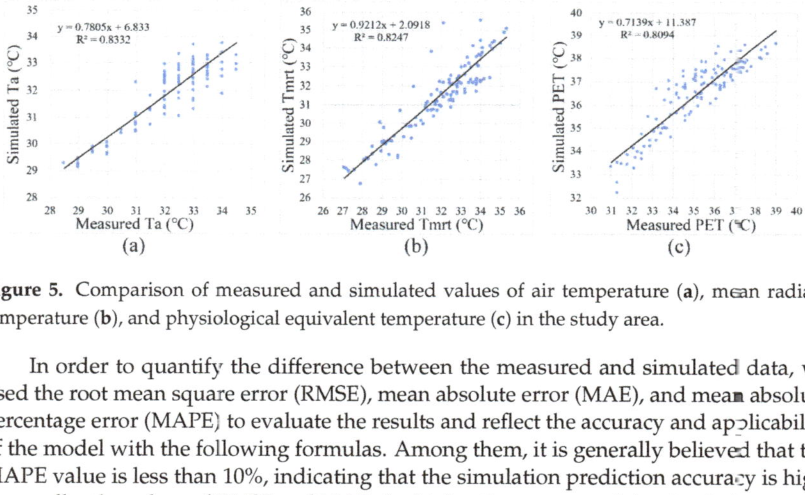

Figure 5. Comparison of measured and simulated values of air temperature (**a**), mean radiant temperature (**b**), and physiological equivalent temperature (**c**) in the study area.

In order to quantify the difference between the measured and simulated data, we used the root mean square error (RMSE), mean absolute error (MAE), and mean absolute percentage error (MAPE) to evaluate the results and reflect the accuracy and applicability of the model with the following formulas. Among them, it is generally believed that the MAPE value is less than 10%, indicating that the simulation prediction accuracy is high; the smaller the values of RMSE and MAE, the higher the accuracy of the simulation results. Moreover, some scholars have found that the RMSE of ENVI-MET simulated values and measured values are approximately between 0.66 and 7.98 °C, and it is currently believed that the RMSE is approximately between 0.52 and 4.30 °C [45], and MAE approximately between 0.27 and 3.67 °C is acceptable.

$$\mathrm{RMSE} = \sqrt{\frac{1}{m}\sum_{i=1}^{m}(Y_{obs,i} - Y_{model,i})^2} \tag{3}$$

$$\mathrm{MAE} = \frac{1}{m}\sum_{i=1}^{m}\left|(Y_{obs,i} - Y_{model,i})\right| \tag{4}$$

$$\mathrm{MAPE} = \frac{1}{m}\sum_{i=1}^{m}\frac{\left|Y_{obs,i} - Y_{model,i}\right|}{Y_{obs,i}} \times 100\% \tag{5}$$

In the formula, RMSE is the root mean square error; MAE is the mean absolute error; MAPE is the mean absolute percentage error; Y_{obs} is the measured value; Y_{model} is the simulated value; m is the number of data samples; and i is the number of samples, i = 1, 2, 3 . . .m.

In Table 5, root mean square error (RMSE), mean absolute error (MAE), and mean absolute percentage error (MAPE) evaluations were performed and analyzed for air temperature, mean radiant temperature, and physiological equivalent temperature in the study area. It can be found that the RMSE, MAE, and MAPE of air temperature in the study area are 0.643, 0.484, and 1.48%, respectively; the RMSE, MAE, and MAPE of mean radiation temperature are 0.970, 0.746, and 2.32%, respectively; and the RMSE, MAE, and MAPE of physiological equivalent temperature are 1.600, 1.387, and 4.04%, respectively.

Table 5. RMSE, MAE, and MAPE of measured and simulated values at monitoring points.

Indicators	Ta	Tmrt	PET
RMSE	0.64 °C	0.97 °C	1.60 °C
MAE	0.48 °C	0.75 °C	1.39 °C
MAPE	1.48%	2.32%	4.04%

The error evaluation metrics, i.e., RMSE value, MAE value, and MAPE value, of all the above-mentioned thermal environment metrics are within the permissible range, indicating that the error between measured and simulated values in this study is small. In this study, the air temperature error is the smallest, the mean radiant temperature is the second largest, and the physiological equivalent temperature error is the last. This is related to the fact that the simulation environment established in ENVI-MET is relatively simple, and the simulation results are more ideal compared with the measured results; in the measured environment, the urban construction environment is more complicated, and the air temperature, relative humidity, wind speed, etc., are easily affected by the surrounding environment and human activities, and the calculation of the physiological equivalent temperature and the mean radiant temperature involves several microclimate parameters; in addition, the data obtained in the actual measurement are instantaneous, and the data obtained in ENVI-MET's simulation are the most accurate. In addition, the data obtained in actual measurements are instantaneous values, while the ENVI-MET simulation environment is set to continuous values. Additionally, for this study, the authors analyzed the stability of the ENVI-MET software 5.0 to ensure the reliability of the results. Based on the existing research results and the calibration experiments in this study, it is still a reliable scientific tool to study the outdoor thermal environment in hot-summer and cold-winter zone, with good modeling accuracy. It can predict the microclimate environment of the study area.

2.4. Data Extraction Point Layout and Data Analysis

In order to comprehensively assess the impact of GPR on outdoor spaces in residential areas, the study categorizes outdoor spaces into street space, courtyard space, and building overhead space based on the function of the space. Street space includes the primary and secondary roads in the residential area, a space with obvious transportation attributes, and is divided into east–west street space and north–south street space in this study. Courtyard space includes enclosed and semi-enclosed courtyards, which are enclosed by buildings with weak openness. Building overhead space refers to the open space layer of a building that is supported only by structural columns without enclosing walls, and it is a semi-indoor, public gray space that is well connected to the outside. For the data extraction of each space, the study refers to the established methods and follows the principle of uniform layout of extraction points to ensure the science and rationality of the extracted data. The layout of data extraction points is shown in Figure 6. Additionally, to investigate the effect of GPR on the thermal environment of the residential area further, the study will analyze the daytime and nighttime hours separately. In contrast to the daytime hours, the nighttime hours are free of solar radiation, and the ground scattering effect dominates.

Figure 6. Layout of simulated data extraction points.

The study extracted data for the air temperature and PET at the pedestrian height of 1.5 m and calculated the urban heat island intensity, which is the difference between the mean air temperature in the city center and the mean air temperature in the surrounding suburbs (countryside). Physiological equivalent temperature (PET) was chosen to quantify thermal comfort. Physiological equivalent temperature integrates air temperature, humidity, wind speed, solar radiation, clothing, and exercise. It is the most accurate and widely used, and its scientific validity has been proved by domestic and international scholars' research [46].

3. Results

3.1. Impact of GPR on Urban Heat Island Intensity in Residential Areas

3.1.1. Impact of GPR on Urban Heat Island Intensity during Daytime

The urban heat island intensity of the residential space is mitigated to varying degrees as the GPR increases. In Figure 7, the urban heat island intensity mitigation occurred in the east–west street space during the GPR increase from 0 to 4.5, with the urban heat island intensity decreasing from 3.91 °C to 2.89 °C. The urban heat island intensity of the north–south street space is reduced by 0.90 °C. The urban heat island mitigation effect of the courtyard space and the building overhead space is the same, with the urban heat island intensity reduced by 0.95 °C and 0.94 °C, respectively. For the north–south street space, the GPR increase from 3.5 to 4.5 still mitigates urban heat island intensity by up to 0.12 °C. However, the GPR increase does not consistently mitigate urban heat island intensity for the other types of spaces. In the courtyard space, the GPR increased from 3.5 to 4.5, with only a 0.02 °C reduction in urban heat island intensity, a 0.04 °C reduction in urban heat island intensity in the east–west street space, and a 0.05 °C reduction in the building overhead space. Figure 8 shows the outdoor air temperature distribution in the residential space at 14:00.

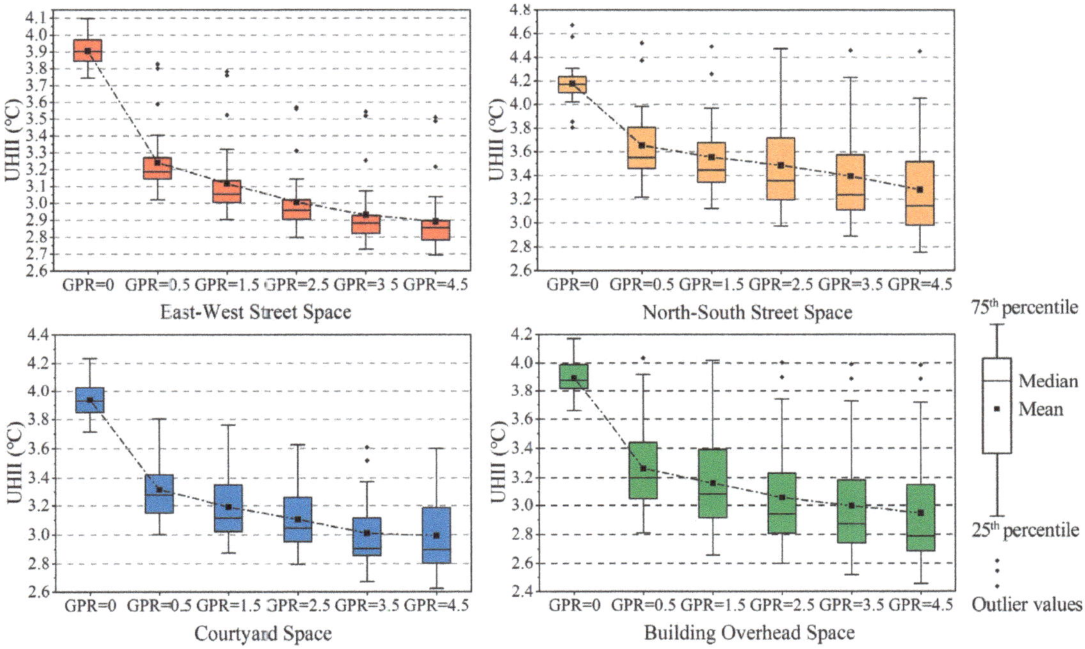

Figure 7. Impact of GPR increase on UHII in four types of spaces at 14:00.

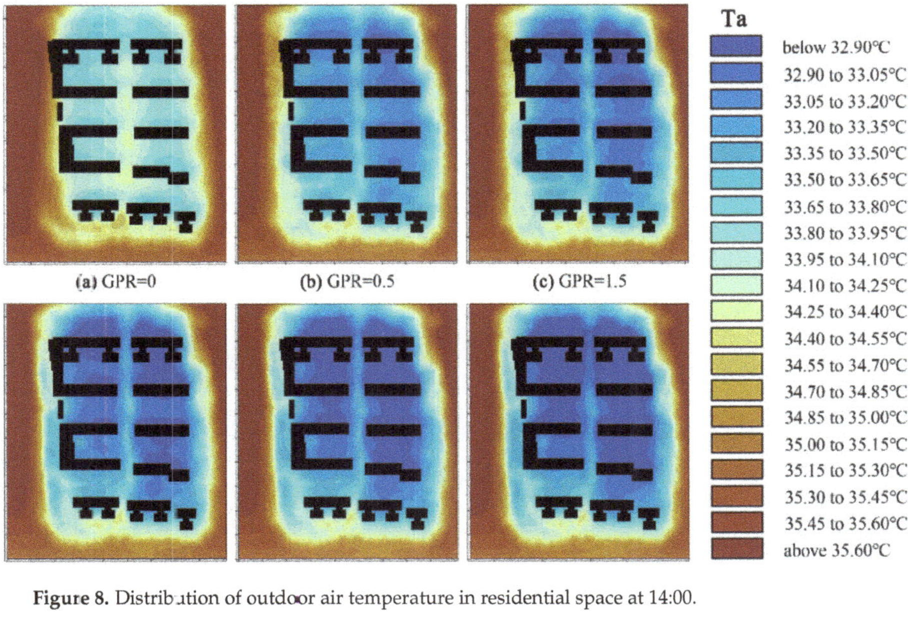

Figure 8. Distribution of outdoor air temperature in residential space at 14:00.

Combined with the urban heat island intensity grading (Table 6), the effects of different GPRs on the four types of space heat intensities vary. Where GPR > 2.5, the urban heat island intensity of the east–west street space is below 3.0, and the urban heat island intensity level is reduced from "strong" to "moderate". For the north–south street space, increasing the GPR mitigates the urban heat island intensity up to 0.90 °C, but the overall intensity level of the space is still in the "strong" category. When the GPR reaches 4.5, the intensity of the urban heat island in the courtyard space and the building overhead space is reduced to the "moderate" level.

Table 6. Urban air temperature urban heat island intensity classification.

UHI Intensity Level	UHII (°C)	Intensity Characteristics
1	UHII ≤ 1	None
2	1 < UHII ≤ 2	Weak
3	2 < UHII ≤ 3	Moderate
4	3 < UHII ≤ 4	Strong
5	UHII > 4	Very strong

The study further analyzes the time-by-time effects of GPR increase on the outdoor air temperature of the four types of spaces. Increasing greenery can reduce the outdoor temperatures of the four types of spaces, and the GPR is increased from 0 to 0.5, and the air temperatures of the four types of spaces are reduced by a maximum of 0.76 °C, 0.69 °C, 0.57 °C, and 0.67 °C, respectively (Figure 9). The GPR is increased from 0.5 to 1.5, and the air temperatures of the four types of spaces are reduced by a maximum of 0.23 °C, 0.12 °C, 0.11 °C, and 0.11 °C, with cooling effects of 0.10 °C or more; the GPR is increased from 1.5 to 2.5, the maximum reductions of the four types of spaces are 0.11 °C, 0.09 °C, 0.08 °C, and 0.10 °C, respectively, i.e., the cooling effect caused by the increase in GPR gradually weakened. The GPR increased from 2.5 to 3.5 and 3.5 to 4.5 in the process of the four types of spaces; the maximum cooling effect of the four types of spaces is less than 0.10 °C.

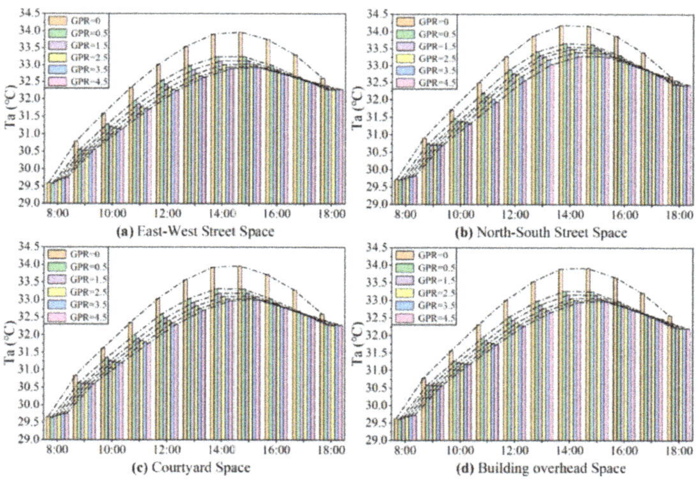

Figure 9. Variation in Ta with the increase in GPR during daytime.

3.1.2. Impact of GPR on Urban Heat Island Intensity during Nighttime

This section analyzes the effect of GPR on the intensity of nighttime urban heat island. Figure 10 illustrates the urban heat island intensity trend with increasing GPR at 22:00. The GPR increases from 0 to 4.5, with the most noticeable increase in the spatial urban heat island intensity for east–west street space at 0.40 °C. The remaining three space urban heat island intensity increase in essentially the same way, at 0.26 °C, 0.25 °C, and 0.26 °C, respectively. Overall, the nocturnal warming effect due to increased GPR is weak.

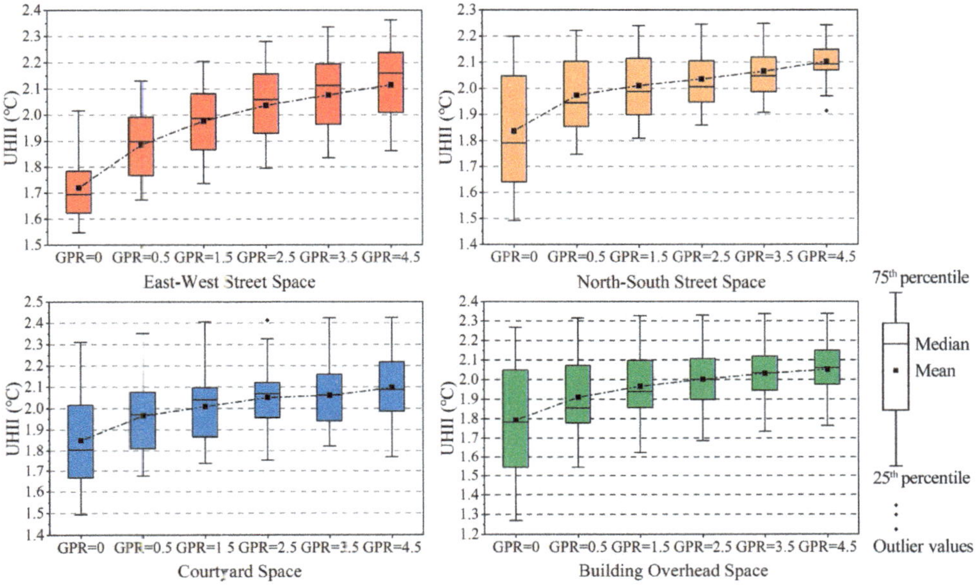

Figure 10. Impact of GPR increase on UHII in four types of spaces at 22:00.

During the nighttime, from 20:00 to 6:00 on the next day, the outdoor air temperature continued to drop (Figure 11). The GPR increase consistently affected outdoor temperatures for all four types of spaces, with outdoor air temperatures increasing with the GPR increase. However, this warming effect diminishes as the GPR increases further.

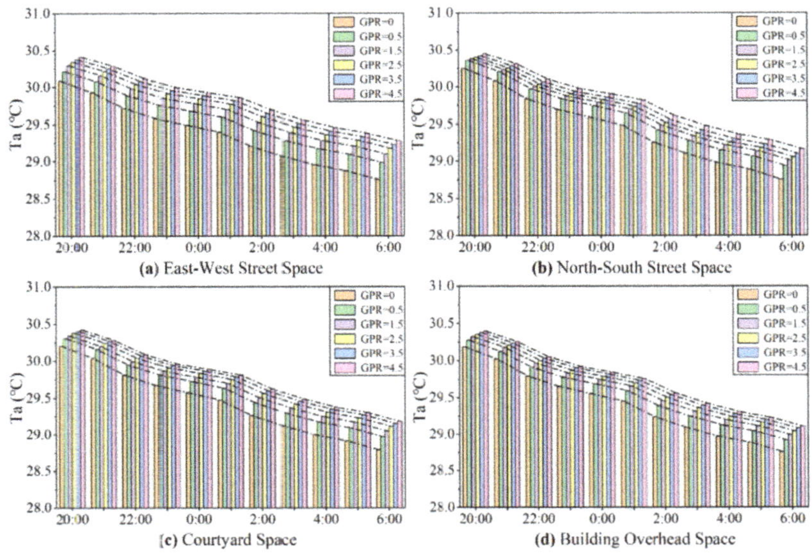

Figure 11. Variation in Ta with the increase in GPR during nighttime.

3.2. Impact of GPR on Outdoor Thermal Comfort in Residential Areas

3.2.1. Impact of GPR on Outdoor Thermal Comfort during Daytime

Figure 12 shows the impact of GPR increase on PET. As the GPR increased from 0 to 4.5, the outdoor PET of four types of spaces had varying degrees of decline during the daytime. The most apparent decline in PET is in the east–west street space, followed by courtyard space, north–south street space, and building overhead space. The PET decreased by 6.61 °C, 6.54 °C, 4.94 °C, and 3.38 °C, respectively. When GPR increased to 0.5, the PET of the building overhead space is reduced to 40.35 °C, and the outdoor thermal comfort transitions from "severe hot" to "hot" according to PET classification in Table 7. When GPR increased to 1.5, the outdoor PET for the north–south street space and the courtyard space is 40.75 °C and 40.86 °C, respectively, and the outdoor thermal comfort is further improved. PET in the east–west street space is obviously improved at GPR above 2.5, with a one-level reduction in human body thermal stress. Notably, the effect of the increase in GPR to improve thermal comfort is not consistently obvious. PET is reduced by 0.38 °C, 0.44 °C, 0.05 °C, and 0.19 °C for the four types of spaces, with the GPR increased from 2.5 to 4.5. Thus, it can be seen that, for courtyard space and building overhead space, the gains in thermal comfort improvement from continued increases in GPR are no longer obvious. Figure 13 shows the outdoor PET distribution at 14:00.

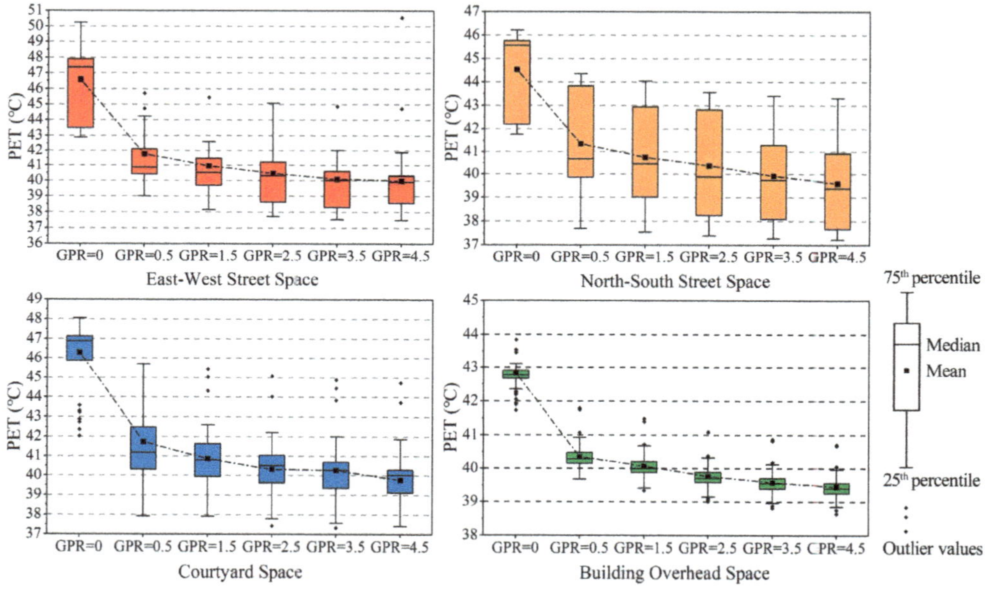

Figure 12. Impact of GPR increase on PET in four types of spaces at 14:00.

Table 7. Thermal Comfort PET Physiological Stress Level Grading Scale for inhabitants [47].

PET	Thermal Sensitivity	Physiological Stress Level
<4 °C	Cold	Strong cold stress
4~8 °C	Cool	Moderate cold stress
8~13 °C	Slightly cool	Slight cold stress
13~23 °C	Comfortable	No thermal stress
23~35 °C	Slightly warm	Slight heat stress
29~35 °C	Warm	Moderate heat stress
35~41 °C	Hot	Strong heat stress
>41 °C	Very hot	Extreme heat stress

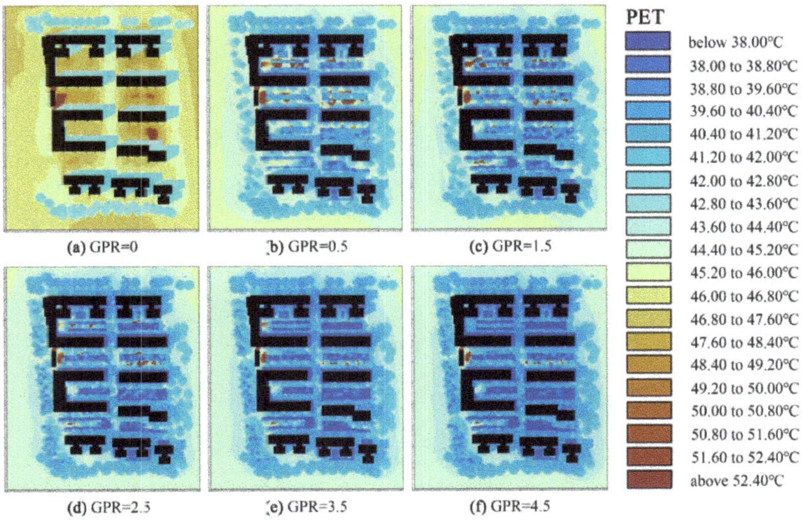

Figure 13. Outdoor PET distribution at 14:00.

Figure 14 illustrates the time-by-time trend of outdoor PET with increased GPR for the four types of spaces. For east–west street space, the spacing between the two curves for GPR = 3.5 and 4.5 varied less, indicating a reduced effect of GPR on the time-by-time effect of PET. In the north–south street space, PET at 18:00 changed with increasing GPR, not showing a trend of higher GPR and lower PET. In the courtyard space, the difference in the effect of increased GPR on PET decreases after 17:00. This is due to the gradual weakening of solar radiation, and the effect of GPR on PET decreases.

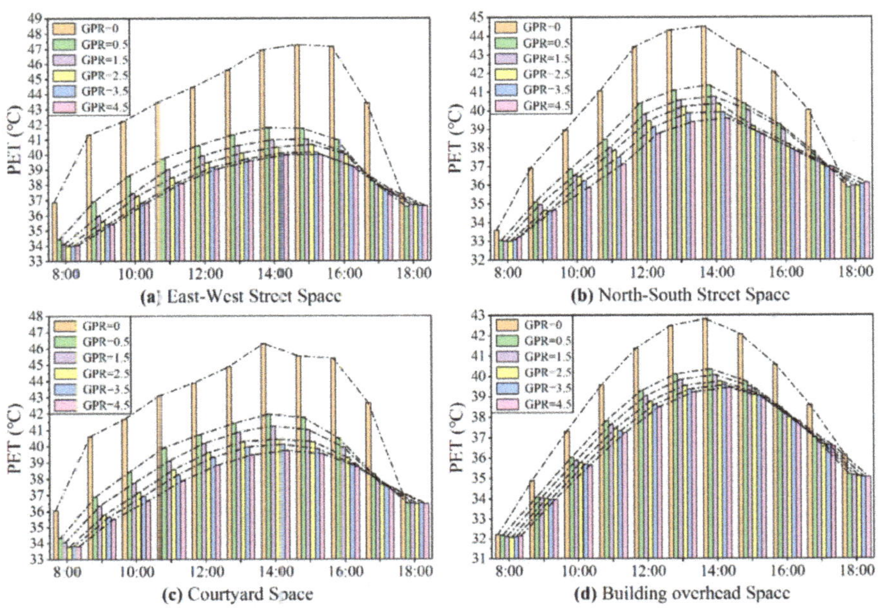

Figure 14. Variation in PET with the increase in GPR during daytime.

3.2.2. Impact of GPR on Outdoor Thermal Comfort during Nighttime

This section demonstrates the impact of GPR on nighttime outdoor thermal comfort. First, the increase in GPR leads to an increase in nighttime PET for all types of spaces in this study (Figure 15). When the GPR was raised from 0 to 4.5, the most noticeable increase in PET was 3.02 °C for the east–west street space, yet the building overhead space only increased by 0.86 °C. Second, the thermal comfort deterioration in the east–west street space was most apparent at night as the GPR continued to increase, and when GPR = 4.5, the outdoor PET in the east–west street space was the highest at 33.57 °C. The effect of GPR on the PET in the courtyard space gradually diminished when GPR > 1.5, as evidenced by an average increment in PET of 0.10 °C.

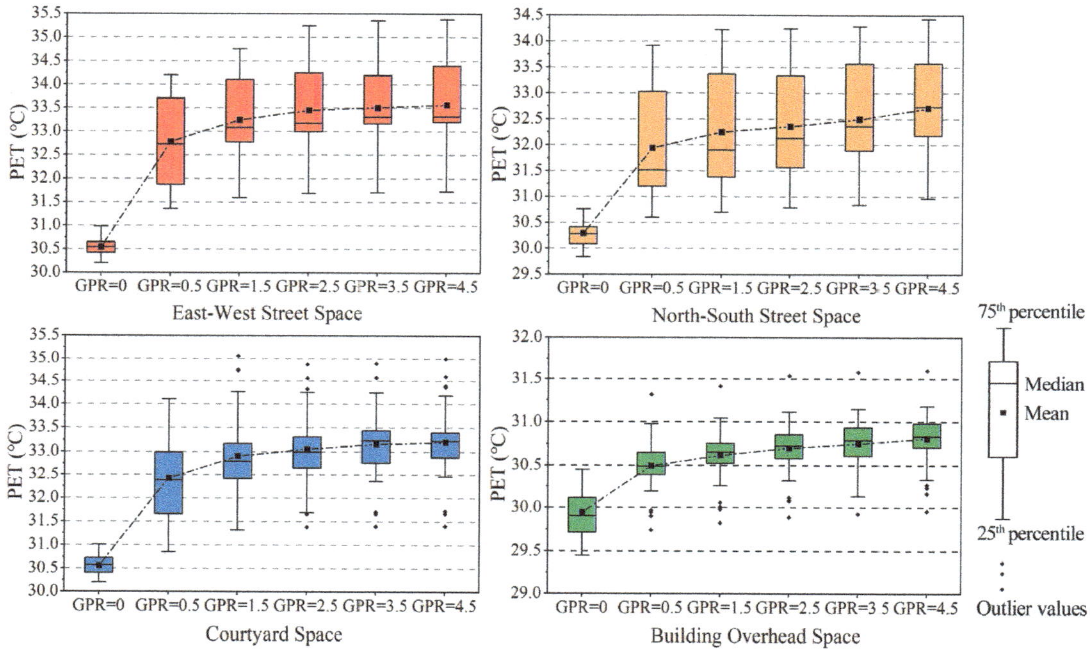

Figure 15. Impact of GPR increase on PET in four types of spaces at 22:00.

Figure 16 shows the night time-by-time variation in outdoor PET with increasing GPR for the four types of spaces. Firstly, nighttime PET shows an overall decreasing trend. Second, among the six GPR scenarios, increasing GPR from 0 to 0.5 has the most noticeable effect on increased outdoor PET in four types of spaces. When GPR is increased from 2.5 to 4.5, the corresponding three survey spacing in Figure 16a is no longer apparent, and when GPR is increased from 3.5 to 4.5, the corresponding curve gap in Figure 16c is also not noticeable. Second, the effect of increasing GPR on the improvement of nighttime outdoor PET in the north–south street space remains evident. The PET in the building overhead space during the nighttime is not as affected by GPR compared with the other three types of spaces obviously.

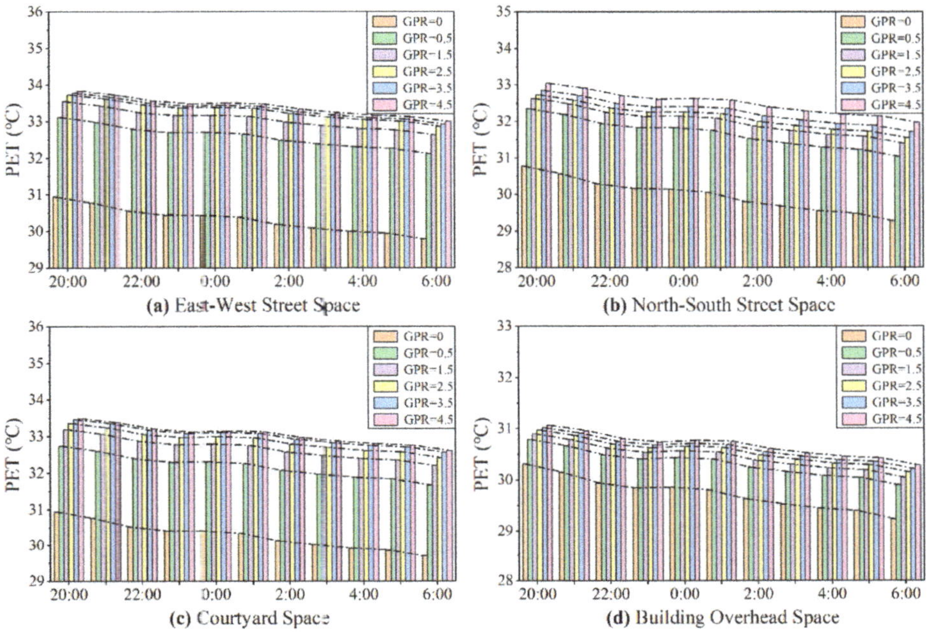

Figure 16. Variation in PET with the increase in GPR during nighttime.

4. Discussion

4.1. Differences in the Effects of GPR on the Intensity of Urban Heat Island of Four Types of Spaces in Residential Areas

The GPR increased from 0 to 4.5, and the daytime urban heat island intensity of the four types of spaces in the residential area decreased by 1.02 °C, 0.95 °C, 0.90 °C, and 0.94 °C, respectively. An increase in GPR represents the increase in the amount of plant greenery in the outdoor space, and as the amount of greenery increases, the overall cooling and humidifying effect of the plants becomes more pronounced [48,49], which effectively reduces air temperature and mitigates the intensity of urban heat islands in settlements. This finding has been confirmed in previous studies [50]. From the analyzed results, it is clear that there are differences in GPR's mitigation of urban heat island intensity for different types of spaces, especially courtyard space. The urban heat island intensity is reduced by only 0.90 °C when the GPR is increased to 4.5, which is due to the more enclosed nature of the courtyard space, resulting in poor air circulation and poor space heat dissipation [51]. Differences in spatial type can have an impact on the effectiveness of GPR in improving daytime urban heat island intensity, which is due to the marked variations in ventilation in different spaces [52], with the north and south street spaces being better ventilated, resulting in an increase in GPR from 3.5 to 4.5 that still reduces the urban heat island intensity by up to 0.12 °C. However, for the remaining three types of spaces, the reduction in urban heat island intensity is less than 0.05 °C.

The GPR increased from 0 to 4.5, and the nighttime urban heat island intensity increased by 0.40 °C, 0.26 °C, 0.25 °C, and 0.26 °C for each of the four types of spaces in the residential area, and a relevant study has confirmed it [53]. This finding means that plants have an "insulating" effect on residential outdoor spaces at night. GPR = 0 represents that there is no planting in the space and the heat received in the space during the daytime will be dissipated at night in the form of longwave radiation [54]. In contrast, with GPR > 0, higher air temperature was observed. Because higher GPR means increased limiting effects of plants on longwave radiation, warm air is trapped under the canopy, with higher air temperature [55].

4.2. Differences in the Effects of GPR on the Thermal Comfort of Four Types of Spaces in Residential Areas

The GPR increased from 0 to 4.5, and the daytime outdoor thermal comfort for the four types of spaces decreased by 7.14 °C (east–west street space), 4.94 °C (north–south street space), 6.56 °C (courtyard space), and 3.38 °C (building overhead space), respectively. Because building overhead space is exposed to less direct sunlight throughout the day, the impact of PET changes due to increased GPR is less apparent [56]. The increase in GPR effectively enhances the cooling and humidifying effect of the plants while increasing the reflection and absorption of solar radiation by the plants, effectively reducing the mean radiant temperature [57], which leads to an obvious decrease in PET. The GPR is less effective in improving thermal comfort in north–south street space than in east–west street space and courtyard space. With limited plant cover and plant greenery, north–south street space is exposed to direct solar radiation for more extended periods throughout the daytime [58]. In addition, a sustained increase in GPR will not constantly improve outdoor thermal comfort. With the increase in the overall green volume of the residential space, the humidification effect of plants is obvious, leading to a further increase in outdoor relative humidity. The conclusion that an increase in relative humidity at high temperatures reduces thermal comfort outdoors has been inferred [59]. The increase in the green volume of plants also impedes the wind speed in the residential area, which is also not conducive to improving thermal comfort [60].

GPR also has a negative effect on nighttime outdoor thermal comfort improvements. The GPR increased from 0 to 4.5, and the thermal comfort index PET for the four space categories increased by 3.02 °C (east–west street space), 2.42 °C (north–south street space), 2.65 °C (courtyard space), and 0.86 °C (building overhead space). While the plant canopy reflects and absorbs solar radiation during the daytime, the plant canopy obstructs long-wave radiation upward from the ground at night [54], resulting in the inability to dissipate heat from space on time and an increase in the mean radiant temperature, thus deteriorating outdoor thermal comfort. The following two points should be focused on for mitigating the impact of greenery on nighttime comfort. The first is to control the volume of greenery, which means keeping the GPR within a specific range. The second is to consider the impact of the greenery layout in residential areas on environmental ventilation because a well-ventilated environment can take away most of the potential heat.

4.3. Proposal of GPR Regulations for Hot-Summer and Cold-Winter Residential Areas Based on Thermal Environment Improvement

In the previous urban planning, two-dimensional indicators such as green space ratio and green coverage ratio were employed to constrain and control the green space in residential areas [61,62]. However, it is not sufficient to consider only from the perspective of two-dimensional indicators. Urban planners have begun to incorporate green plot ratio as one of the evaluation indicators of green space quality to assess the effectiveness of green space from the perspective of three-dimensional spatial green volume. Therefore, this paper studies the effect of the three-dimensional green volume of plants on urban heat island intensity and thermal comfort in urban residential areas. Based on the analyzed results, we provide a reasonable green plot ratio value interval for urban residential spaces according to local conditions, considering the ecological benefits of greenery plants and the economic benefits of environmental construction.

- From the perspective of urban heat island intensity mitigation in urban residential areas, the study recommends that the GPR be set at a value of not less than 3.5 because among the four types of spaces in residential areas, when the GPR is above 3.5, the urban heat island intensity in the east–west street space, the courtyard space, and the building overhead space is below 3.0 °C, with a "moderate" urban heat island intensity.
- From the perspective of thermal comfort improvement in urban residential areas, the study suggests that the GPR should be higher than 1.5 because then the thermal

comfort of the four types of spaces in the residential areas will be reduced from "extreme heat stress" to "strong heat stress". However, it should be noted that the GPR should not be higher than 3.5, because the improvement of outdoor thermal comfort after the GPR is higher than 3.5 is weak, and a very high GPR will deteriorate the outdoor thermal comfort at night to a certain extent. For the mitigation of urban heat island intensity and the optimization of thermal comfort in different types of spaces in residential areas, GPR can be included as a control indicator.

5. Conclusions

This paper investigated the effect of green plot ratio (GPR) on urban heat island and thermal comfort in residential areas. We selected Tong-tai residence as a typical residential area in Changsha based on the research. According to the calculation method of GPR, we constructed several sets of scenarios and conducted simulations to analyze the effect of GPR on the thermal environment of four types of spaces in residential areas. The conclusions of this study are mainly as follows:

- The greenery can obviously improve the outdoor thermal comfort in residential areas. The east–west street space had the most effective PET reduction, with a PET reduction of 7.14 °C, and the building overhead space had the least effective PET reduction, with a PET reduction of 3.38 °C.
- Increasing GPR will not always have a positive effect. When the GPR exceeds 3.5 in the residential area, the mitigation of urban heat island intensity and the improvement of thermal comfort are not obvious. At the same time, with the increase in GPR, outdoor thermal comfort at night deteriorates.
- Combining the analysis of daytime and nighttime, we suggest that the GPR value should be controlled in a specific range, which can facilitate the cooling effect of greenery during the daytime while improving the thermal comfort at night.
- We need to be fully aware of the nocturnal warming effect of plants near the ground. In our research, we have been concerned that a very high GPR leads to some increase in outdoor air temperature at night since energy stored in the ground is released to the outside world in the form of longwave radiation, but the plants in the high GPR scenarios impede this process. Based on this study, we concluded that it is not better to grow more plants. More attention should be paid to the study of the negative impact of greenery on nighttime thermal comfort in subsequent studies, rational plant arrangement, and good ventilation environment creation, all of which should be carefully explored.

In the practical aspect, this paper proposes the recommended values of GPR for urban heat island intensity mitigation and thermal comfort improvement in residential areas. From a theoretical aspect, it complements the existing studies on green space rate and green coverage and further explores the influence of GPR on the thermal environment of urban space. Admittedly, this paper has some limitations for subsequent research. First, due to the diversity of plant species in the living space, to simplify the model, the study temporarily considered five numerically dominant species of trees and one each of grasses and shrubs in the plant modeling. Second, this study is currently only exploring typical summer weather days, and how GPR affects outdoor comfort in residential spaces in winter can be further explored in subsequent studies. Finally, the GPR planning and control indexes proposed in this paper are for the hot-summer and cold-winter climate zone, and whether their values are suitable for other regions needs to be studied further.

The quality and effectiveness of urban green space should be evaluated more in the context of the change from "incremental development" to "quality enhancement and optimization" of urban development and construction. The study analyzes the impact of GPR on urban heat island intensity and thermal comfort in urban residential areas from the perspective of the amount of plant greenery in residential areas, and based on this, the study proposes the planning control values of GPR, which makes up for the current lack of thermal environment planning indexes in urban residential areas, and provides feasible index suggestions for urban planning practice.

Author Contributions: Conceptualization, B.Z. and J.Z.; methodology, B.Z.; software, Z.L. validation, J.Z., B.Z. and Z.L.; formal analysis, Z.L.; investigation, J.Z.; resources, J.Z. and B.Z.; data curation, J.Z.; writing—original draft preparation, Z.L.; writing—review and editing, J.Z.; visualization, Z.L.; supervision, B.Z.; project administration, B.Z.; funding acquisition, B.Z. and J.Z. All authors have read and agreed to the published version of the manuscript.

Funding: This research was funded by the Hunan Provincial Natural Science Foundation, grant numbers 2023JJ30693; the Changsha University of Science and Technology New Teachers Research Funding, grant number 2023008809; and the Hunan Provincial Innovation Foundation for Postgraduate, grant number CX20230144.

Data Availability Statement: The data presented in this study are available on request from the corresponding authors.

Conflicts of Interest: The authors declare no conflicts of interest.

References

1. Ren, Z.; Fu, Y.; Dong, Y.; Zhang, P.; He, X. Rapid urbanization and climate change obviously contribute to worsening urban human thermal comfort: A national 183-city, 26-year study in China. *Urban Clim.* **2022**, *43*, 101154. [CrossRef]
2. Ye, C.; Schroder, P.; Yang, D.; Chen, M.; Cui, C.; Zhuang, L. Toward healthy and liveable cities: A new framework linking public health to urbanization. *Environ. Res. Lett.* **2022**, *17*, 064035. [CrossRef]
3. Wang, Z.; Zhu, P.; Zhou, Y.; Li, M.; Lu, J.; Huang, Y.; Deng, S. Evidence of relieved urban heat island intensity during rapid urbanization through local climate zones. *Urban Clim.* **2023**, *49*, 101537. [CrossRef]
4. Debbage, N.; Shepherd, J.M. The urban heat island effect and city contiguity. *Comput. Environ. Urban Syst.* **2015**, *54*, 181–194. [CrossRef]
5. Leng, H.; Liang, S.; Yuan, Q. Outdoor thermal comfort and adaptive behaviors in the residential public open spaces of winter cities during the marginal season. *Int. J. Biometeorol.* **2020**, *64*, 217–229. [CrossRef]
6. Rui, L.; Buccolieri, R.; Gao, Z.; Gatto, E.; Ding, W. Study of the effect of green quantity and structure on thermal comfort and air quality in an urban-like residential district by ENVI-met modelling. *Build. Simul.* **2019**, *12*, 183–194. [CrossRef]
7. Gidlof-Gunnarsson, A.; Ohrstrom, E. Noise and well-being in urban residential environments: The potential role of perceived availability to nearby green areas. *Landsc. Urban Plan.* **2007**, *83*, 115–126. [CrossRef]
8. Yang, Y.; Zhou, D.; Wang, Y.; Meng, X.; Gu, Z.; Xu, D.; Han, X. Planning method of centralized greening in high-rise residential blocks based on improvement of thermal comfort in summer. *Sustain. Cities Soc.* **2022**, *80*, 103802. [CrossRef]
9. Kubes, J.; Ourednicek, M. Functional types of suburban settlements around two differently sized Czech cities. *Cities* **2022**, *127*, 103742. [CrossRef]
10. Chen, X.; Zhao, P.; Hu, Y.; Ouyang, L.; Zhu, L.; Ni, G. Canopy transpiration and its cooling effect of three urban tree species in a subtropical city-Guangzhou, China. *Urban For. Urban Green.* **2019**, *43*, 126368. [CrossRef]
11. Richards, D.R.; Fung, T.K.; Belcher, R.N.; Edwards, P.J. Differential air temperature cooling performance of urban vegetation types in the tropics. *Urban For. Urban Green.* **2020**, *50*, 126651.
12. Chen, H.; Liu, R.; Zhang, Y. The Impact of Vegetation Canopy on the Outdoor Thermal Environment in Cold Winter and Spring. *Sustainability* **2023**, *15*, 12818. [CrossRef]
13. Zhang, L.; Zhan, Q.; Lan, Y. Effects of the tree distribution and species on outdoor environment conditions in a hot summer and cold winter zone: A case study in Wuhan residential quarters. *Build. Environ.* **2018**, *130*, 27–39. [CrossRef]
14. Yang, Y.; Zhou, D.; Wang, Y.; Ma, D.; Chen, W.; Xu, D.; Zhu, Z. Economical and outdoor thermal comfort analysis of greening in multistory residential areas in Xi'an. *Sustain. Cities Soc.* **2019**, *51*, 101730. [CrossRef]
15. Sabrin, S.; Karimi, M.; Nazari, R. The cooling potential of various vegetation covers in a heat-stressed underserved community in the deep south: Birmingham, Alabama. *Urban Clim.* **2023**, *51*, 101623. [CrossRef]
16. Lai, D.; Liu, Y.; Liao, M.; Yu, B. Effects of different tree layouts on outdoor thermal comfort of green space in summer Shanghai. *Urban Clim.* **2023**, *47*, 101398. [CrossRef]
17. Yang, J.; Zhao, Y.; Zou, Y.; Xia, D.; Lou, S.; Liu, W.; Ji, K. Effects of Tree Species and Layout on the Outdoor Thermal Environment of Squares in Hot-Humid Areas of China. *Buildings* **2022**, *12*, 1867. [CrossRef]
18. Liu, S.; Zhao, J.; Xu, M.; Ahmadian, E. Effects of landscape patterns on the summer microclimate and human comfort in urban squares in China. *Sustain. Cities Soc.* **2021**, *73*, 103099. [CrossRef]
19. Li, H.; Meng, H.; He, R.; Lei, Y.; Guo, Y.; Ernest, A.-A.; Jombach, S.; Tian, G. Analysis of Cooling and Humidification Effects of Different Coverage Types in Small Green Spaces (SGS) in the Context of Urban Homogenization: A Case of HAU Campus Green Spaces in Summer in Zhengzhou, China. *Atmosphere* **2020**, *11*, 862. [CrossRef]

20. Kotharkar, R.; Bagade, A.; Singh, P.R. A systematic approach for urban heat island mitigation strategies in critical local climate zones of an Indian city. *Urban Clim.* **2020**, *34*, 100701.
21. Sayad, B.; Alkama, D.; Ahmad, H.; Baili, J.; Aljahdaly, N.H.; Menni, Y. Nature-based solutions to improve the summer thermal comfort outdoors. *Case Stud. Therm. Eng.* **2021**, *28*, 101399. [CrossRef]
22. Meili, N.; Acero, J.A.; Peleg, N.; Manoli, G.; Burlando, P.; Fatichi, S. Vegetation cover and plant-trait effects on outdoor thermal comfort in a tropical city. *Build. Environ.* **2021**, *195*, 107733. [CrossRef]
23. Huang, Z.; Wu, C.; Teng, M.; Lin, Y. Impacts of Tree Canopy Cover on Microclimate and Human Thermal Comfort in a Shallow Street Canyon in Wuhan, China. *Atmosphere* **2020**, *11*, 588. [CrossRef]
24. Abdulkarim, K.H.; Abd Ghafar, A.; Lai, L.Y.; Said, I. Effects of Vegetation Covers for Outdoor Thermal Improvement: A Case Study at Abubakar Tafawa Balewa University, Bauchi, Nigeria. *Pertanika J. Sci. Technol.* **2021**, *29*, 2125–2147. [CrossRef]
25. Li, Y.; Lin, D.; Zhang, Y.; Song, Z.; Sha, X.; Zhou, S.; Chen, C.; Yu, Z. Quantifying tree canopy coverage threshold of typical residential quarters considering human thermal comfort and heat dynamics under extreme heat. *Build. Environ.* **2023**, *233*, 110100. [CrossRef]
26. Wu, Z.; Kong, F.; Wang, Y.; Sun, R.; Chen, L. The Impact of Greenspace on Thermal Comfort in a Residential Quarter of Beijing, China. *Int. J. Environ. Res. Public Health* **2016**, *13*, 1217. [CrossRef]
27. Chen, H.; Ooka, R.; Harayama, K.; Kato, S.; Li, X.F. Study on outdoor thermal environment of apartment block in Shenzhen, China with coupled simulation of convection, radiation and conduction. *Energy Build.* **2004**, *36*, 1247–1258. [CrossRef]
28. Hong, Z.; Xu, W.; Liu, Y.; Wang, L.; Ou, G.; Lu, N.; Dai, Q. Retrieval of Three-Dimensional Green Volume in Urban Green Space from Multi-Source Remote Sensing Data. *Remote Sens.* **2023**, *15*, 5364. [CrossRef]
29. Ong, B.L. Green plot ratio: An ecological measure for architecture and urban planning. *Landsc. Urban Plan.* **2003**, *63*, 197–211. [CrossRef]
30. *GB 50180-2018*; Urban Residential Planning and Design Standards Building Code. Ministry of Housing and Urban-Rural Development of the People's Republic of China: Beijing, China, 2018.
31. Luo, X.; Yu, C.W.; Zhou, D.; Gu, Z. Challenges and adaptation to urban climate change in China: A viewpoint of urban climate and urban planning. *Indoor Built Environ.* **2019**, *28*, 1157–1161. [CrossRef]
32. Zain, Z.M.; Taib, M.N.; Baki, S.M.S. Hot and humid climate: Prospect for thermal comfort in residential building. *Desalination* **2007**, *209*, 261–268. [CrossRef]
33. Hunan Provincial Department of Housing and Urban-Rural Development. *Design Standard for Green Building in Hunan Province*; China Construction Industry Publishing House: Beijing, China, 2018.
34. *DB32/T 4174-2021*; Standard for Greening of Urban Residential Area and Companies. Jiangsu Provincial Department of Housing and Urban-Rural Development, Jiangsu Provincial Department of Housing and Urban-Rural Development: Nanjing, China, 2021.
35. *DBJ51/168-2021*; Design Standard for Residential Buildings in Sichuan Province. Sichuan Provincial Department of Housing and Urban-Rural Development, Southwest Jiaotong University Publishing House: Chengdu, China, 2021.
36. *SJG 47-2018*; Assessment Standard for Green Building. Shenzhen Municipal Bureau of Housing and Construction, Shenzhen Municipal Bureau of Housing and Construction: Shenzhen, China, 2018.
37. *DGJ 08-2139-2018*; Green Design Standard for Residential Building. Shanghai Municipal Commission of Housing and Urban-Rural Development, Shanghai Municipal Commission of Housing and Urban-Rural Development: Shanghai, China, 2020.
38. *DBJ 13-197-2017*; Design Standard for Green Buildings in Fujian. Fujian Provincial Department of Housing and Urban-Rural Development, Fujian Provincial Department of Housing and Urban-Rural Development: Quanzhou, China, 2018.
39. Chen, D.; Chen, H.W. Using the Köppen classification to quantify climate variation and change: An example for 1901–2010. *Environ. Dev.* **2013**, *6*, 69–79. [CrossRef]
40. Liu, X.; Chen, X.; Shahrestani, M. Optimization of insulation thickness of external walls of residential buildings in hot summer and cold winter zone of China. *Sustainability* **2020**, *12*, 1574. [CrossRef]
41. Liu, X.; Wang, W.; Wang, Z.; Song, J.; Li, K. Simulation Study on Outdoor Wind Environment of Residential Complexes in Hot-Summer and Cold-Winter Climate Zones Based on Entropy-Based TOPSIS Method. *Sustainability* **2023**, *15*, 12480. [CrossRef]
42. Liu, J.; Tang, H.; Zheng, B. Simulation study of summer microclimate in street space of historic conservation areas in China: A case study in Changsha. *Front. Environ. Sci.* **2023**, *11*, 1146801. [CrossRef]
43. Qiu, X.; Kil, S.-H.; Jo, H.-K.; Park, C.; Song, W.; Choi, Y.E. Cooling Effect of Urban Blue and Green Spaces: A Case Study of Changsha, China. *Int. J. Environ. Res. Public Health* **2023**, *20*, 2613. [CrossRef]
44. Xiong, Y.; Zhang, F. Effect of human settlements on urban thermal environment and factor analysis based on multi-source data: A case study of Changsha city. *J. Geogr. Sci.* **2021**, *31*, 819–838. [CrossRef]
45. Chow, W.T.L.; Pope, R.L.; Martin, C.A.; Brazel, A.J. Observing and modeling the nocturnal park cool island of an arid city: Horizontal and vertical impacts. *Theor. Appl. Climatol.* **2011**, *103*, 197–211. [CrossRef]
46. Zhang, S.; Zhang, X.; Niu, D.; Fang, Z.; Chang, H.; Lin, Z. Physiological equivalent temperature-based and universal thermal climate index-based adaptive-rational outdoor thermal comfort models. *Build. Environ.* **2023**, *228*, 109900. [CrossRef]
47. Lin, T.-P.; Matzarakis, A.; Hwang, R.-L. Shading effect on long-term outdoor thermal comfort. *Build. Environ.* **2010**, *45*, 213–221. [CrossRef]
48. Jáuregui, E. Influence of a large urban park on temperature and convective precipitation in a tropical city. *Energy Build.* **1990**, *15*, 457–463. [CrossRef]

49. Cohen, P.; Potchter, O.; Matzarakis, A. Daily and seasonal climatic conditions of green urban open spaces in the Mediterranean climate and their impact on human comfort. *Build. Environ.* **2012**, *51*, 285–295. [CrossRef]
50. Fu, J.; Dupre, K.; Tavares, S.; King, D.; Banhalmi-Zakar, Z. Optimized greenery configuration to mitigate urban heat: A decade systematic review. *Front. Archit. Res.* **2022**, *11*, 466–491. [CrossRef]
51. Forouzandeh, A. Comparative analysis of sol-air temperature in typical open and semi-closed courtyard spaces. *Build. Simul.* **2022**, *15*, 957–973. [CrossRef]
52. Erlwein, S.; Zoelch, T.; Pauleit, S. Regulating the microclimate with urban green in densifiying cities: Joint assessment on two scales. *Build. Environ.* **2021**, *205*, 108233. [CrossRef]
53. Li, Y.; Fan, S.; Li, K.; Zhang, Y.; Kong, L.; Xie, Y.; Dong, L. Large urban parks summertime cool and wet island intensity and its influencing factors in Beijing, China. *Urban For. Urban Green.* **2021**, *65*, 127375. [CrossRef]
54. Chen, G.; Wang, D.; Wang, Q.; Li, Y.; Wang, X.; Hang, J.; Gao, P.; Ou, C.; Wang, K. Scaled outdoor experimental studies of urban thermal environment in street canyon models with various aspect ratios and thermal storage. *Sci. Total Environ.* **2020**, *726*, 138147. [CrossRef]
55. Wujeska-Klause, A.; Pfautsch, S. The Best Urban Trees for Daytime Cooling Leave Nights Slightly Warmer. *Forests* **2020**, *11*, 945. [CrossRef]
56. Weng, J.; Luo, B.; Xiang, H.; Gao, B. Effects of Bottom-Overhead Design Variables on Pedestrian-Level Thermal Comfort during Summertime in Different High-Rise Residential Buildings: A Case Study in Chongqing, China. *Buildings* **2022**, *12*, 265. [CrossRef]
57. Lachapelle, J.A.; Krayenhoff, E.S.; Middel, A.; Coseo, P.; Warland, J. Paper Maximizing the pedestrian radiative cooling benefit per street tree. *Landsc. Urban Plan.* **2023**, *230*, 104608. [CrossRef]
58. Zaki, S.A.; Toh, H.J.; Yakub, F.; Saudi, A.S.M.; Ardila-Rey, J.A.; Muhammad-Sukki, F. Effects of Roadside Trees and Road Orientation on Thermal Environment in a Tropical City. *Sustainability* **2020**, *12*, 1053. [CrossRef]
59. Du, C.; Li, B.; Liu, H.; Li, C. Effect of air humidity on human acceptable thermal environments and evaluation. *Chin. Sci. Bull.-Chin.* **2020**, *65*, 311–324. [CrossRef]
60. Li, J.; Zheng, B.; Ouyang, X.; Chen, X.; Bedra, K.B. Does shrub benefit the thermal comfort at pedestrian height in Singapore? *Sustain. Cities Soc.* **2021**, *75*, 103333. [CrossRef]
61. Zhu, R.; Liu, Y.; Yan, B.; Zhang, X.; Yuan, L.; Wang, Y.; Pan, Y. Effects of district parameters, green space and building density on thermal comfort- a case study of Badaguan District in Qingdao. *Case Stud. Therm. Eng.* **2023**, *42*, 102705. [CrossRef]
62. Luo, Y.; He, J. Evaluating the heat island effect in a planned residential area using planning indicators. *J. Build. Eng.* **2021**, *43*, 102473. [CrossRef]

forests

MDPI

Article

A Simulation Study of the Impact of Urban Street Greening on the Thermal Comfort in Street Canyons on Hot and Cold Days

Junyou Liu, Bohong Zheng and Fan Yang *

School of Architecture and Art, Central South University, Changsha 410075, China; 204801023@csu.edu.cn (J.L.); zhengbohong@csu.edu.cn (B.Z.)
* Correspondence: 207103@csu.edu.cn

Abstract: The urban heat island effect has become a widely concerning issue worldwide. Many researchers have made great efforts to improve the summer thermal comfort of urban street canyons by optimizing street greening. Relatively less research has focused on how to improve the thermal comfort of street canyons by optimizing street greening during cold days. Many researchers have proposed strategies to improve the summer thermal comfort of street canyons using road greening. This may have a significant negative impact on the winter thermal comfort of street canyons due to the lack of consideration of the impact on hot and cold days simultaneously, especially when the road green space is mainly composed of evergreen tree species. We aimed to explore the impacts of urban street greening on thermal comfort on hot and cold days at the same time. We used Zhutang West Road in Changsha, China, as an example and built six different models to explore the impacts of the street vegetation types, number of street trees, tree heights, crown widths, and Leaf Area Index on the thermal comfort of the street canyon. In addition, we also considered the impact of different building features and wind directions on the thermal comfort of the street canyon. We employed ENVI-met (version 5.5.1) to simulate different urban street greening models. The results show that the model with a high tree canopy density, tall trees, large and dense crowns, and sufficient building shade has good thermal comfort on hot days (the average physiological equivalent temperature (PET) is 31.1 °C for the study period) and bad thermal comfort on cold days (the PET is 13.3 °C) when it is compared with the other models (the average PETs are 36.2 °C, 31.5 °C, 41.5 °C, 36.2 °C, and 35.5 °C, respectively, on hot days and for other models). In addition, the model with a very large number of short hedges has a positive impact on thermal comfort during hot days (the PET is 31.1 °C). The PET value of another comparable model which does not have hedges is 31.5 °C. Even if the model with a small building area has good ventilation, the small building shade area in the model has a more obvious impact and the model has relatively good thermal comfort during cold days (the PET is 14.2 °C) when it is compared to models with bigger building areas (the PET is 13.9 °C). In summer, when the wind is parallel to the direction of the street canyon, the wind speed in the street canyon is high and the model has relatively good thermal comfort (the PET is 35.5 °C) compared with another model which has different wind direction and lower wind speed at the street canyon (the PET is 36.2 °C). In winter, when the wind is perpendicular to the direction of the street canyon, buildings and trees have a strong windproof effect and this is beneficial to the improvement of thermal comfort (the PET is 15.3 °C for this model and 13.9 °C for another comparable model). This research lays a solid foundation and encourages people to think about the impact of building and tree composition and configuration on the thermal comfort of street canyons during hot and cold days simultaneously.

Keywords: urban street greening; tree canopy; leaf area density; thermal comfort; simulation; physiological equivalent temperature; hot and cold days

1. Introduction

Urban heat island effects and global warming have obvious negative impacts on thermal comfort and public health [1–3]. The Climate Change 2023 Synthesis Report of the Intergovernmental Panel on Climate Change (IPCC) indicates that from 2011 to 2020, the global surface temperature was 1.1 °C above that from 1850 to 1900. This includes a 1.59 °C increase over land and a 0.88 °C increase over the ocean [4]. Some researchers have pointed out that approximately 5 million deaths worldwide are associate with non-optimal temperatures per year. This accounts for 9.43% of the total number of deaths [5]. Planting trees is considered one of the most effective ways to improve the thermal comfort of urban street canyons during hot summer days [6]. However, this may not be a good way to improve thermal comfort in winter [7]. For better urban planning, planners should have a better understanding of the impact of street trees and buildings on the thermal comfort in street canyons [3].

Some researchers have studied the impact of street trees and buildings on the thermal comfort in street canyons during hot summer conditions [3,6–12]. Street trees can significantly improve the thermal comfort of the street canyon during hot days [6]. Li et al. (2023) considered the impact of street trees on thermal comfort and air pollution in street canyons on hot days. They found that a higher Leaf Area Index (LAI) and planting density could increase the concentration of $PM_{2.5}$ but improve the thermal comfort in street canyons [8]. Cheng et al. (2022) explored the impact of tree canopy characteristics on human thermal comfort in street canyons. They found that for the same type of tree species, trees with higher LAI and crown width can be much more effective in improving thermal comfort. For the selection of tree species, LAI is an important factor to consider. Those trees with higher LAI tend to be more effective in improving thermal comfort [10]. Huang et al. (2020) found that tree canopies block solar radiation and influence wind speed. This can improve the summer thermal comfort [11]. Narimani et al. (2022) considered the impact of street orientation and tree species on thermal comfort on the streets. They pointed out that street orientation has less effect on improving the thermal comfort than tree cover [9]. Cui et al. (2023) found that if the street orientation and vegetation distribution are the same, when the aspect ratio (a ratio of building height to street width) [13] is 0.9, the thermal comfort of the street canyon is the best. Their research indicates that the same road greening features may have different effects on thermal comfort in different street canyon environments [13]. Cao et al. (2022) found that the impact of green space on the thermal comfort of streets was affected by the street orientation and aspect ratio [7].

The abovementioned aspects are more related to the component of street greening and buildings on thermal comfort. In addition, the spatial configuration of green space and buildings on thermal comfort has also been widely focused on [12–15]. For example, Ren et al. (2022) studied the impact of different tree coverage on thermal comfort and public health in street canyons. They found that the streets with a higher street tree coverage had a lower physiological equivalent temperature (PET) and better thermal comfort. In addition, they found people in areas with a higher street tree coverage had lower systolic blood pressure, diastolic blood pressure, and pulse rate during hot days [3]. Yang et al. (2018) found that if the street canyon is narrow, wind has a weak impact on thermal comfort. In this case, the shadows of buildings and trees can have a more significant impact on thermal comfort. Under the same lighting conditions, the shade generated by buildings and trees in east–west streets has a different effect on the thermal comfort of street canyons compared to the shade generated by buildings and trees in north–south streets. This is mainly because the city they studied, Xi'an, is located in the northern hemisphere, north of the Tropic of Cancer. Affected by the specific location of shadows generated by buildings and trees at different times throughout the day, the buildings and trees on the south side of the street can play a quite active role in improving the thermal comfort of the street canyon at many different times of the day, especially around noon. Thus, on the east–west streets, planting street trees near buildings had the best effect for the improvement of thermal comfort in street canyons. For north–south streets, the thermal comfort is better if the trees were planted in the middle of one side of the street. This is mainly because the buildings' and trees' shade improves the thermal comfort of the street canyon at many different times of the day [14]. Zheng et al. (2018) explored the thermal comfort of roads under different street configurations and tree distributions. They found that denser tree canopies were better for cooling and the interval between the trees was observed to have an optimal effect when it was equal to the canopy width [12]. Morakinyo and Lam (2016) explored the impact of a street's tree layout on the thermal comfort in street canyons. They found the thermal comfort of a layout of two rows of trees is better than that of a layout of street trees in the center; proper ventilation is important for the improvement of the thermal comfort of roads; and vegetation is more effective in improving thermal comfort under calm vertical wind than strong skew wind [15]. Yilmaz et al. (2018) found that vegetation, public space layouts, street orientation, and the height and shape of the buildings can have a significant impact on the thermal comfort of street canyons. These factors need to be well considered when improving the thermal comfort of street canyons [16].

In winter, vegetations and the buildings can still have a significant impact on the thermal comfort of street canyons [17–19]. Cohen et al. (2012) pointed out that plants can reduce the air temperature during winter [18]. Cao et al. (2022) pointed out that street trees on pedestrian streets have an improvement effect on the thermal comfort of street canyons in summer and a deterioration effect on the thermal comfort of street canyons in winter [7]. Yilmaz et al. (2018) believe that deciduous trees should be given priority for improving the thermal comfort of street canyons because they can effectively block sunlight and fully utilize transpiration effects in the summer. In winter, due to the shedding of their leaves, their canopy no longer blocks sunlight from shining onto the street canyons. This is also beneficial for the thermal comfort of street canyons in winter [16]. Fan and Dong (2021) studied the thermal comfort of narrow urban street canyons in Shanxi, China. They found a "single-arch" tree-lined street canyon space had a good effect on summer thermal comfort. In addition, they found the street space can achieve good thermal comfort in winter due to reasonable coordination between the street trees and the building layout [19]. This indicates that not only the summer thermal comfort can be improved by optimizing the spatial configuration of the street canyon but also the winter thermal comfort. It is impossible to implement two different renovation plans simultaneously in the same street canyon area. In this case, a renovation plan that is beneficial for improving the summer thermal comfort of street canyons taking into account measures such as building shadows and the transpiration effect of street trees while also improving the winter thermal comfort of the street canyons considering measures like the use of buildings' and street trees' wind-blocking effects is very important.

Compared with the widely discussed urban summer thermal comfort challenges, people have focused less on urban winter thermal comfort improvement [20–22]. In addition, if people propose urban planning strategies for the improvement of the thermal comfort of street canyons only based on a single season, their strategies may not be good from the whole year's perspective. We attempted to study the impacts of street tree configuration on the thermal comfort of street canyons during hot and cold days simultaneously. We tried to explore the impact of the urban street canyon layout on the thermal comfort. We will focus on the impact of street tree quantity, height, and canopy features (like leaf area density (LAD), crown width, crown height and canopy density) on the thermal comfort of the street canyon during hot and cold days. The LAD is defined as the total one-sided leaf area per unit volume [23–25]. In addition, we will also consider whether the impact of street trees on the thermal comfort of the street canyon may be different under different building features, wind speeds, and wind directions.

2. Material and Methods

2.1. Selection of the Study Time and Area

The scope of Changsha is 27°51′~28°40′ north latitude and 111°53′~114°15′ east longitude. It is located on the western margin of the Changliu Basin and far from the sea. Changsha is an important central city in the middle reaches of the Yangtze River in China and the capital of the Hunan province. By the end of 2021, the total permanent population of Changsha reached 10.2393 million [26]. According to the Köppen–Geiger climate classification updated by Beck et al. (2018), Changsha is located in Cfa of the climate classes, that is, temperate, hot summers without a dry season [27]. Its climate is a subtropical monsoon climate. It is hot in summer and cold in winter. In summer, it is often sunny continuously for several days. The sun shines fiercely and evaporation is strong. The average temperature is above 25 °C. The winter is short and cold, with an average temperature of 5–12 °C [28]. Evergreen is the main street tree type in Changsha [29]. We will use Changsha as a typical case to explore how to optimize the spatial form of street canyons in areas with hot summer and cold winter climate characteristics and evergreen trees as the main street trees. We aim to achieve the most favorable thermal comfort for hot and cold days. In this study, a section of Zhutang West Road within the scope of the Third Ring Road of Changsha was studied. The street trees in the selected road section are camphor trees. These are one of the most common street tree types in Changsha. The central urban area of Changsha is mainly composed of multi-story and high-rise buildings. The road section we selected has both high-rise and multi-story buildings on both sides, which are also very representative. Figure 1 shows the study area. We can see from the picture below that the sidewalk in this section is narrow, and the canopy of the street trees is large. The distance between the canopies of two adjacent street trees is relatively small and the street trees block the sun well.

Figure 1. Picture of the study area.

In China, the lunar definition of four seasons is widely used [30]. Based on the lunar defining of the four seasons, the beginning of spring, summer, autumn, and winter have been dated consistently on 3–4 February; 5–6 May; 7–8 August; and 7–8 November of each year, respectively, for the last 10 years [31]. Thus, 13 May 2023 is a hot summer day and 16 December 2022 is a cold winter day. We conducted meteorological data collection using a Kestrel 5500 (Nielsen-Kellerman, Upper Chichester, PA, USA) in the Third Ring Road of Changsha City over 2 days, and used the measured data to represent the meteorological data within the Third Ring Road of Changsha City and further applied it to the research.

According to the China Meteorological Administration, a comfortable air temperature for people in summer is between 19 and 24 °C [32,33]. The air temperature we measured on 13 May 2023 from 8:00 to 18:00 was mostly higher than the highest comfortable air temperature for the human body, 24 °C. The weather was sunny and the air temperature was hot on 13 May 2023. It is necessary to fully utilize the cooling effect of road green spaces to effectively improve the thermal comfort of street canyons. This day belongs to a representative hot summer day. According to the China Meteorological Administration, the comfortable air temperature range for winter is between 12 and 22 °C [32,33]. The air temperature we measured from 10:00 to 19:00 on 16 December 2022 was below the minimum comfortable air temperature, 12 °C. The morning was relatively sunny, and the noon and afternoon were cloudy and windy. This day belongs to a representative cold winter day. In this case, measures to increase the air temperature of the street canyon can enhance the comfort of the street canyon.

2.2. Data Collection and Selection of a Thermal Comfort Index

We selected a Kestrel 5500 to collect data on the air temperature, relative humidity, wind direction, and speed in the study area. The Kestrel 5500 is a weather meter designed and produced by Nielsen-Kellerman in Boothwyn, PA, USA [34]. The accuracy of the Kestrel 5500 for the measurement of meteorological parameters is shown in Table 1 [35].

Table 1. Measurement accuracy of Kestrel 5500.

Meteorological Parameters	Air Temperature	Relative Humidity	Wind Speed	Wind Direction
Accuracy	0.5 °C	2% RH	Larger than 3% of reading, least significant digit, or 20 ft/min	5°

We collected the air temperature, relative humidity, wind direction, and wind speed data from fixed observation points located within the Third Ring Road of Changsha City from 8:00 to 18:00 on 13 May 2023, and from 10:00 to 19:00 on 16 December 2022. We used them as the meteorological data within the Third Ring Road of Changsha City during the research period. The instrument records data every minute. For the air temperature and relative humidity, we average the data measured within each hour and used it as the data for that hour. We used the average wind direction and speed throughout 8:00 to 18:00 on 13 May 2023, and 10:00 to 19:00 on 16 December 2022, respectively, as the input wind directions and speeds for the two different days. We used ENVI-met version 5.5.1 to conduct the simulation study since the full forcing mode in ENVI-met can report errors when the hourly wind speed and direction change considerably. This study chose the simple forcing mode for simulation due to the fact that the hourly change in wind direction and speed were relatively large in the measured data. The wind direction and speed throughout the entire research period were uniformly set as 225° and 0.42 ms^{-1} on 13 May 2023 and as 47.9° and 3 ms^{-1} on 16 December 2022. The air temperature and relative humidity data on 13 May 2023 are shown in Table 2. The air temperature and relative humidity on 16 December 2022 are shown in Table 3.

Table 2. Summary of measured meteorological data on 13 May 2023.

Time	Air Temperature (°C)	Relative Humidity (%)
08:00:00	20.8	76.7
09:00:00	22.4	67.5
10:00:00	23.9	61.6
11:00:00	25.7	56.8
12:00:00	26.1	54.0
13:00:00	27.1	52.0
14:00:00	27.2	49.4
15:00:00	26.9	50.3
16:00:00	27.0	52.0
17:00:00	26.6	53.4
18:00:00	25.8	58.3

We judged the canopy diameter of trees using a Deli-EDL330060 mechanical roller diastimeter (Deli, Ningbo, China). The error between the measured and actual values of the roller diastimeter is less than 0.5% of the actual value [36]. We determined two points on the ground that could represent the positions of the two endpoints of the canopy and measured the diameter of the tree. We judged the crown width of trees on this basis. We randomly chose 12 trees in the study area. The canopies diameters of the 12 trees were measured and averaged as 9 m. The canopies diameter of the trees were uniformly set as 9 m in the simulation. We measured the heights of the 12 street trees using the direct reading height finder CGQ-1 from Huaguang Forests, Harbin, China. We judged that the height of most street trees in the study area were around 15 m. We uniformly set the street trees as 15 m in the simulation study. We analyzed the LAI of the 12 street trees using the HemiView plant canopy analyzer from Delta-T in Cambridge, the UK. The LAI is defined as the one-sided green leaf area per unit ground surface area [25,37]. We found that the average LAI of most street trees ranged from 1.8 to 3.6 m^2m^{-2}. The street tree canopies are sparse. Referring to the relevant sparse tree canopy information in ENVI-met and the canopy characteristics of the street canyon, the LAD of the street canyon was determined as 0.225 m^2m^{-3}.

Table 3. Summary of Measured Meteorological Data on 16 December 2022.

Time	Air Temperature (°C)	Relative Humidity (%)
10:00	8.7	69.9
11:00	8.2	68.5
12:00	7.7	70.2
13:00	7.8	65.3
14:00	7.7	62.4
15:00	7.6	60.5
16:00	7.5	57.0
17:00	7.5	55.7
18:00	7.7	54.9
19:00	8.0	52.5

PET was uniformly used to reflect thermal comfort in this study. It is defined as the corresponding air temperature when the skin temperature and internal temperature of the human body reach the same thermal state as the typical indoor environment in an indoor or outdoor environment [38]. We simulated the relevant microclimate data (mean radiant temperature, air temperature, relative humidity, wind direction and speed, etc.) in the study area using ENVI-met. We calculated the PET values using the BIO-met module in ENVI-met [39]. Elraouf et al. (2022) have verified the PET simulation results of ENVI-met are valid in their study [40]. This laid a solid foundation for future researchers to use ENVI-met to simulate PET values. The PET values for 13 May 2023 were calculated based on the default male summer clothing (male, 35 years old, height: 1.75 m, weight: 75 kg, outdoor: 0.50 clo, pre. speed: 1.34 ms^{-1}, total metabolic rate: 86.21 W/m^2). The PET value calculation for 16 December 2022 is based on the default male average clothing (male, 35 years old, height: 1.75 m, weight: 75 kg, outdoor: 0.90 clo, pre. speed: 1.34 ms^{-1}, total metabolic rate: 86.21 W/m^2). The heat balance equation below is its main basis [38]:

$$M + W + R + C + E_D + E_{Re} + E_{Sw} + S = 0 \quad (1)$$

M is the metabolic rate; W represents the physical work output; R is the net radiation of body; C is the convective heat flow; E_D is the latent heat flow to evaporate water into water vapor diffusing through skin; and E_{Re} is the sum of heat flows for heating and humidifying the inspired air. E_{Sw} is the heat generated by sweat evaporation. S represents stored heat.

By combining the following Formulas (2) and (3) with Formula (1), the core temperature of the human body, skin surface temperature, and clothing surface temperature can be calculated. These three indicators are the main indicators that determine the thermal sensation of the human body. After ascertaining these three indicators, people can know the human thermal sensation further [41,42].

$$F_{CS} = v_b \times \rho_b \times c_b \times (T_c - T_{sk}) \quad (2)$$

F_{CS} is the heat flow from the core of the body to the skin, v_b is the blood flow from the core of the body to the skin, ρ_b represents the density of the blood, c_b represents the specific heat capacity of the blood, T_c represents the core body temperature, and T_{sk} represents the skin surface temperature.

$$F_{SC} = (1/I_{cl}) \times (T_{sk} - T_{cl}) \quad (3)$$

F_{SC} is the heat flow from the skin surface to the clothing surface, I_{cl} represents the clothing thermal resistance, T_{sk} represents the skin surface temperature, and T_{cl} represents the clothing surface temperature.

2.3. Model Building and Simulation Analysis

From the perspective of a single tree, its height, crown width, and LAD have been highly valued by previous researchers when exploring the impact of trees on environmental thermal comfort [5,7,10]. From the perspective of the thermal comfort of the whole street canyon area, vegetation types, tree quantity, canopy density, ventilation, open space, building shadow, and wind direction and speed are some influencing factors on thermal comfort that have been highly valued by previous researchers [7–10]. In the past, researchers usually focused on the impact of these features on environmental thermal comfort only in hot or cold weather [3–10]. We explored the different effects of these factors on thermal comfort on hot and cold days at the same time by constructing different scenario models and conducting simulation analysis.

We used a Mavic Air 2 unmanned aerial vehicle designed and produced by DJI in Shenzhen, China. to take aerial pictures of the research area. We spliced the pictures using DJI Terra to form an orthograph. This can correct the geometric deformation of the pictures caused by sensors, external orientation elements, and terrain fluctuations. The shape of the orthograph is consistent with the actual situation. The scale of the different areas of the orthophoto map is consistent. The relevant orientations of each part in the orthophoto remain unchanged [43,44]. We first traced the contours of buildings and roads in AutoCAD 2008 and determined the positions of all kinds of plants in the picture. We imported the plan into SketchUp 2020 and built a three-dimensional model of the study area. Model 1 was built according to the actual conditions of the research area. The canopy of trees plays an important role in improving thermal comfort [45–47]. In order to better explore the different effects of tree canopy characteristics on thermal comfort, LAD, as an indicator of the leaf area per unit area, is explored to consider the different situations of tree canopy [48]. This study attempts to explore the differences in comfort level of street canyons under different LAD conditions. Street tree species is uniformly set as camphor tree for the model 1. This is one of the most common street trees in Changsha.

Xue and Li (2017) pointed out that the LADs of broadleaf trees usually range from 0.2 to 1.1 m^2m^{-3} [49]. The studies of Turhan (2023) [50], Katul et al. (2004) [51], Chen et al. (2022) [45], Zhou et al. (2020) [52], Zhao et al. (2014) [53], Wu et al. (2016) [54], and Li (2012) [55] show the LADs of different trees are within this range. Among these studies, Zhao et al. (2014) [53]; Wang et al. (2018) [54]; and Li (2012) [55] explored the canopy features of *Cinnamomum camphora* and *Platanus hispanica*. For some studies which only provide LAI values, we refer to Lalic et al. (2004) [25] to determine the LAD based on the LAI. Based on the measured data and reference to other relevant studies [56–58], when the LAD of the street tree is 0 225 m^2m^{-3}, the canopy of the street tree is relatively sparse. Referring to the common range of broadleaf tree LAD values pointed out by Xue and Li (2017) [49], and combined with our actual measurement of the street tree LAI in different areas of Changsha, we ultimately determined that the LAD of a dense street tree model is uniformly set as 0.825 m^2m^{-3}. We built Model 2, which had the same feature as Model 1 except the LAD of street trees on both sides of the road was 0.825 m^2m^{-3}.

Due to the tight arrangement of street trees on the road section we are studying, it is difficult to further add street trees 15 m high and with a 9 m crown width on this basis. We selected street trees with a height of 5 m and a crown width of 3 m to construct Model 3. The trees are small. We arranged 5 m street trees whose number was near twice the number of those in Model 1 and built Model 3. These are the only differences between Models 1 and 3.

Different plant species can also have different impacts on the thermal comfort of the surrounding environment [59]. We constructed Model 4 with dense canopy trees and hedges. The hedge species was *Ligustrum lucidum*. This is a common hedge species in Changsha. Models 2 and 4 are almost the same except that Model 4 has hedges and Model 2 does not.

We considered rebuilding some low-rise buildings within Model 1 for the purpose of decreasing the floor area of the buildings. The building height remained unchanged. This can add air ducts. We named the amended model Model 5. We could explore the impact of the improvement of ventilation on thermal comfort in street canyons using Models 1 and 5.

We set different wind directions from Model 1 and constructed Model 6 based on it. We set the wind direction of Model 6 as 90° on 13 May 2023. This wind direction was parallel to the street direction. We set the wind direction of Model 6 as 180° on 16 December 2022. This wind direction was perpendicular to the direction of the street. Model 6 had different wind directions on summer and winter days. This was beneficial for the comparison of the thermal comfort of the street canyon under different wind directions.

The canopy width of all 15 m high street trees was set to 9 m, and the trunk height was set to 4 m. All 5 m high street trees had a uniform canopy width of 3 m and a trunk height of 1 m. When simulating based on meteorological conditions on 13 May 2023, the main differences among the six models are shown in Figure 2 and Table 4. When simulating based on meteorological conditions on 16 December 2022, the main differences among the six models are shown in Figure 3 and Table 5.

Table 4. A table which compares the differences of the six models on 13 May 2023.

	Model 1	**Model 2**	**Model 3**	**Model 4**	**Model 5**	**Model 6**
Street tree height (m)	15	15	5	15	15	15
Crown width (m)	9	9	3	9	9	9
Street tree numbers in the study area	36	36	71	36	36	36
Leaf area density (m^2m^{-3})	0.225	0.825	0.225	0.825	0.225	0.225
Hedge area (m^2)	0	0	0	40	0	0
Building area (m^2)	4060	4060	4060	4060	3672	4060
Wind direction	225.0°	225.0°	225.0°	225.0°	225.0°	90.0°

Table 5. A table which compares the differences between the six models on 16 December 2022.

	Model 1	**Model 2**	**Model 3**	**Model 4**	**Model 5**	**Model 6**
Street tree height (m)	15	15	5	15	15	15
Crown width (m)	9	9	3	9	9	9
Street tree numbers in the study area	36	36	71	36	36	36
Leaf area density (m^2m^{-3})	0.225	0.825	0.225	0.825	0.225	0.225
Hedge area (m^2)	0	0	0	40	0	0
Building area (m^2)	4060	4060	4060	4060	3672	4060
Wind direction	47.9°	47.9°	47.9°	47.9°	47.9°	180°

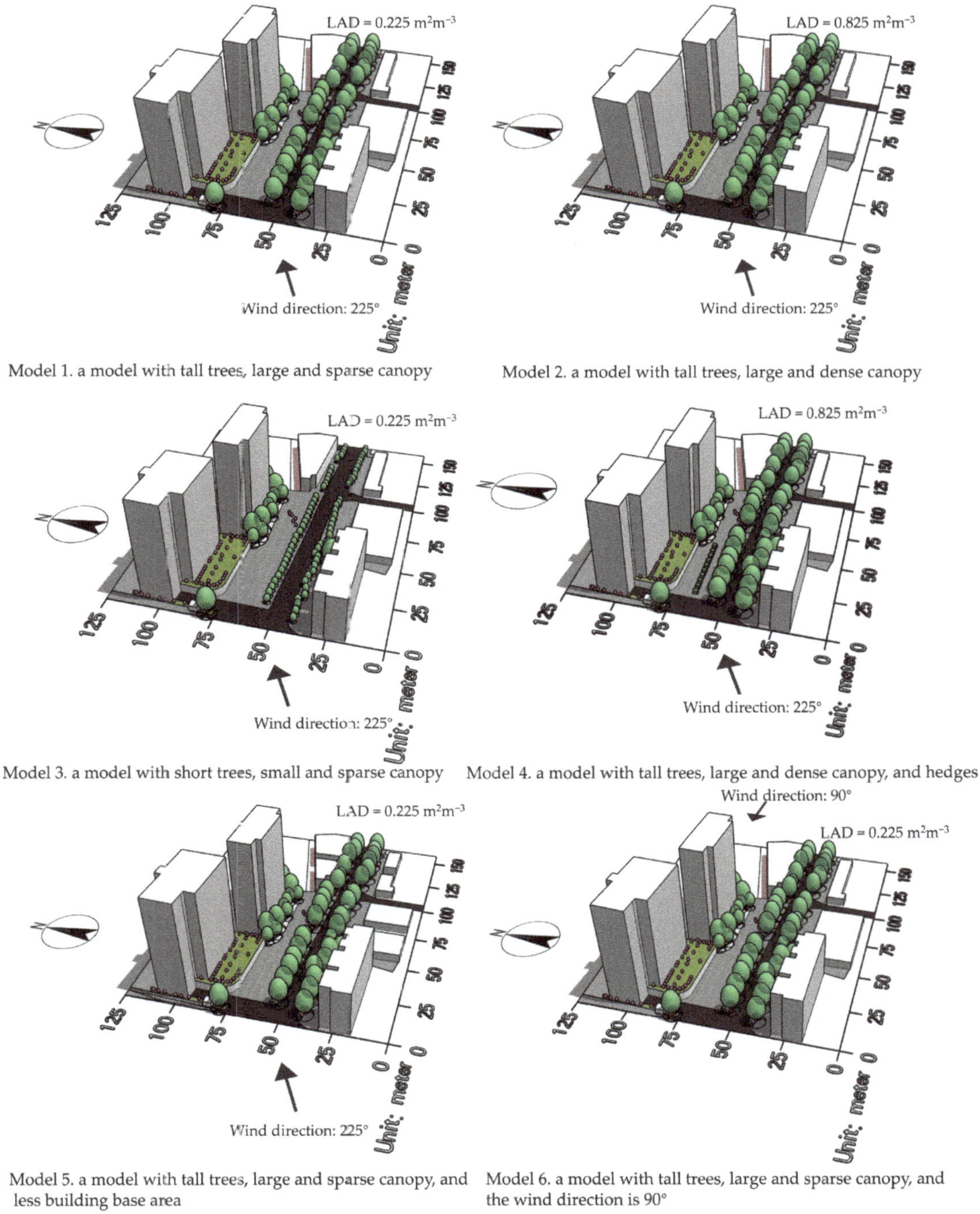

Figure 2. A schematic diagram to compare the differences between Models 1–6 on 13 May 2023.

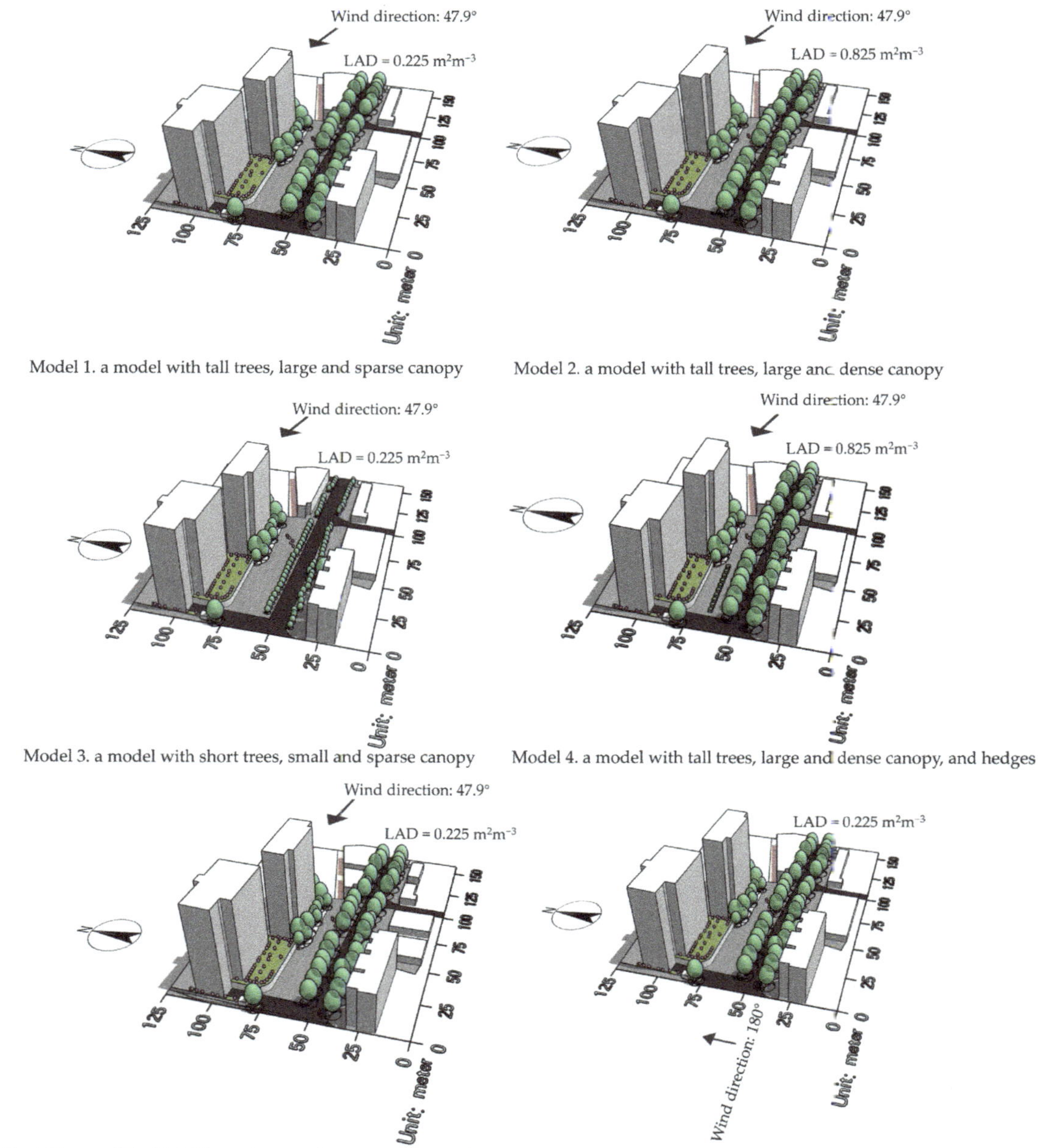

Model 1. a model with tall trees, large and sparse canopy

Model 2. a model with tall trees, large and dense canopy

Model 3. a model with short trees, small and sparse canopy

Model 4. a model with tall trees, large and dense canopy, and hedges

Model 5. a model with tall trees, large and sparse canopy, and less building base area

Model 6. a model with tall trees, large and sparse canopy, and the wind direction is 180°

Figure 3. A schematic diagram to compare the differences between Models 1–6 on 16 December 2022.

After the three-dimensional models of the study area were made in SketchUp, we used the INX plugin in SketchUp to determine the geographical coordinates of the study area, materials of the ground and buildings, and information on the plants in the area. We imported the models made in SketchUp into ENVI-met using the INX plugin. We entered the measured air temperature and relative humidity into ENVI-met. We used ENVI-met to simulate the microclimate conditions and BIO-met to simulate the PET.

2.4. Error Analysis

We analyzed the measured and simulated air temperature and relative humidity to explore the accuracy of the simulation. On 13 May 2023, the measurement point selected by us was a fixed point. The corresponding coordinate of the measurement point is (42, 43) in Model 1. After completing the simulation, we extracted the simulated air temperature and relative humidity of the observation point (42, 43) over 10 h. The height selected for reading the value of the simulation was 1.4 m. This was roughly consistent with the height of the Kestrel 5500 in the process of the actual measurement. The indicators selected by us were the coefficient of determination (R^2), root mean square error (RMSE), and index of agreement (d) [60–63]:

$$R^2 = 1 - \left[\sum_{i=1}^{N}(Pi - Oi)^2 / \sum_{i=1}^{N}(\overline{O} - Oi)^2\right] \quad (4)$$

$$\text{RMSE} = \sqrt{\frac{1}{N}\sum_{i=1}^{N}(Pi - Oi)^2} \quad (5)$$

$$d = 1 - \left[\sum_{i=1}^{N}(Pi - Oi)^2 / \sum_{i=1}^{N}\left(|Pi - \overline{O}| + |Oi - \overline{O}|\right)^2\right] \quad (6)$$

in which Pi is the i-th simulated value, Oi is the i-th measured value, $\overline{O}$ is the average value of the measured values, and N is the number of measurements.

These three indicators have been widely used to analyze the accuracy of the simulation results in urban microclimate research [61,64–66]. The measured and simulated air temperature and relative humidity on 13 May 2023 are shown in Table 6. We made scatterplots to analyze the errors between the simulated values and measured values in Excel. The R^2 between the measured and the simulated air temperature was 0.97, the RMSE was 0.55, and the d was 0.98. The R^2 between the measured and simulated humidity was 0.96, the RMSE was 1.62, and the d was 0.99.

Table 6. Measured and simulated air temperature values on 13 May 2023.

Time	Measured Air Temperature (°C)	Simulated Air Temperature (°C)	Measured Relative Humidity (%)	Simulated Relative Humidity (%)
08:00	20.8	20.7	76.7	75.5
09:00	22.4	23.0	67.5	67.8
10:00	23.9	23.9	61.6	62.6
11:00	25.7	24.7	56.8	59.2
12:00	26.1	25.5	54.0	56.1
13:00	27.1	26.2	52.0	53.5
14:00	27.2	26.6	49.4	51.6
15:00	26.9	26.7	50.3	50.9
16:00	27.0	26.6	52.0	51.6
17:00	26.6	26.2	53.4	52.9
18:00	25.8	25.5	58.3	55.4

Except for using measured and simulated data at a height of 1.4 m for the error analysis, we chose a unified 1 m height for thermal comfort analysis in the study area. This height is relatively close to adults' center of gravity height [67,68]. Creating a comfortable wind and heat environment involves attributing great importance to the thermal comfort of various parts of the human body at corresponding heights. This 1 m height for research can consider the feelings of various parts of the human body better [69,70].

3. Results and Discussions

3.1. Results

We further explored the impact of street greening on the thermal comfort of street canyons on hot and cold days based on existing research [3,6–12]. Some previous studies have focused on how one factor can maximize the improvement of thermal comfort in street canyons. For example, Huang et al. (2020) conducted research on the impact of different tree canopy covers on the thermal comfort of street canyons [11]. Our study considers many different ways to improve the thermal comfort of the street canyon simultaneously. These include using densely covered street trees to replace sparsely covered street trees, further increasing the number of street trees by selecting street trees with smaller tree heights and crown widths in limited street canyon spaces, and increasing the green space rate of the street canyon by adding low plants such as hedges or shrubs in narrow street canyons, to improve the ventilation effect of the street canyon. We tried to compare which method was more likely to significantly improve the thermal comfort of the street canyon. In addition, we also considered whether street greening would have different impact under different wind directions. Compared to previous studies that only focused on the impact of street greening on the thermal comfort of street canyons during hot days or cold days [12,20,71], this study considers the impacts of street greening on street canyons during both hot and cold days. This can help to improve the thermal comfort of street canyons more comprehensively. Compared to previous studies that only focused on thermal comfort indices such as the PET, Predicted Mean Vote (PMV), and Universal Thermal Climate Index (UTCI), or only focused on air temperature and relative humidity [7,72,73], this study considers the PET, mean radiant temperature, air temperature, relative humidity, and wind speed in the street canyon simultaneously. This helps to better explore the different impacts of different road greening characteristics on street canyons.

Deng et al. (2022) pointed out that in Changsha, when the PET value is between 24.7–27.8 °C, people will neither feel cold nor hot and the metabolic level of the body will be at its lowest level [70]. The average PET values of each model in each period are shown in Table 7 below. Figure 4 shows the average value of the PET in each model in the street canyon area.

Table 7. The average PET values (°C) of each model in the street canyon area at different times on 13 May 2023.

Time	Model 1	Model 2	Model 3	Model 4	Model 5	Model 6
09:00	33.5	29.2	39.5	28.9	33.6	33.0
10:00	36.1	30.4	42.0	30.1	36.2	35.5
11:00	38.7	32.9	43.5	32.5	38.7	37.9
12:00	40.1	34.2	45.1	33.9	40.1	39.3
13:00	41.3	34.8	46.9	34.4	41.3	40.5
14:00	41.3	34.8	47.5	34.4	41.4	40.6
15:00	39.8	34.9	46.5	34.5	39.9	39.1
16:00	35.4	31.8	42.8	31.3	35.5	34.7
17:00	30.2	28.0	34.7	27.4	30.3	29.6
18:00	25.3	23.9	26.2	23.7	25.3	24.8

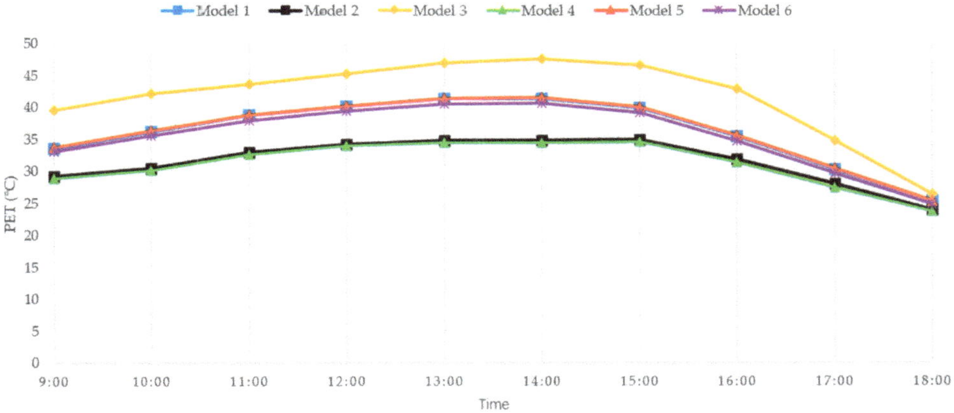

Figure 4. The average PET values in the street canyon area of each model at different times on 13 May 2023.

The tree canopy is sparse (0.225 m^2m^{-3}) in Model 1, while the tree canopy is dense (0.825 m^2m^{-3}) in Model 2. This is the only difference between the two models. The average PET value in the street canyon in Model 2 is 4.6 °C lower than that in Model 1 during the 10 h. This can be attributed to a denser canopy being beneficial for fully utilizing the effect of transpiration, ventilation, and shade. This can improve the thermal comfort of street canyons in hot summer conditions [71]. This is consistent with the findings of Huang et al. (2020) on tree canopy research [11]. A dense tree canopy is beneficial for blocking sunlight [72]. The mean radiant temperature and air temperature of the tree shading area can be very low. Street trees can convert sensible heat into latent heat and release water vapor in to the air via transpiration and reduce air temperature [73]. Dense canopy trees can have a stronger transpiration effect than sparse canopy trees [74]. Dense canopy trees have better effects on increasing the relative humidity than sparse canopy trees. This is the main reason why the average relative humidity of the street canyon area of Model 2 was higher than that of Model 1. Street trees can achieve the effect of blocking wind through their canopy. Dense canopies have a stronger effect on blocking wind. When the wind speed is low, heat dissipation by the wind is also weak [75]. Although the wind speed in Model 1 is higher than Model 2, the mean radiation temperature and air temperature in Model 2 are both lower than in Model 1. In addition, the relative humidity of Model 2 is higher than Model 1. The mean radiation temperature and air temperature are believed to be the two most influential factors on human thermal sensation [72,75]. This can explain why Model 2 has a better thermal comfort than Model 1 even if the average wind speed of Model 1 is higher. Thus, improving the LAD characteristics of trees should be given priority in increasing summer thermal comfort.

The number of street trees in Model 3 is around twice that of Model 1. However, the trees in Model 3 have lower heights and crown heights than in Model 1. The canopy of the street tree in Model 3 is significantly smaller than that in Model 1. The canopy density of Model 3 is smaller than that in Model 1. In these 10 h, the average hourly PET value in Model 3 is 5.3 °C higher than that in Model 1. The PET values have the greatest difference at 16 o'clock, which is 7.4 °C. PET values have the smallest difference at 18 o'clock, which is 0.9 °C. The shadow area formed by trees is relatively small in Model 3. This results in a larger area of direct solar radiation. In addition, the transpiration effect is also weaker in Model 3 due to a short crown height and smaller canopy density. Trees with a lower canopy height have a better effect in blocking wind within the height range of 0–2 m [76]. There is an unfavorable heat dissipation environment in Model 3. Therefore, the average mean radiation temperature and air temperature in the street canyon area of Model 3 are higher than that of Model 1. The average relative humidity and wind speed in Model 3 are lower than those in Model 1. These contribute to the poor human thermal comfort in this model. Equally, 15 m street trees improve the thermal comfort in street canyons more significantly than 5 m street trees. This indicates that from the perspective of the canopy structure, tall plants with large canopy volumes have a much stronger effect on improving summer thermal comfort in street canyons than short plants with small canopy volumes. This is consistent with the noticeable negative correlation between crown width and air temperature pointed out by Huang et al. (2022) [10]. This also inspires urban designers to reasonably lay out tall street trees in street canyons based on the transpiration and shade characteristics of tall street trees.

We added a certain number of hedges in an open area on the basis of Model 2 and built Model 4. We can see from Table 7 and Figure 4 that Model 4 has the lowest PET value and the best thermal comfort. In Model 4, hedges block sunlight in the open space. This reduces the area affected by direct solar radiation. This can reduce the mean radiant temperature and air temperature in the open space area. Hedges can also increase the relative humidity due to transpiration. Low hedges can effectively block wind [76]. The average wind speed of Model 4 is smaller than that in Model 2. However, the hedges have weak effects on the wind due to their size and the low wind speed. Model 4 increased the relative humidity more efficiently than Model 1. This indicates that we can improve the thermal comfort in street canyons by adding hedges in the open area, but the effect is not obvious. The PET value in the street canyon area of Model 4 is only less than 1 °C lower than that in Model 2 each hour. The average PET value in the 10 hours is 0.38 °C lower. In a street canyon where the street trees are closely arranged, it is difficult for people to further increase the number of tall street trees. Setting low hedges in open areas can achieve a certain improvement in summer thermal comfort, which is an option to consider. However, its effect on improving summer thermal comfort is weak and will occupy a certain amount of open space. Whether to adopt this method to improve the summer thermal comfort of street canyons needs to be carefully considered based on the actual situation.

We considered demolishing some short buildings based on Model 1. We formed Model 5 with the same building heights and a smaller floor area than Model 1. The hourly average PET in the street canyon area in Model 5 slightly increased on the basis of Model 1. However, the difference between the corresponding values in the two models in each period was within 1 °C. The thermal comfort in the street canyon is deteriorated by reconstructing some large buildings into several small buildings even if this improves the ventilation. This is different from the findings found by some researchers that improving ventilation can achieve the effect of improving thermal comfort in hot summer conditions [77–79]. This does not mean that increasing ventilation will have a negative impact on the thermal comfort. On the one hand, due to the low average wind speed during the research period, the effect achieved by increased ventilation is not noticeable. On the other hand, demolishing some buildings can also lead to a reduction in the shadow area of the buildings. Some researchers have pointed out that thermal comfort can be significantly improved due to the shading effect of buildings [19,80,81]. Building shading is beneficial for reducing the mean radiation temperature [81]. The building area in Model 5 is relatively small. The shaded area formed by the buildings in the street canyon is also relatively small. This has a negative impact on the mean radiation temperature in the street canyon area. This is the main reason why the average mean radiation temperature of the street canyon area of Model 5 is higher than that of Model 1. The cooling effect achieved by demolishing some buildings to improve ventilation is sometimes weaker than the cooling effect achieved by blocking out sunlight and forming shaded areas [70]. The air temperature in shaded areas can be significantly reduced [74,81]. In addition, this can promote the flow of surrounding hot air into the street canyon [82]. This results in the relative humidity of the street canyon area in Model 5 being slightly lower than that in Model 1. Although the average wind speed in the street canyon area of Model 5 is slightly higher than in Model 1, the average mean radiation temperature and air temperature in the street canyon area of Model 1 are both lower than Model 5. The average relative humidity of Model 1 is higher than Model 5. The mean radiation temperature and air temperature are considered the two most influential factors that can affect human thermal sensation [72,75]. This can explain why the average PET value of Model 1 is lower than in Model 5. Therefore, the decrease in the building shadow area may have a negative impact on the thermal comfort of the street canyon.

We constructed Model 6 in the same way as Model 1 except for wind direction. The wind direction of Model 1 and Model 6 are 225° and 90°, respectively. It can be seen from Table 7 and Figure 4 that the average PET value in the street canyon area of Model 6 is 0.67 °C lower than the average PET value of Model 1 during the 10 h. This can be attributed to the fact that the wind blowing toward the roadway area in Model 1 will be blocked by buildings and roadside trees to a serious extent. This is not beneficial for reducing the air temperature [82]. The wind is parallel to the road direction in Model 6. The wind speed is high. A higher wind speed is also beneficial for increasing the relative humidity of the street canyon [82]. This is similar to what some researchers have found under suitable direction and speed winds [15,83–85].

We compared the simulation results of each model at 12 o'clock. Figure 5 is a plan view of the PET simulation results at a height of 1 m for each model. Each point within the simulation area has a specific PET value. The specific PET value of each point can be known based on the color of each point and the explanation of the PET value range in the legend section. The black area in the figure is the area where the building is located, so there is no specific PET value.

From the comparison of the simulation results in Model 1 and Model 2 in the picture below, we can see that the thermal comfort of the area under the trees is good compared with the surrounding environment. The PET under the street trees of Model 2 with a denser canopy is much better than that in Model 1. This is attributed to the fact that plants with a higher LAD can generate more significant transpiration and shade, improving the thermal comfort of street canyons in hot summer conditions [10].

The number of street trees in Model 3 is almost twice that in Model 2, but the canopy size of the street trees in Model 3 is significantly smaller than that in Model 2. Many small areas with good thermal comfort under the tree are formed in Model 3, but these areas are very small and the PET values are often 35.60–38.80 °C, which are much higher than those under the tree in Model 1 (29.20–35.60 °C) and Model 2 (below 29.20 °C).

The hedges in Model 4 are shown in green in Figure 5. This is the only difference between the two models. We can see from the picture that the thermal comfort in the area where the hedges are located is relatively poor. The PET values are often 42–48.4 °C, which is much higher than that (below 29.20 °C) in the area under the trees in Model 4. The effect of the hedges on thermal comfort is very limited. Tochaiwat et al. (2023) have already pointed out that tall trees always have a greater impact on the thermal comfort of the surrounding environment than shrubs [86]. This shows consistency even if they studied shrubs and we studied hedges.

The positions of the buildings in Model 1 and Model 5 are shown in the black area in the figure. By comparing the black areas in the simulation results of Model 1 and Model 5 in Figure 5, we can see that some buildings in Model 5 are relatively small, and more open spaces can help with the ventilation of street canyons. However, from Table 4 and Figure 5, we can see that the thermal comfort in Model 5 with more buildings with smaller floor areas is worse than that in Model 1. This street is an east–west street. At this time, the architectural shadow is on the north side of the building. By comparing the conditions of Model 1 and Model 5 in Figure 5, we can see that an area with good thermal comfort is formed in the area where the building shadow is located in the two models. In Model 5, the floor area of buildings on both sides of the road is smaller than that in Model 1 and the area with good thermal comfort in street canyons formed due to the building shadow is also relatively small. Rodríguez Algeciras and Matzarakis (2016) compared the effects of using ventilation and shadows to improve thermal comfort and found that using shadows to improve thermal comfort was more effective [87]. Our findings are consistent with this.

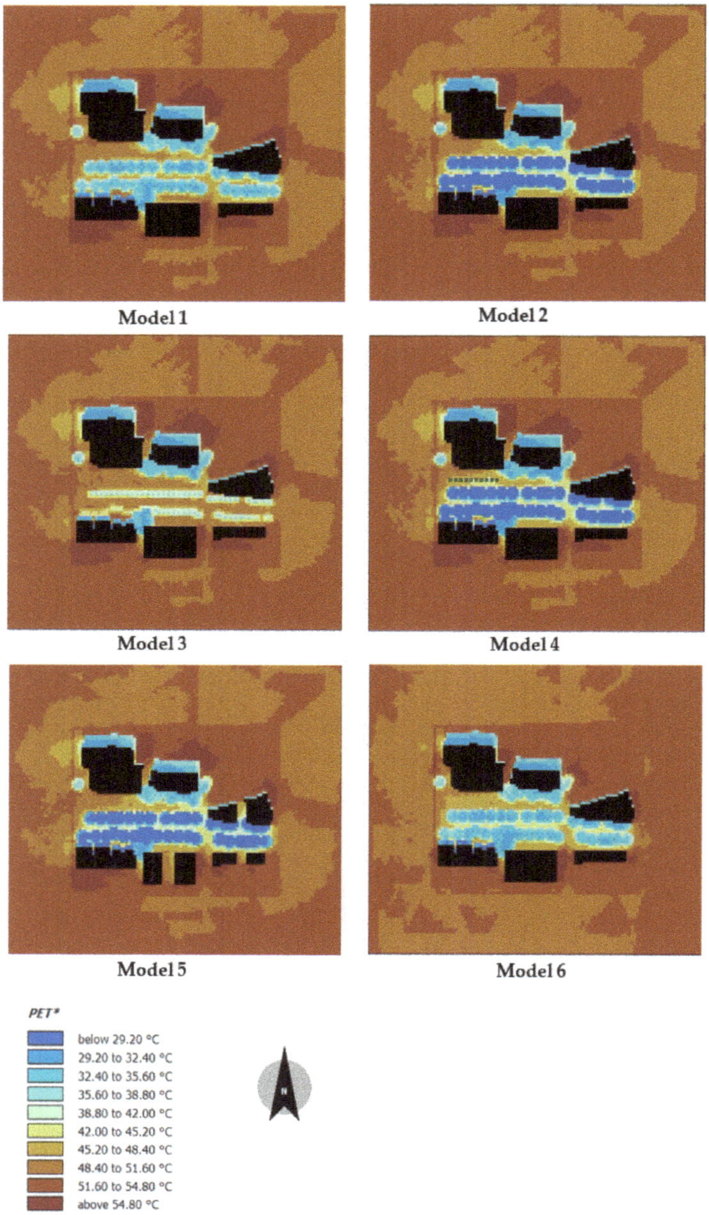

Figure 5. Simulation results of PET values in each model at 12:00, 13 May 2023 (In the figure, the black area represents the area occupied by buildings, for which a PET value is not available).

A wind direction of 90° was selected for the simulation of Model 6. The wind direction of Model 1 is 225°. This is the only difference between the two models. Compared with the impact of the LAD, crown width, and canopy density, the impact of the wind on the thermal comfort of street canyons is not quite as obvious. In Model 6, the wind blowing from the east (90°) can benefit the thermal comfort of the street canyon more due to a higher wind speed and lower blockage. In Model 1, the southeast wind (225°) is less effective in improving the thermal comfort of the street canyon, influenced by the barrier formed by buildings and trees to the wind blowing through the street canyon.

We found that the hourly average PET (Table 8 and Figure 6) in all six models was lower than the neutral PET range in Changsha (24.7–27.8 °C) pointed out by Deng et al. (2022) [70]. In this case, the closer the PET is to the neutral range, the weaker the feeling of cold in the human body.

Table 8. The average PET values (°C) of each model in the street canyon area at different times on 16 December 2022.

Time	Model 1	Model 2	Model 3	Model 4	Model 5	Model 6
11:00	17.7	16.3	20.7	16.3	18.1	19.5
12:00	17.8	16.3	20.8	16.3	18.1	19.6
13:00	17.1	15.7	20.1	15.7	17.5	18.9
14:00	15.8	14.7	18.5	14.7	16.1	17.4
15:00	14.2	13.6	15.7	13.6	14.4	15.6
16:00	12.1	12.0	12.6	11.9	12.3	13.2
17:00	10.3	10.4	10.3	10.4	10.4	11.2
18:00	10.2	10.3	10.2	10.3	10.3	11.0
19:00	10.2	10.2	10.1	10.3	10.2	11.0

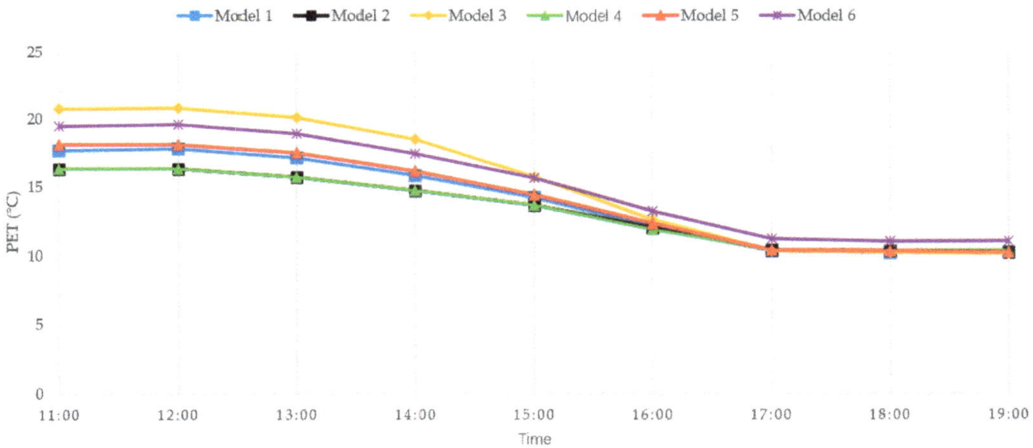

Figure 6. The average PET values (°C) in the street canyon area of each model at different times on 16 December 2022.

The average PET value of Model 2 is 0.66 °C lower than Model 1 during the 9 h. The LAD is the only variable between the two models. We can find that the PET value in Model 2 is lower than that in Model 1 from 11 o'clock to 16 o'clock. This can be attributed to sufficient sunlight during the day, and the mean radiation temperature and air temperature of the sunlit area are generally higher than that of the shaded area formed by building and vegetation coverage [88]. From 17:00 to 19:00, it is almost dark. The tree canopy inhibits longwave radiative cooling [11]. This leads to a higher average mean radiation temperatures in the street canyon area in Model 2 than in Model 1 even if the average air temperature of Model 2 still lower than Model 1. The average PET of Model 2 is higher than Model 1 between 17 o'clock and 19 o'clock. The model with high LADs has a higher average relative humidity in the street canyon area [48]. Reducing the LADs of street trees is beneficial for improving the thermal comfort of street canyons during the daytime in winter. Our findings are consistent with the findings by Yang et al. (2022) that the higher the LAD, the more effectively the air temperature is reduced [14]. Model 2 with a dense tree canopy has a stronger wind resistance effect than Model 1. The average wind speed in the canyon area of Model 2 is lower. The change trend reflected by the PETs of Model 1 and Model 2 is closest to the change trend reflected by the mean radiation temperature of the two models. This also indicates that the mean radiation temperature is one of the most influential factors on human thermal sensation [72].

The average PET value of Model 3 is 1.51 °C higher than Model 1 during the 9 h. The PET in Model 3 is closest to the neutral PET range, and the human comfort is the best. This is opposite to the worst thermal comfort in Model 3 in summer. In winter, plants can increase the water vapor in the air through transpiration and block sunlight. This can make street canyons colder [71]. The transpiration caused by the smaller canopy of 5 m roadside trees is relatively weak, and there are relatively few areas that block sunlight. The lower canopy height is beneficial for blocking wind near the ground. This can explain why Model 3 has the lowest overall average PET. In colder weather conditions in winter, an increased air temperature is needed in the street canyon to improve the comfort level [75]. Model 3, with the lowest canopy density, has the highest average mean radiation temperature and air temperature in the street canyon area. This can be attributed to the mean radiation temperature and air temperature of the sunlit area being generally higher than that of the shaded area formed by building and vegetation coverage during the day [88]. In addition, short trees and a low canopy density have a weak transpiration effect [73]. Although the crowns of the street trees in Model 3 are smaller, the starting height of the crowns of the street trees in Model 3 are shorter. This is beneficial for blocking wind at the height of 0–2 m. We found that the average wind speed at a height of 1 m in the street canyon area of Model 3 is lower than that of Model 1. The high PET value of Model 3 can be attributed to its high mean radiant temperature and air temperature and low relative humidity and wind speed.

The impact of hedges on the thermal comfort of street canyons can be explored by comparing the simulation results of Models 2 and 4. The average PET value of Model 4 is almost the same as Model 2. Although the hedge in Model 4 can block out sunlight, this results in a smaller area of direct solar radiation. This can result in a lower mean radiation temperature [72]. The average mean radiant temperature of Model 4 is higher than that of Model 2 due to hedges inhibiting longwave radiative cooling from the surface [11]. Hedges can reduce the air temperature of the surrounding environment during winter days due to them blocking sunlight and having a transpiration effect [74]. The average relative humidity of Model 4 is higher than that of Model 2 due to the hedges also increasing the relative humidity due to transpiration [74]. Near the ground, 2 m hedges have a good wind blocking effect. The average wind speed of Model 4 is lower than that of Model 2. Reducing the wind speed is beneficial for the thermal comfort of the street canyon in winter [74]. In addition, the hedges are small in size, and this makes their influence weak. Although the street canyon area in Model 4 shows good thermal comfort during summer compared to Model 2, in winter, the hedges in Model 4 do not have a noticeable impact on the comfort level of the street canyon area.

The average PET value of Model 5 is 0.22 °C higher than Model 1 during the 9 h. Model 5 has a smaller building area than Model 1. This is more conducive to ventilation. In addition, Model 5 also has smaller shaded areas. Model 5 shows higher average PET values than Model 1. The mean radiation temperature and air temperature of the sunlit area is generally higher than that of the shaded area formed by the buildings [88]. The mean radiant temperature and air temperature of Model 5 is higher than those in Model 1. The average relative humidity of Model 5 is lower than Model 1. A larger area of direct sunlight can contribute to a lower relative humidity in Model 5 [89]. In cold winter, when there is strong wind in the area, their narrowness generates a more significant effect when the wind blows through street canyons. This leads to a rapid increase in wind speed in the street canyons [90]. Adding an open space at this time is beneficial for reducing the effect of the narrow space. This can be used to explain why the average wind speed in the street canyon area of Model 5 is lower than that of Model 1. The average mean radiant temperature and air temperature of the street canyon area of Model 5 is higher than those of Model 1. The relative humidity and wind speed of Model 5 are lower than Model 1. These can explain why the PET value of Model 5 is lower than Model 1. The overall comfort of Model 5 is relatively good in winter. Habib et al. (2015) also found that building shadows can have a significant impact on the thermal comfort of street canyons during hot and cold days. This also supports our viewpoint [91].

The average PET value of Model 6 is 1.33 °C higher than Model 1 during the 9 h. The only difference between Model 1 and Model 6 is the wind direction. The wind direction is 47.9° in Model 1 and 180° in Model 6. The simulated wind speeds in Models 1 and 6 are both 3 ms^{-1}. This is much higher than the simulated wind speed on 13 May 2023. The better thermal comfort of Model 6 can be attributed to the good blocking effect of buildings on both sides of the street on the cold air blown. In Model 6, the wind direction is perpendicular to the street direction. The buildings and roadside trees on both sides of the street have a good wind resistance effect. The average mean radiation temperature and air temperature of the street canyon area of Model 6 re lower than in Model 1. This is consistent with the significant impact of the wind environment on the mean radiation temperature pointed out by Sangkertadi and Syafriny (2016) [92]. The effect of water evaporation is weak [93]. The average relative humidity in the street canyon area in this model is relatively low. The average wind speed of the street canyon area of Model 6 is lowest among the six models since buildings and trees play quite positive roles in blocking the wind in this model. This means that fully utilizing the wind resistance effect of buildings and trees is beneficial for improving the comfort level of street canyons in winter.

Figure 7 shows a plan view of the simulation results of each model within the study area at 17:00, 16 December 2022. On a cold winter day, it is necessary to increase the PET value of the street canyons to achieve the goal of improving thermal comfort. In general, the enclosed area formed by buildings and trees shows a relatively better thermal comfort than open areas at the outskirts of the research area. This shows the importance of buildings and trees in blocking the wind during cold and windy winter days. From the following figure, it can be seen that the areas with higher PETs in each model appear on both sides of the building. This is due to wind resistance. The areas where the road green space is located show a relatively lower PET value than the surrounding environment. Based on the analysis of the average PETs of the street canyon areas in the six models, we did not find warming effects of the green space in winter as pointed out by some researchers [94–96]. Although street trees can decrease winter thermal comfort by blocking sunlight and transpiration [57,58], their effect on blocking the wind should not be neglected. Buildings can effectively block wind blowing from the south (180°) into the street canyon. The thermal comfort in the street canyon of Model 6 is better than Model 1 with a wind direction of 47.9°. Model 6 shows a larger area of relatively good thermal comfort (PET 12–13.6 °C) than Model 1. This indicates to us that we can think more about how to improve the winter thermal comfort of the street canyon by optimizing the building layout as well as the building–tree arrangement.

3.2. Discussion

Table 9 shows the advantages and disadvantages of the building and tree features in relation to the thermal comfort of the street canyon area during hot and cold days. Although the dense distribution of road green spaces may effectively block sunlight and increase moisture on hot days, this type of green space layout may worsen the thermal comfort of the street canyon due to its good cooling effect on cold days. The buildings and green spaces in the street canyon area need to be considered simultaneously to form an optimal street canyon spatial form and achieve good thermal comfort throughout the year.

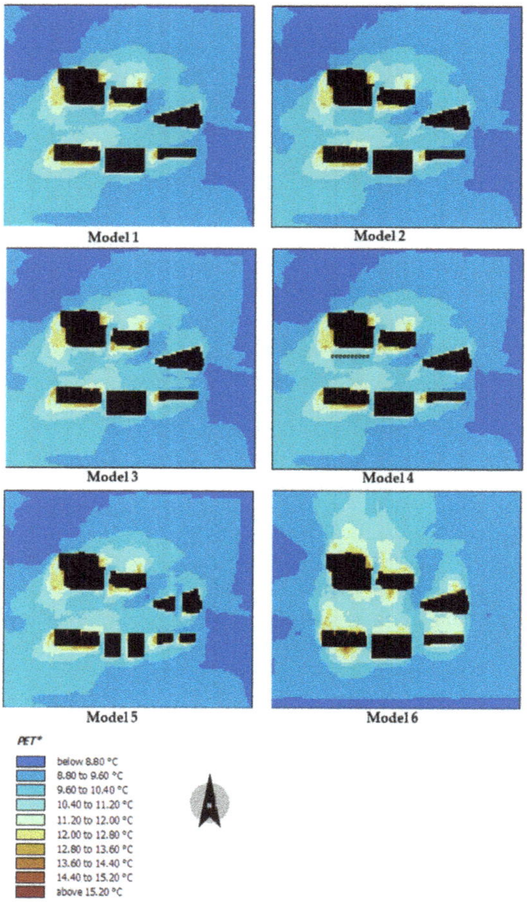

Figure 7. Simulation results of PET values in each model at 17:00, 16 December 2022.

Table 9. A comparison between different tree and building features on the thermal comfort of the street canyon during hot and cold days.

	Advantages	**Disadvantages**
Tall trees, large crown high canopy density, and LADs	Beneficial for decreasing air temperature and improving thermal comfort during hot days.	Play a quite active role in the reduction in air temperature and deteriorates thermal comfort during cold days.
Short and sparse hedges	Reduced air temperature and increased moisture due to hedges are beneficial for hot day thermal comfort improvement. Wind-resistant effects of hedges are beneficial for winter thermal comfort improvement.	Even if they are quite active in wind proofing and this is beneficial for thermal comfort during cold days, their transpiration effect can reduce the air temperature.

Table 9. *Cont.*

	Advantages	Disadvantages
Buildings	Form building shadows and this is quite helpful for thermal comfort improvement during hot days. The wind-proofing effect of buildings can make the thermal comfort of the street canyon area better than other open spaces, which are heavily influenced by winds.	Even if buildings have a good wind-proofing effect and make the thermal comfort of the street canyon generally good during cold days, the building shadow area still has a lower temperature than areas with direct sunlight in the street canyon.
Areas with direct sunlight and without building or tree shadows	Good sunlight conditions are beneficial for the improvement of thermal comfort in the street canyon area during cold days.	Sunlit areas can have worse thermal comfort than shadowed areas during hot days.
Wind environment	On hot days, wind can play an active role in thermal comfort improvement. Thus, if the wind direction is parallel to the street and the wind speed is high, this is beneficial for the thermal comfort of the street canyon.	On cold days, wind can the reduce air temperature and deteriorate thermal comfort. Thus, if the wind direction is perpendicular to the street and the wind speed is low, this is beneficial for the thermal comfort of the street canyon.

Whether in hot or cold weather, street trees have a certain cooling effect. Street trees with high tree heights and large and dense canopies can have a good cooling effect in summer and are beneficial for summer thermal comfort. However, such street trees are not conducive to improving winter thermal comfort in the street canyon. When the street trees in the street canyon are short, with small and sparse tree crowns, the street canyon has relatively good thermal comfort in winter, while the street canyon has poor thermal comfort in summer. Given that evergreen trees can produce good cooling effects in both summer and winter, prioritizing deciduous trees for street trees in hot summer and cold winter areas is beneficial for improving the comfort of hot and cold weather.

A reasonable arrangement of a certain number of short hedges can achieve cooling effects in summer by blocking sunlight and transpiration. In winter, low hedges can provide excellent wind protection at pedestrian heights. The wind-proofing effect of hedges can also weaken the cooling effect of the street canyon area affected by the environment. We suggest arranging low hedges in open areas to fully utilize their potential.

The existence of certain open spaces between buildings can facilitate ventilation. A well-ventilated environment is conducive to cooling. The design of tall buildings with large building base areas is more conducive to forming building shadows to block sunlight in summer and improve thermal comfort. In cold weather, the building can generally produce good wind resistance, making the thermal comfort of the street canyon area superior to other open spaces. However, having a certain amount of open space between buildings is also beneficial for increasing the area with direct sunlight, which may further increase the temperature of the street canyon area and improve the thermal comfort of the street canyon in winter.

Considering this, we should reduce the air temperature of the street canyon on hot days to improve the thermal comfort, while increasing the air temperature in winter to improve the thermal comfort of street canyons. We believe that there is no spatial pattern of street canyons that can achieve the goal of reducing the temperature as much as possible in summer and increasing the temperature as much as possible in winter. We believe that improving the thermal comfort of the street canyon on hot and cold days by improving the green space and buildings on both sides is more of a trade-off, that is, trying to have a greater positive impact on the thermal comfort of hot or cold days while minimizing the negative impact on the thermal comfort of cold or hot days. Our research is beneficial for guiding future researchers to consider how to fully utilize various cooling and heating conditions and make the overall thermal comfort of the street canyon better throughout the year. Due to differences in conditions like annual average temperature, dominant wind direction, street direction, street width, building height, and street tree species among different street canyons, there is no single good street canyon spatial form that can suit all street canyon areas.

4. Conclusions

It is important to improve the thermal comfort of urban street areas which many people pass by every day. Many researchers have conducted research on how to improve the thermal comfort of street canyons [3,6–12,19–21]. Many studies only focus on improving the summer or winter thermal comfort of street canyons. We selected a section of Zhutang West Road in the central urban area of Changsha for our research. We used the microclimate simulation software ENVI-met 5.5.1 to construct different models for comparative simulation research. Camphor trees as an evergreen tree species was selected as the main street trees for the research. We selected PET as the thermal comfort indicator. In addition, we also analyzed the corresponding mean radiant temperature, air temperature, relative humidity, and wind speed. We selected a hot day (13 May 2023) and a cold day (16 December 2022) to study. We found that on the hot day, Model 2, which had tall trees, large canopies, and dense leaves, could effectively cool down the street canyon by blocking sunlight and having a transpiration effect. The average PET value of this model is 31.5 °C. This is lower than the PET value of Model 1 with tall trees and large and sparse canopies (36.2 °C) and Model 3 with a large number of short street trees with small crowns, and sparse leaves (31.1 °C). Short trees have a certain cooling effect, but this is weaker. Our simulation did not achieve the expected goal of adding open spaces to improve ventilation to enhance summer thermal comfort in the street canyon. Model 5 with a smaller building base area and better ventilation has almost the same thermal comfort condition as the model with a greater building base area and poor ventilation. The average PETs of the two models are both 36.2 °C. This is due to the decrease in the building shadow area for Model 5 and this is not beneficial for the thermal comfort. This also indicates that when exploring the improvement of the spatial structure of the street canyon to improve thermal comfort, building shadows and ventilation effects are two factors that need to be considered simultaneously. The wind parallel to the direction of the street canyon promotes wind flow through the street canyon due to lower barriers. This is beneficial for the thermal comfort of the street canyon in summer. This can be used to explain why Model 6 with a wind direction parallel to the street can have a lower PET value than Model 1 with a wind direction of 225°. The average PET value of Model 6 during the study period on 13 May 2023 was 35.5 °C and the value was 36.2 °C for Model 1. In winter, tall trees, large canopies, and dense leaves still have a significant cooling effect in Model 2. This cannot achieve the goal of improving the thermal comfort of street canyons. On the contrary, the short, small, and sparse roadside trees in Model 3 have a weaker transpiration effect. This results in smaller shadow areas that block sunlight. In addition, tree crowns also have a certain degree of wind resistance. The average PET value of Model 2 during the study period was 13.3 °C and the value was 15.4 °C for Model 3. The overall thermal comfort of the models with small trees and hedges in the street canyons were relatively good. By demolishing some buildings in the street canyon, open spaces can be added. This is beneficial for reducing the shadow area of buildings and improving the thermal comfort of the street canyon in winter. In addition, increasing the open space on both sides of the road can be beneficial for weakening the effect of narrow, reducing wind speed, and improving the thermal comfort of the street canyon. Blocking the wind blowing into the street canyon through building spaces is beneficial for reducing the wind speed and improving the thermal comfort of the street canyon in winter. These factors can be used to explain why the average PET value of Model 5 (14.2 °C) was higher than the average PET value of Model 1 (13.9 °C).

Our comparative research confirms that street canyon green space and building layouts that have good cooling effects on hot days also produce good cooling effects in cold days.

This results in relatively poor thermal comfort in the street canyon in winter. We argue it is quite important to highly value the impact of building and tree layouts on the thermal comfort of street canyons on hot and cold days at the same time, especially for those cities with hot summers and cold winters. We should make trade-offs among different street canyon spatial patterns based on the thermal comfort of the street canyon during hot and cold days at the same time. We should try to maximize the advantages and minimize the disadvantages. Since different street canyon environments have different specific situations, like dominant wind directions and speeds, building heights, street widths, tree canopy densities, and so on, no one optimization strategy can suit all different environment. We suggest following research highlighting the thermal comfort on hot and cold days at the same time and formulating an optimization strategy based on the specific situations for different street canyons.

Author Contributions: Conceptualization, J.L.; methodology, J.L. and B.Z.; software, J.L.; validation, J.L.; formal analysis, J.L. and B.Z.; investigation, J.L.; resources, J.L. and F.Y.; data curation, J.L. and F.Y.; writing—original draft preparation, J.L.; writing—review and editing, J.L., B.Z. and F.Y.; visualization, J.L.; supervision, F.Y. and B.Z.; project administration, J.L. and F.Y.; funding acquisition, B.Z., F.Y. and J.L. All authors have read and agreed to the published version of the manuscript.

Funding: This research was funded by the Natural Science Foundation of Hunan Province (2023JJ30693), Hunan Provincial Innovation Foundation for Postgraduates (CX20230091), and for the Postgraduate Independent Exploration and Innovation Program of Central South University (2023ZZTS0001).

Data Availability Statement: The data presented in this study are available on request from the corresponding author.

Conflicts of Interest: The authors declare no conflict of interest.

References

1. Wu, J. Urban ecology and sustainability: The state-of-the science and future directions. *Landsc. Urban Plan.* **2014**, *125*, 209–221. [CrossRef]
2. Hwang, R.L.; Lin, C.Y.; Huang, K.T. Spatial and temporal analysis of urban heat island and global warming on residential thermal comfort and cooling energy in Taiwan. *Energy Build.* **2017**, *152*, 804–812. [CrossRef]
3. Ren, Z.B.; Zhao, H.B.; Fu, Y.; Xiao, L.; Dong, Y.L. Effects of urban street trees on human thermal comfort and physiological indices: A case study in Changchun city, China. *J. For. Res.* **2022**, *33*, 911–922. [CrossRef]
4. IPCC. 2023: Sections. In *Climate Change 2023: Synthesis Report. Contribution of Working Groups I, II and III to the Sixth Assessment Report of the Intergovernmental Panel on Climate Change*; Core Writing Team, Lee, H., Romero, J., Eds.; IPCC: Geneva, Switzerland, 2023; pp. 35–115. [CrossRef]
5. Zhao, Q.; Guo, Y.; Ye, T.; Gasparrini, A.; Tong, S.; Overcenco, A.; Urban, A.; Schneider, A.; Entezari, A.; Vicedo-Cabrera, A.M.; et al. Global, regional, and national burden of mortality associated with non-optimal ambient temperatures from 2000 to 2019: A three-stage modelling study. *Lancet Planet Health* **2021**, *5*, e415–e425. [CrossRef] [PubMed]
6. Martini, A.; Biondi, D.; Batista, A.C. Thermal comfort provided by street trees in cities. *Arboric. J.* **2020**, *42*, 153–164. [CrossRef]
7. Cao, L.S.; Xu, H.; Li, H. Effect of Greening Trees on Thermal Comfort of the Pedestrian Streets in Hot Summer and Cold Winter Regions in China. *Nat. Environ. Pollut. Technol.* **2022**, *21*, 1543–1552. [CrossRef]
8. Li, Z.; Zhang, H.; Juan, Y.-H.; Lee, Y.-T.; Wen, C.-Y.; Yang, A.-S. Effects of urban tree planting on thermal comfort and air quality in the street canyon in a subtropical climate. *Sustain. Cities Soc.* **2023**, *91*, 104334. [CrossRef]
9. Narimani, N.; Karimi, A.; Brown, R.D. Effects of street orientation and tree species thermal comfort within urban canyons in a hot, dry climate. *Ecol. Inform.* **2022**, *69*, 101671. [CrossRef]
10. Cheng, Y.; Huang, Z.; Guo, Y.; Cheng, W.; Wu, C. Tree Canopy Characteristics Affect Street Canyon's Microclimate Conditions and Human Thermal Comfort in Hot-Humid Climate. In *Climate Change and Environmental Sustainability. Advances in Science, Technology & Innovation*; He, B.J., Prasad, D., Pignatta, G., Jupesta, J., Eds.; Springer: Cham, Switzerland, 2022. [CrossRef]
11. Huang, Z.; Wu, C.; Teng, M.; Lin, Y. Impacts of Tree Canopy Cover on Microclimate and Human Thermal Comfort in a Shallow Street Canyon in Wuhan, China. *Atmosphere* **2020**, *11*, 588. [CrossRef]
12. Zheng, B.; Bernard Bedra, K.; Zheng, J.; Wang, G. Combination of Tree Configuration with Street Configuration for Thermal Comfort Optimization under Extreme Summer Conditions in the Urban Center of Shantou City, China. *Sustainability* **2018**, *10*, 4192. [CrossRef]
13. Cui, P.; Jiang, J.J.; Zhang, J.; Wang, L. Effect of street design on uhi and energy consumption based on vegetation and street aspect ratio: Taking harbin as an example. *Sust. Cities Soc.* **2023**, *92*, 29. [CrossRef]

14. Yang, Y.J.; Zhou, D.A.; Gao, W.J.; Zhang, Z.H.; Chen, W.; Peng, W.C.Y. Simulation on the impacts of the street tree pattern on built summer thermal comfort in cold region of China. *Sustain. Cities Soc.* **2018**, *37*, 563–580. [CrossRef]
15. Morakinyo, T.E.; Lam, Y.F. Simulation study on the impact of tree-configuration, planting pattern and wind condition on street-canyon's micro-climate and thermal comfort. *Build. Environ.* **2016**, *103*, 262–275. [CrossRef]
16. Yilmaz, S.; Mutlu, E.; Yilmaz, H. Quantification of thermal comfort based on different street orientation in winter months of urban city Dadaşkent (Conference Paper). *Acta Hortic.* **2018**, *1215*, 67–72. [CrossRef]
17. Chan, S.Y.; Chau, C.K. On the study of the effects of microclimate and park and surrounding building configuration on thermal comfort in urban parks. *Sust. Cities Soc.* **2021**, *64*, 13. [CrossRef]
18. Cohen, P.; Potchter, O.; Matzarakis, A. Daily and seasonal climatic conditions of green urban open spaces in the Mediterranean climate and their impact on human comfort. *Build. Environ.* **2012**, *51*, 285–295. [CrossRef]
19. Fan, Y.; Dong, L. Analysis of Winter Microclimate Test of "Single Arch" Tree-lined Street Valley Space in Xi'an City Street. *Chinese Landsc.* **2022**, *38*, 115–120. [CrossRef]
20. Huang, H.; Peng, M. The outdoor thermal comfort of urban square: A field study in a cold season in Chongqing. In *IOP Conference Series: Earth and Environmental Science*; IOP Publishing: Bristol, UK, 2020; Volume 467, p. 012215. [CrossRef]
21. He, X.C.; Gao, W.J.; Wang, R.; Yan, D.Y. Study on outdoor thermal comfort of factory areas during winter in hot summer and cold winter zone of China. *Build. Environ.* **2023**, *228*, 12. [CrossRef]
22. Ebrahimabadi, S.; Nilsson, K.L.; Johansson, C. The problems of addressing microclimate factors in urban planning of the subarctic regions. *Environ. Plan. B-Plan. Des.* **2015**, *42*, 415–430. [CrossRef]
23. Chen, Q.; Baldocchi, D.; Gong, P.; Kelly, M. Isolating individual trees in a savanna woodland using small footprint lidar data. Photogramm. *Eng. Remote Sens.* **2006**, *72*, 923–932.
24. Rouzbeh Kargar, A.; MacKenzie, R.; Asner, G.P.; van Aardt, J. A Density-Based Approach for Leaf Area Index Assessment in a Complex Forest Environment Using a Terrestrial Laser Scanner. *Remote Sens.* **2019**, *11*, 1791. [CrossRef]
25. Lalic, B.; Mihailovic, D.T. An empirical relation describing leaf-area density inside the forest for environmental modeling. *J. Appl. Meteorol.* **2004**, *43*, 641–645. [CrossRef]
26. Changsha Municipal Government. Overview of Changsha. 2022. Available online: http://www.changsha.gov.cn/xfzs/zjmlzs/zsgl/200907/t20090727_5686409.html?ivk_sa=1024320u (accessed on 13 May 2023).
27. Beck, H.E.; Zimmermann, N.E.; McVicar, T.R.; Vergopolan, N.; Berg, A.; Wood, E.F. Present and future Köppen-Geiger climate classification maps at 1-km resolution. *Sci. Data* **2018**, *5*, 180214. [CrossRef] [PubMed]
28. Xiang, J.; Liu, H. Study on Optimal Design of Exterior Sunshade for Residential Buildings in Hot Summer and Cold Winter Zone. *Build. Sci.* **2016**, *32*, 80–84+126. [CrossRef]
29. Wang, X. Investigation, Analysis and Evaluation of Street Trees in Changsha City. In Proceedings of the Third China Forestry Academic Conference, Fuzhou, China, 12–13 September 2013.
30. Jiang, H.; Zhu, M.; Li, J.; Zhong, L.; Su, X.; Feng, Z.; Qi, Y.; Zhang, X. Data Mining Based on Chinese Traditional Calendar in the Han Dynasty Yang Mausoleum Museum. *Appl. Sci.* **2019**, *9*, 5442. [CrossRef]
31. Ren, X. *Chinese Folk Perpetual Calendar*; China Commercial Publishing House: Beijing, China, 2012.
32. The Central People's Government of the people's Republic of China. Meteorological Science Popularization: Why Does Comfort Feel Different at the Same Temperature. 2007. Available online: https://www.gov.cn/govweb/fwxx/kp/2007-04/06/content_573324.htm (accessed on 30 September 2023).
33. China Meteorological Administration. What Is a High-Temperature Heat Wave? 2011. Available online: https://www.cma.gov.cn/2011qxfw/2011qqxkp/2011qkpdt/201110/t20111026_124192.html#:~:text=%E4%BA%BA%E7%9A%84%E6%AD%A3%E5%B8%B8%E4%BD%93%E6%B8%A9%E5%A4%A7%E7%BA%A6,%E5%BF%8D%E8%80%90%E6%9E%81%E9%99%90%EF%BC%8C%E9%80%A0%E6%88%90%E6%AD%BB%E4%BA%A1%E4%BA%8B%E6%95%85%E3%80%82 (accessed on 30 September 2023).
34. Kestrel Instruments. Company History. 2023. Available online: https://kestrelinstruments.com/company-history (accessed on 13 May 2023).
35. Kestrel Instruments. Kestrel 5500 Weather Meter. Available online: https://kestrelinstruments.com/kestrel-5500-weather-meter (accessed on 6 July 2023).
36. Deli. Product Details. 2023. Available online: https://www.deli-tools.com/productinfo/info/53938.html (accessed on 13 May 2023).
37. Watson, D.J. Comparative physiological studies on the growth of field crops: I. Variation in net assimilation rate and leaf area between species and varieties and within and between years. *Ann. Bot.* **1947**, *11*, 41–76. [CrossRef]
38. Höppe, P. The physiological equivalent temperature—A universal index for the biometeorological assessment of the thermal environment. *Int. J. Biometeorol.* **1999**, *43*, 71–75. [CrossRef]
39. ENVI-met. Microclimate Model Calculations: Exploring the Comprehensive Analysis. 2023. Available online: https://www.envi-met.com/zh-hans/%E8%BD%AF%E4%BB%B6%E4%BA%A7%E5%93%81/ (accessed on 5 May 2023).
40. Elraouf, R.A.; ELMokadem, A.; Megahed, N.; Eleinen, O.A.; Eltarabily, S. Evaluating urban outdoor thermal comfort: A validation of ENVI-met simulation through field measurement. *J. Build. Perform. Simul.* **2022**, *15*, 268–286. [CrossRef]
41. Ji, T.; Yuan, W.; Wang, K.; Ji, L. Perspective on thermal sensation in sports under thermal environment. *China Sport Sci. Technol.* **2022**, *4*, 73–80.

42. Krüger, E.; Rossi, F.; Drach, P. Calibration of the physiological equivalent temperature index for three different climatic regions. *Int. J. Biometeorol.* **2017**, *61*, 1323–1336. [CrossRef]
43. Lin, J.; Cang, G. *Photogrammetry*; National Defense Industry Press: Beijing, China, 2006.
44. Liu, K.; Zhou, J.; Liu, Y.; Zeng, P.; Xin, W.; Liu, K.; Zhao, Z. Accuracy Analysis of Multitemporal Image-free Orthographic of Consumer Aerial Survey. *J. Univ. South China (Sci. Technol.)* **2022**, *36*, 71–78.
45. Chen, Y.; Liu, J.; Yao, X.; Deng, Y.; Hao, Z.; Lin, L.; Wu, N.; Yu, K. Effect of layer thickness and voxel size inversion on leaf area density based on the voxel-based canopy profiling method. *J. For. Res.* **2022**, *33*, 1545–1558. [CrossRef]
46. Middel, A.; Selover, N.; Hagen, B.; Chhetri, N. Impact of shade on outdoor thermal comfort—A seasonal field study in Tempe, Arizona. *Int. J. Biometeorol.* **2016**, *60*, 1849–1861. [CrossRef] [PubMed]
47. Ornelas, A.; Cordeiro, A.; Lameiras, J.M. Thermal Comfort Assessment in Urban Green Spaces: Contribution of Thermography to the Study of Thermal Variation between Tree Canopies and Air Temperature. *Land* **2023**, *12*, 1568. [CrossRef]
48. Kang, G.; Kim, J.J. Effects of Vertical Forests on Air Quality in Step-up Street Canyons. *Sustain. Cities Soc.* **2023**, *94*, 14. [CrossRef]
49. Xue, F.; Li, X. The impact of roadside trees on traffic released PM10 in urban street canyon: Aerodynamic and deposition effects. *Sustain. Cities Soc.* **2017**, *30*, 195–204. [CrossRef]
50. Turhan, C.; Atalay, A.S.; Gokcen Akkurt, G. An Integrated Decision-Making Framework for Mitigating the Impact of Urban Heat Islands on Energy Consumption and Thermal Comfort of Residential Buildings. *Sustainability* **2023**, *15*, 9674. [CrossRef]
51. Katul, G.G.; Mahrt, L.; Poggi, D.; Sanz, C. ONE- and TWO-Equation Models for Canopy Turbulence. *Bound. Layer Meteorol.* **2004**, *113*, 81–109. [CrossRef]
52. Zhou, M.; Gao, F.; Ma, K. Research on Species Optimization of Typical Road Trees in Beijing Based on ENVI-met. *Chin. Landsc. Archit.* **2020**, *36*, 141–144.
53. Zhao, Y.; Kan, L.; Che, S. A Preliminary Study about Common Garden Plants' Effect of Carbon Fixation and Oxygen Release in Shanghai's Community and Optimal Arrangement. *J. Shanghai Jiaotong Univ. Agric. Sci.* **2014**, *32*, 45–53.
54. Wu, Z.; Kong, F.; Wang, Y.; Sun, R.; Chen, L. The Impact of Greenspace on Thermal Comfort in a Residential Quarter of Beijing, China. *Int. J. Environ. Res. Public Health* **2016**, *13*, 1217. [CrossRef]
55. Li, B. The Study on Common Street Trees Fixing Carbon and Releasing Oxygen Sluggish Dust Benefits in Changsha. Master's Thesis, Central South University of Forestry and Technology, Changsha, China, 2012.
56. Bréda, N.J.J. Ground-based measurements of leaf area index:: A review of methods, instruments and current controversies. *J. Exp. Bot.* **2003**, *54*, 2403–2417. [CrossRef] [PubMed]
57. Segura, R.; Krayenhoff, E.S.; Martilli, A.; Badia, A.; Estruch, C.; Ventura, S.; Villalba, G. How do street trees affect urban temperatures and radiation exchange? Observations and numerical evaluation in a highly compact city. *Urban Clim.* **2022**, *46*, 101288. [CrossRef]
58. Teng, R.; Liu, G.; Chen, X.; Li, C.; Wang, Z. Study on the effect of solar radiation on the mean radiant temperature and PMV in winter. *J. Qingdao Univ. Technol.* **2023**, *41*, 95–102.
59. Sayad, B.; Alkama, D.; Rebhi, R.; Menni, Y.; Ahmad, H.; Inc, M.; Sharifpur, M.; Lorenzini, G.; Azab, E.; Elnaggar, A.Y. Outdoor Thermal Comfort Optimization through Vegetation Parameterization: Species and Tree Layout. *Sustainability* **2021**, *13*, 11791. [CrossRef]
60. Chicco, D.; Warrens, M.J.; Jurman, G. The coefficient of determination R-squared is more informative than SMAPE, MAE, MAPE, MSE and RMSE in regression analysis evaluation. *Peer J Comput. Sci.* **2021**, *7*, e623. [CrossRef]
61. Tsoka, S.; Tsikaloudaki, A.; Theodosiou, T. Analyzing the ENVI-met microclimate model's performance and assessing cool materials and urban vegetation applications—A review. *Sustain. Cities Soc.* **2018**, *43*, 55–76. [CrossRef]
62. Willmott, C.J. On the validation of models. *Phys. Geogr.* **1981**, *2*, 184–194. [CrossRef]
63. Willmott, C.J.; Robeson, S.M.; Matsuura, K. A refined index of model performance. *Int. J. Climatol.* **2012**, *32*, 2088–2094. [CrossRef]
64. Zhu, S.J.; Causone, F.; Gao, N.P.; Ye, Y.; Jin, X.; Zhou, X.; Shi, X. Numerical simulation to assess the impact of urban green in-frastructure on building energy use: A review. *Build. Environ.* **2023**, *228*, 20. [CrossRef]
65. Lam, C.K.C.; Lee, H.; Yang, S.R.; Park, S. A review on the significance and perspective of the numerical simulations of outdoor thermal environment. *Sust. Cities Soc.* **2021**, *71*, 30. [CrossRef]
66. Liu, J.; Tang, H.; Zheng, B. Simulation study of summer microclimate in street space of historic conservation areas in China: A case study in Changsha. *Front. Environ. Sci.* **2023**, *11*, 1146801. [CrossRef]
67. Chi, M.; Dan, J. Chinese Grow in Height Rankings. 2016. Available online: https://www.chinadaily.com.cn/china/2016-07/27/content_26242503.htm (accessed on 7 March 2023).
68. NCD Risk Factor Collaboration. A century of trends in adult human height. *eLife* **2016**, *5*, e13410. [CrossRef]
69. Lu, G.; Hu, Y.; Yang, Z.; Zhang, Y.; Lu, S.; Gong, S.; Li, T.; Shen, Y.; Zhang, S.; Zhuang, H. Geographic latitude and human height statistical analysis and case studies from China. *Arab. J. Geosci.* **2022**, *15*, 335. [CrossRef]
70. Deng, Y.; Gan, D.; Tang, N.; Cai, Z.; Li, X.; Chen, S.; Li, X. Research on Outdoor Thermal Comfort and Activities in Residential Areas in Subtropical China. *Atmosphere* **2022**, *13*, 1357. [CrossRef]
71. Boukhabl, M.; Alkam, D. Impact of vegetation on thermal conditions outside, thermal modeling of urban microclimate, case study: The street of the Republic, Biskra. *Energy Procedia* **2012**, *18*, 73–78. [CrossRef]
72. Liu, J.; Tang, H.; Zheng, B.; Sun, Z. A Study on the Summer Microclimate Environment of Public Space and Pedestrian Commercial Streets in Regions with Hot Summers and Cold Winters. *Appl. Sci.* **2023**, *13*, 5263. [CrossRef]

73. Berardi, U.; Jandaghian, Z.; Graham, J. Effects of greenery enhancements for the resilience to heat waves: A comparison of analysis performed through mesoscale (WRF) and microscale (Envi-met) modeling. *Sci. Total. Environ.* **2020**, *747*, 141300. [CrossRef]
74. Wang, Z.; Li, Y.; Song, J.; Wang, K.; Xie, J.; Chan, P.W.; Ren, C.; Di Sabatino, S. Modelling and optimizing tree planning for urban climate in a subtropical high-density city. *Urban CLim.* **2022**, *43*, 101141. [CrossRef]
75. Darbani, E.S.; Rafieian, M.; Parapari, D.M.; Guldmann, J.M. Urban design strategies for summer and winter outdoor thermal comfort in arid regions: The case of historical, contemporary and modern urban areas in Mashhad, Iran. *Sustain. Cities Soc.* **2023**, *89*, 104339. [CrossRef]
76. Li, J.Y.; Zheng, B.H.; Ouyang, X.; Chen, X.; Bedra, K.B. Does shrub benefit the thermal comfort at pedestrian height in Singapore? *Sustain. Cities Soc.* **2021**, *75*, 10. [CrossRef]
77. Hong, B.; Lin, B.R. Numerical studies of the outdoor wind environment and thermal comfort at pedestrian level in housing blocks with different building layout patterns and trees arrangement. *Renew. Energy* **2015**, *73*, 18–27. [CrossRef]
78. Müller, N.; Kuttler, W.; Barlag, A.B. Counteracting urban climate change: Adaptation measures and their effect on thermal comfort. *Theor. Appl. Climatol.* **2014**, *115*, 243–257. [CrossRef]
79. Jang, G.; Kim, S.; Lee, J. Planning scenarios and microclimatic effects: The case of high-density riverside residential districts in Seoul, South Korea. *Build. Environ.* **2022**, *223*, 109517. [CrossRef]
80. Goldberg, V.; Kurbjuhn, C.; Bernhofer, C. How relevant is urban planning for the thermal comfort of pedestrians? Numerical case studies in two districts of the City of Dresden (Saxony/Germany) (Article). *Meteorol. Z.* **2013**, *22*, 739–751. [CrossRef]
81. Acero, J.A.; Koh, E.J.Y.; Tan, Y.S.; Norford, L.K. Quantifying the Effect of Building Shadowing and Cloudiness on Mean Radiant Temperature in Singapore. *Atmosphere* **2021**, *12*, 1012. [CrossRef]
82. Heaviside, C.; Cai, X.M.; Vardoulakis, S. The effects of horizontal advection on the urban heat island in birmingham and the west midlands, united kingdom during a heatwave. *Q. J. R. Meteorol. Soc.* **2015**, *141*, 1429–1441. [CrossRef]
83. Zhang, C.; Wang, Q.; Chan, P.; Ren, C.; Li, Y. The effect of background wind on summertime daily maximum air temperature in Kowloon, Hong Kong. *Build. Environ.* **2022**, *210*, 108693. [CrossRef]
84. Al-Obaidi, I.; Rayburg, S.; Półrolniczak, M.; Neave, M. Assessing the Impact of Wind Conditions on Urban Heat Islands in Large Australian Cities. *J. Ecol. Eng.* **2021**, *22*, 1–15. [CrossRef]
85. Lopes, A.; Alves, E.; Alcoforado, M.J.; Machete, R. Lisbon Urban Heat Island Updated: New Highlights about the Relationships between Thermal Patterns and Wind Regimes. *Adv. Meteorol.* **2013**, *2013*, 487695. [CrossRef]
86. Tochaiwat, K.; Phichetkunbodee, N.; Suppakittpaisarn, P.; Rinchumphu, D.; Tepweerakun, S.; Kridakorn Na Ayutthaya, T.; Sittisom, P. Eco-Efficiency of Green Infrastructure on Thermal Comfort of Outdoor Space Design. *Sustainability* **2023**, *15*, 2566. [CrossRef]
87. Rodríguez Algeciras, J.A.; Matzarakis, A. Quantification of thermal bioclimate for the management of urban design in Mediterranean climate of Barcelona, Spain. *Int. J. Biometeorol.* **2016**, *60*, 1261–1270. [CrossRef] [PubMed]
88. Lindberg, F.; Holmer, B.; Thorsson, S.; Rayner, D. Characteristics of the mean radiant temperature in high latitude cities-implications for sensitive climate planning applications. *Int. J. Biometeorol.* **2014**, *58*, 613–627. [CrossRef] [PubMed]
89. Wu, H.; Cao, J.; Miao, M.; LI, C. Characteristics of sunshine duration in ecological conservation area of West Beijing and its influencing factors. *J. Atmos. Environ. Opt.* **2022**, *17*, 195.
90. Wang, K.; Liang, H.; Shi, P.; Zhao, M. A comparative study of wind environment measurement and CFD simulation in a compact urban open space based on wind sense. *Acta Ecol. Sin.* **2019**, *39*, 6051–6057.
91. Habib, F.; Barzegar, Z.; Cheshmehghasabani, M. Adjustment comparison of veranda effect on building shadow area in semi-arid and moderate climates. *Ksce J. Civ. Eng.* **2015**, *19*, 1256–1264. [CrossRef]
92. Sangkertadi, S.; Syafriny, R. Pair influence of wind speed and mean radiant temperature on outdoor thermal comfort of humid tropical environment. *J. Urban Environ. Eng.* **2016**, *10*, 177–185. [CrossRef]
93. Hathway, E.A.; Sharples, S. The interaction of rivers and urban form in mitigating the Urban Heat Island effect: A UK case study. *Build. Environ.* **2012**, *58*, 14–22. [CrossRef]
94. Thompson, K.L.; Pauli, J.N.; Erker, T.; Kucharik, C.J.; Schatz, J.; Townsend, P.A.; Zuckerberg, B. Urban greenspaces promote warmer soil surface temperatures in a snow-covered city. *Landsc. Urban Plan.* **2022**, *227*, 12. [CrossRef]
95. Yang, J.; Zhao, Y.; Zou, Y.; Xia, D.; Lou, S.; Liu, W.; Ji, K. Effects of Tree Species and Layout on the Outdoor Thermal Environment of Squares in Hot-Humid Areas of China. *Buildings* **2022**, *12*, 1867. [CrossRef]
96. Huttner, S. Further Development and Application of the 3D Microclimate Simulation ENVI-Met. Ph.D. Thesis, Johannes Gutenberg University Mainz, Mainz, Germany, 2012.

forests

MDPI

Article

Disentangling the Cooling Effects of Transpiration and Canopy Shading: Case Study of an Individual Tree in a Subtropical City

Zhe Shi [1], Chunhua Yan [2,3], Weiting Hu [1], Zifan Luo [1] and Guo Yu Qiu [1,*]

1 School of Environment and Energy, Peking University, Shenzhen 518055, China; shiz@pku.edu.cn (Z.S.); huwt@pku.edu.cn (W.H.)
2 Shenzhen Campus of Sun Yat-sen University, Shenzhen 518107, China; yanchh@mail.sysu.edu.cn
3 School of Ecology, Sun Yat-sen University, Guangzhou 510275, China
* Correspondence: qiugy@pkusz.edu.cn

Abstract

Transpiration and canopy shading are the main ways that trees cool urban environments; this is crucial to human survival and improving urban livability in the context of global warming and rapid urbanization. So far, most studies focus on the combined cooling effect of transpiration and canopy shading, but their individual contributions have not been widely explored. Therefore, a quantitative framework was developed by carrying out a long-term field experiment and microenvironment simulations to investigate the cooling effect of a single *Ficus concinna*. The results show that the annual mean cooling effects of shading and transpiration are 0.17 ± 0.27 °C and 0.30 ± 0.13 °C, accounting for 21.2 ± 51.6% and 44.7 ± 26.3% of total cooling, respectively. Shade cooling demonstrates strong radiative dependence, reaching a peak of 0.63 °C with a cooling contribution of 77.1% during summer at noon due to solar radiation interception. In contrast, nighttime and winter conditions revealed shading-induced temperature increases up to 0.52 °C via longwave radiation reflection. By contrast, transpiration cooling demonstrated temperature dependence, which increased with air temperature and peaked at 1.03 °C (contributing 70.0% to the total cooling) before stomata closing. This mechanistic analysis quantitatively reveals that *F. concinna* provides cooling effects through a dynamic complementarity between transpiration and shading. These findings could offer a biophysically grounded basis for optimizing urban greening strategies and contribute to the theoretical advancement of nature-based urban climate solutions.

Keywords: cooling effect; transpiration; canopy shading; tree; urban

1. Introduction

Subtropical cities are among the areas most severely affected by the urban heat island effect and heat waves [1] due to their higher air temperature, denser population, and increasing impermeable materials [2–4]. This situation could become more severe with ongoing global warming and the increasing frequency of heat waves [5–7]. Elevated urban temperatures not only reduce thermal comfort but also pose serious public health risks, notably elevated mortality rates in vulnerable groups such as older adults [8–10]. In response, policymakers and urban planners have sought strategies to mitigate urban heat island effects [11] and green infrastructures have been recognized as an effective approach [12–15].

Trees are one of the most important components of urban green infrastructures and play a significant role in thermal environment regulation. Extensive studies have shown that urban areas with tree cover exhibit lower temperatures compared to treeless urban green spaces [16–18]. Trees regulate temperatures in two major ways [19]: a well-known process is shading, where tree canopies intercept solar radiation to reduce heat absorption on underlying surfaces and lower temperatures [20,21]. In addition to shading, trees can cool the atmosphere through transpiration, a process in which water is released from leaf surfaces. Transpiration can utilize nearly all available net radiation through latent heat, thereby decreasing sensible heat and cooling both the canopy and the surrounding air [22,23]. Collectively, shading and transpiration enable trees to intercept approximately 50%–90% of incoming shortwave radiation [24], reduce air temperatures by an average of 0.5–3.0 °C [25], and lower ground surface temperatures by as much as 3–20 °C between shaded and exposed asphalt [26].

Although the biophysical cooling processes of trees are well understood, a quantitative study on the individual cooling potential of shading and transpiration remains limited. Most research focuses on their combined effects, with relatively few studies attempting to isolate and evaluate their individual contributions [27–29]. Furthermore, the tree cooling effect is influenced by many factors such as tree-specific traits, background climate conditions, and green space configuration [30,31]. Crucially, shading and transpiration exhibit divergent responses to the same environmental variables. For instance, increased green cover enhances canopy shading but may reduce transpiration rates [32,33], which means that the dominant cooling mechanism varies with environmental conditions. Separate quantification of tree shading and transpiration cooling effects can deepen our understanding of tree cooling mechanisms, allowing for more economical and effective strategies for mitigating urban heat stress.

So far, several investigations have applied experimental or modeling approaches to understand these effects. For example, Tan et al. [34] assessed the role of transpiration by severing plant roots to inhibit water uptake; Pace et al. [35] used an energy budget analysis to estimate the energy expended through shading and transpiration; and Manickathan et al. [36] employed a computational fluid dynamics model to investigate the cooling effects of tree transpiration and shading. These studies offer innovative, valuable methodological advancements for quantifying the cooling contributions of shading and transpiration. However, most of the studies identify shading as the dominant mechanism in tree-induced cooling, while the temperature regulation of transpiration processes received limited attention. As a matter of fact, these findings are often based on short-term measurements conducted during sunny summer days or heatwave periods [37,38], when transpiration may be constrained by stomatal closure under high temperature conditions [39], whereas shading effects are amplified due to heightened solar radiation. Such temporal limitations can lead to an incomplete understanding of the cooling mechanisms of urban trees.

In this study, we propose the following hypotheses for a tree in a subtropical city: (1) transpiration would be the dominant cooling mechanism with ample water availability, and (2) the relative contributions of shading and transpiration would shift dynamically with meteorological conditions, exhibiting a complementary relationship. To test these hypotheses, we conducted a year-long field experiment in Shenzhen, a subtropical city characterized by intense solar radiation and abundant precipitation, which create favorable conditions for canopy shading and vegetation transpiration [40,41]. The climatic conditions in this city offer a unique opportunity to examine the effects of tree transpiration and canopy shading across different environmental conditions. To disentangle the intertwined effects of transpiration and shading—which is challenging through field measurements alone—we employed the ENVI-met microclimate model (version 5.1.1). This tool allows for the creation of controlled, idealized scenarios (e.g., simulating a tree without transpiration) that are infeasible in real-world experiments. By integrating continuous in situ monitoring with ENVI-met microclimate simulations, we independently quantified the cooling contributions of tree shading and transpiration and analyzed their temporal dynamics over an annual cycle. The objectives of this study are to (1) investigate the combined cooling effects across various temporal scales, (2) qualify the relative contributions of tree transpiration and canopy shading, and (3) analyze the response mechanisms of tree cooling processes to environmental changes.

2. Materials and Methods

2.1. Study Area and Field Experiments

This study was conducted in Shenzhen, China (22°35′ N, 113°58′ E, Figure 1a), a subtropical coastal city characterized by a mild winter and hot summer, with average temperatures of 14.9 °C in January (coldest month) and 28.6 °C in July (hottest month). The study area is located in a university campus, with the coverage of buildings, paved surfaces, grasslands and woodlands being about 25.68%, 4.65%, 24.43% and 30.72% (Figure 1b). The campus vegetation is representative of urban green spaces in subtropical South China, dominated by species such as the evergreen *Ficus concinna* (the primary street tree), *Terminalia neotaliala*, *Araucaria cunninghamii*, and *Albizia julibrissin*. *F. concinna* is a native and universal tree species that is commonly cultivated in subtropical areas. Its prevalence and use in numerous urban ecological studies make it a representative and ecologically significant subject for investigating tree cooling mechanisms. In this study, an individual 12-year-old *F. concinna* was chosen for detailed observation. The tree has a height of 5 m, a diameter at the breast (DBH) of 0.32 m, and a crown area of 29 m^2 (Figure 1d). It is situated on an open lawn with no surrounding structures or nearby trees, ensuring minimal external interference with microenvironmental measurements. To account for the influence of lawn evapotranspiration beneath tree canopies, an adjacent lawn area within the same campus was selected as the control site (Figure 1c).

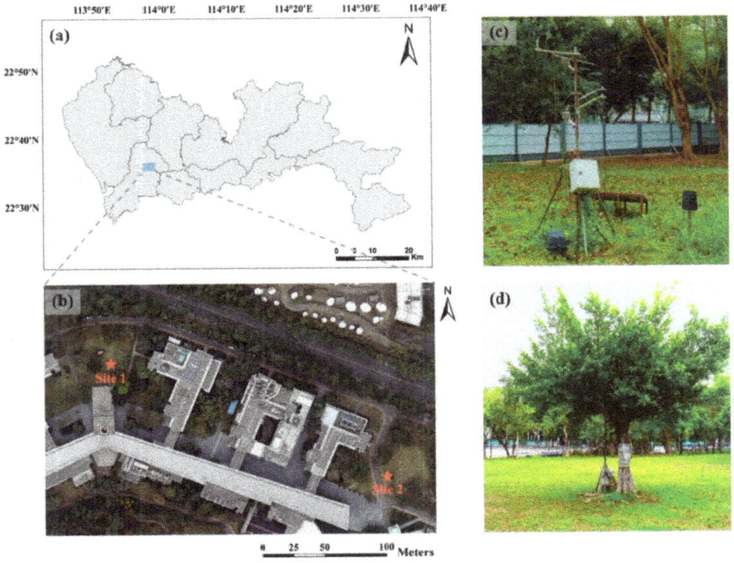

Figure 1. (**a**) Location of the study area in Shenzhen City; (**b**) plan view of the study area and the location of the two experiment sites; (**c**) Site 1, an urban lawn (*Zoysia matrella*); and (**d**) Site 2, the experimental target tree (*Ficus concinna*).

In this study, the air temperature at pedestrian height was used as an indicator to evaluate the cooling effect of the tree. Air temperature and relative humidity were measured simultaneously in both direct sunlight and beneath the tree canopy using sensors (225-050YA, Novalynx, Grass Valley, CA, USA) installed at a height of 2 m above the ground at both experiment sites. Shortwave and longwave radiation were measured at both sites using radiometers (PYP-PA, Apogee, Santa Monica, CA, USA) mounted at a height of 2 m. These instruments captured radiation conditions under both sunlight and shaded areas. Wind speed and direction were monitored using an anemometer (03002L, Campbell Scientific, Logan, UT, USA) mounted 3 m above the ground at the lawn experiment site. In addition, we conducted transpiration measurements of the *F. concinna* using a sap flow system (SF-G type probes, Ecomatic, Munich, Germany). All measurements were recorded at 1 min intervals, averaged over 10 min periods, and stored using a data logger (CR1000, Campbell Scientific, Logan, UT, USA).

This study conducted a year-long continuous observation, collecting meteorological data from January to December in 2022, to analyze the dynamic variations in the tree cooling effect. Sap flow data were continuously collected from July to November.

2.2. Microclimate Simulation

In existing studies, the cooling effects of transpiration and shading are usually separated through experimental approaches or model simulations. Experimental approaches often require direct intervention to suppress the transpiration process, such as halting water supply or severing root systems, which can cause detrimental damage to the plants. In contrast, microclimate models provide a more flexible and non-invasive means of simulating idealized scenarios. Among them, Computational Fluid Dynamics (CFD) is widely used to simulate heat transfer and airflow dynamics across spatial scales ranging from street canyons to entire urban districts [42].

A commonly adopted CFD model is ENVI-met, which is specifically designed to simulate surface–plant–air interactions in urban environments at high spatial and temporal resolutions. ENVI-met employs the numerical solution of a set of governing physical equations to derive key microclimatic variables, such as air temperature, surface temperature, wind speed, relative humidity, and various thermal comfort indices [43]. ENVI-met has been proven to be reliable for simulating air temperature and relative humidity, showing high agreement with measured data [44]. Its capacity to capture fine-scale environmental variables allows for detailed assessments of urban design impacts on local climate conditions and human thermal comfort [45]. ENVI-met facilitates the creation of "ideal scenarios" through its advanced modeling capabilities, allowing researchers to quantify the benefits of different green infrastructure and surface material strategies on microclimate and thermal comfort, ultimately informing sustainable urban design [46].

In this study, ENVI-met version 5.1.1 was employed to simulate the cooling capacity of an individual tree. The ENVI-met simulation was conducted following a systematic workflow: spatial scenario design, parameterization, model execution, and result extraction. The model domain consisted of 40 × 40 × 20 three-dimensional grid cells, with a spatial resolution of 0.5 × 0.5 × 0.5 m. The physical properties of surface covers, including albedo, thermal conductivity, heat capacity, and initial soil moisture, were parameterized based on field measurements from our previous study [47]. A custom-designed *F. concinna* tree model was incorporated at the center of the domain. Its height and leaf area density (LAD) were assigned based on field measurements, while its root depth and density were referenced from studies on the same species in a similar climatic zone [27,48]. Finally, the meteorological data (air temperature, relative humidity, wind speed/direction, and solar radiation) measured over the lawn area (Section 2.1) were used as driving forces for the simulation. The simulations were initialized using fully forced lateral boundary conditions to maintain realistic meteorological inputs. For the analysis of the tree's cooling effect, the simulated air temperature at a height of 1.75 m from the grid cell nearest to the observation point was extracted and used for subsequent calculation.

Due to the high sensitivity of the ENVI-met model to input driving data, excessively high wind speeds can cause model divergence, leading to computational instability. While rainfall introduces an additional source of evaporation, it complicates the isolation of the transpiration cooling effect. Therefore, data from days with severe winds (wind speeds exceeding 8 m/s) and rainy conditions were excluded from the analysis. Meteorological data from 195 days in 2022 with complete observational records were used as input for the simulations. During the study period, air temperatures ranged from 6.9 °C to 38.7 °C, and relative humidity varied between 32.1% and 100%. Each simulation was run for a 24 h period (from 00:00 to 24:00), with model outputs generated at hourly intervals.

2.3. Designing the Simulation Scenarios

To assess the individual effects of tree transpiration and shading on air temperature, four idealized scenarios were established within the ENVI-met model framework (Figure 2): (a) transpiring tree; (b) transpiring lawn; (c) non-transpiring tree; (d) non-transpiring lawn. Scenarios (a) and (b) represent actual conditions of a tree and lawn under normal physiological states and were used to replicate real-world microclimatic conditions. Simulated temperature outputs ($T_{c,s}$,$T_{l,s}$) from the two scenarios were compared with field measurements ($T_{c,m}$,$T_{l,m}$) to validate the model's performance, where subscript "*c*" refers to the canopy, "*l*" to the lawn, "*s*" to the simulated results, and "*m*" to the measured data. Scenario (c), which limits the transpiration process of the tree in the model, was designed to isolate the shading effect of tree transpiration on temperature. By comparing the temperature under the tree with and without transpiration (scenario a and scenario c), the specific cooling effect from transpiration could be determined. Scenario (d), a lawn without transpiration, served as the baseline for assessing the shading-induced cooling. The comparison between scenarios (c) and (d) allowed for quantification of the net cooling effect attributable to tree shading alone.

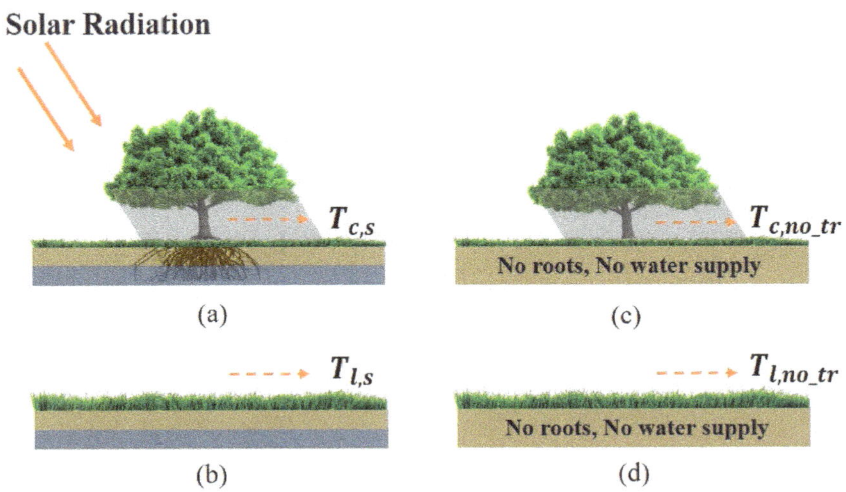

Figure 2. Schematic diagram of the 4 designed simulated scenarios: (**a**) transpiring tree with normal roots and water supply; (**b**) transpiring lawn; (**c**) non-transpiring tree without roots and water supply; (**d**) non-transpiring lawn. The subscript "no_tr" refers to the simulation results under non-transpiring conditions.

As plant transpiration in ENVI-met is a function of stomatal conductance and soil moisture availability, direct control over this process is not an inherent model feature [49]. To simulate non-transpiring conditions for comparative analysis, a methodological approach was employed to induce an extreme hydrological limitation. This was implemented by setting the volumetric water content of the entire 0–50 cm soil profile to 0%, in conjunction with specifying a Root Area Density (RAD) of zero for the target tree. This parameter combination eliminates both soil water availability and the plant's capacity for water uptake, thereby precluding transpiration. All other biophysical parameters (e.g., leaf area density, albedo) were held at a constant relative to the baseline scenarios. Thus, the resultant differences in microenvironmental variables are unequivocally attributable to the suppression of transpiration cooling.

2.4. Estimation of the Cooling Effect of Tree Transpiration and Canopy Shading

In this study, air temperature (T_a) differences are used to evaluate the cooling effects of the individual tree. The measured temperature differences between the shade ($T_{c,m}$) and the lawn ($T_{l,m}$) are the combined cooling effect of the tree due to shading and transpiration (ΔT_c):

$$\Delta T_c = T_{l,m} - T_{c,m} \tag{1}$$

The four simulated temperatures are utilized to calculate the temperature variations induced by tree shading and transpiration. The shade cooling effect (ΔT_{Sh}) is determined by simulating the temperature differential between the non-transpiring lawn (T_{l,no_tr}) and the non-transpiring tree (T_{c,no_tr}):

$$\Delta T_{Sh} = T_{l,no\_tr} - T_{c,no\_tr} \tag{2}$$

The tree transpiration cooling effect (ΔT_{Tr}) is quantified by the temperature differences between the simulated non-transpiring trees (T_{c,no_tr}) and the transpiration tree ($T_{c,s}$). In scenario c, the soil moisture content was set at 0, which inhibits tree transpiration as well as the transpiration effect of the grass, potentially resulting in an overestimation of the cooling effect attributed to tree transpiration. To minimize this effect, the temperature difference between the non-transpiring lawn (T_{l,no_tr}) and the transpiration lawn ($T_{l,s}$) is incorporated in the calculation of tree transpiration cooling:

$$\Delta T_{Tr} = T_{c,no\_tr} - T_{c,s} - (T_{l,no\_tr} - T_{l,s}) \tag{3}$$

The contributions of the two processes of shading and transpiration to the combined cooling effect of the tree are calculated by Equations (4) and (5). Considering that the impact of individual processes on temperature may not always be a cooling effect, their absolute values are used when calculating the sum of their effects:

$$Con_{\mathrm{Sh}} = \Delta T_{Sh} / (|\Delta T_{Sh}| + |\Delta T_{Tr}|) \tag{4}$$

$$Con_{Tr} = \Delta T_{Tr} / (|\Delta T_{Sh}| + |\Delta T_{Tr}|) \tag{5}$$

The simulated transpiration cooling contribution (Equation (5)) is further compared against an observation-based metric. This metric is defined as the proportion of latent heat (LE) attributable to tree transpiration within the net radiation (Rn), which was calculated from measured radiation and sap flow data:

$$Con_{Tr,EB} = \mathrm{LE}/\mathrm{Rn} \tag{6}$$

3. Results

3.1. Characteristics of Meteorological Conditions During the Experiment Period

Daily average air temperature and precipitation at the two observation sites over the entire measurement period are shown in Figure 3. During the observation period, the study area exhibited distinct seasonal climatic patterns: winter (December to February) and spring (March to May) were characterized by pronounced temperature variability, while summer experienced frequent high-temperature events and concentrated precipitation. The mean annual air temperature recorded was 23.6 °C, with the maximum daily temperature of 32.7 °C on day 205 (July 24) of the observation year. In total, 32 days of daily temperatures above 30 °C were recorded. The cumulative annual precipitation was 1653.6 mm, with most of the rainfall occurring between May and September. Notably, August alone experienced 20 rainy days, which contributed to significantly lower air temperatures compared to July and September.

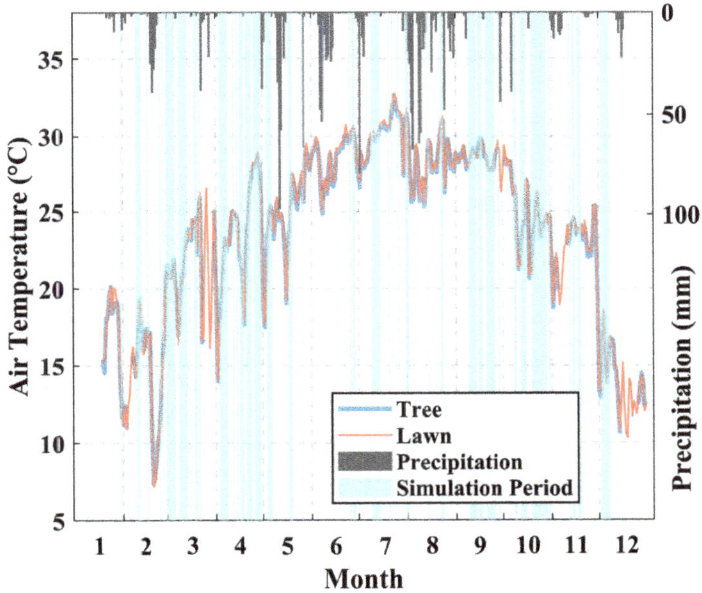

Figure 3. Air temperatures and precipitation of the study area during the observation period from January to December 2022, the blue shaded areas indicate the days that are used for ENVI-met simulations.

3.2. Characteristics of the Combined Cooling Effects of Tree Transpiration and Shading

Field observations provide a robust foundation for quantifying the cooling effects of the tree. In this study, the combined cooling effect of trees was determined by calculating the temperature differences between in situ measurements taken beneath the tree canopy and those in the adjacent open lawn area. Figure 4 illustrates the annual and diurnal dynamics of ΔT_c during 2022, with an annual mean value of 0.31 ± 0.46 °C. Daily cooling intensity exhibited substantial intra-annual variability (Figure 4a), demonstrating enhanced cooling performance during summer months (May–September). The maximum daily cooling effect reached 0.47 °C in late July. In contrast, negative cooling values predominated during winter and early spring (December–March), with the minimum daily value of −0.41 °C recorded in early March. This suggests that trees may contribute to localized warming under certain seasonal conditions.

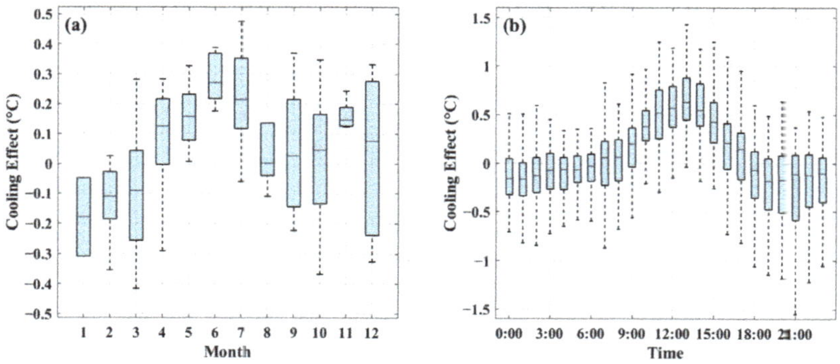

Figure 4. Combined cooling effect of the individual tree. (**a**) Annual variations in the tree cooling effect on a daily scale. (**b**) Intraday variations in the tree cooling effect on an hourly scale. The boxplots for each time illustrate the 25th, 50th, and 75th quantiles of the data.

The diurnal cooling pattern followed a characteristic unimodal distribution (Figure 4b), with the peak cooling effect occurring at 13:00, reaching a maximum instantaneous value of 1.43 °C and a mean cooling intensity of 0.64 °C at noon. Positive cooling effects persisted throughout the daytime (09:00–16:00), while negative values, indicative of a warming effect, prevailed during the nighttime (17:00–08:00). The most pronounced nocturnal warming effect was recorded at 21:00 on April 3rd, with a temperature increase of 1.71 °C under the tree canopy relative to the open lawn.

Overall, the cooling effect of the tree exhibited pronounced thermal dependence, with enhanced cooling efficiency observed during hotter periods, particularly in summer and around midday. Conversely, the cooling capability was diminished during winter and nighttime, when the canopy covers frequently contributed to localized warming. Additionally, the cooling performance of the tree was strongly modulated by weather variability. For example, in March, cold air incursions led to substantial temperature fluctuations, with the daily average cooling effect ranging from −0.41 °C to 0.28 °C. In August, the occurrence of typhoons and frequent precipitation events significantly reduced the tree's cooling capacity compared to the more stable and hotter conditions observed in July.

3.3. Model Validation of the Simulated Temperature

The observational experiment conducted across the tree canopy and lawn effectively quantified the cooling effects of the tree and established a robust data foundation for partitioning transpiration- and shading-induced cooling effects. To ensure the reliability of the ENVI-met model in replicating real-world microclimatic conditions, two representative simulation scenarios were configured: Scenario a (canopy-shading zones) and Scenario c (unobstructed lawn areas), both aligned with corresponding field observation sites. The model simulated hourly thermal conditions for 104 sunny days and 111 cloudy days throughout the study period. After excluding anomalous outputs, primarily induced by abrupt wind speed/direction changes that can cause numerical instability, 1718 valid hourly data points were retained for analysis.

Comparative validation against field measurements (Figure 5) demonstrated exceptional model performance, with coefficients of determination (R^2) between simulated and observed temperatures exceeding 0.98 (0.986–0.997) in both shaded and unshaded areas, the slope coefficients approaching 1.0 (0.974–0.992), indicating strong agreement. Slightly higher discrepancies were observed during daytime hours on cloudy days, likely due to rapid and transient fluctuations in solar irradiance that are challenging to capture in simulations. Additionally, a systematic underestimation in simulated temperatures was attributed to the vertical resolution constraints of ENVI-met: observational data were recorded at 2 m above ground, whereas model outputs corresponded to the nearest vertical layer at 1.75 m within the 0.5 m grid resolution.

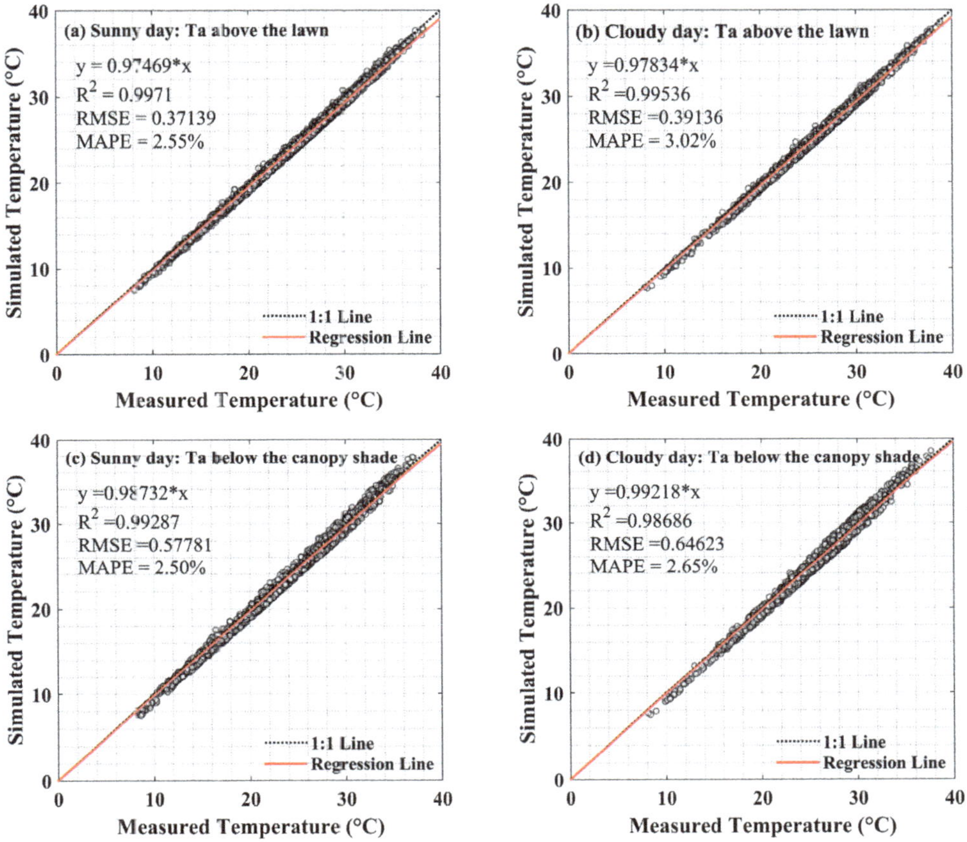

Figure 5. Comparison of the measured and simulated air temperatures. (**a**) Lawn on sunny days; (**b**) lawn on cloudy days; (**c**) canopy shade on sunny days; (**d**) canopy shade on cloudy days. RMSE: root mean square error; MAPE: mean absolute percentage error.

Overall, the detailed ENVI-met modeling accurately replicated air temperature dynamics and successfully achieved realistic representations of lawn and tree cover. This validation provides a critical foundation for the use of scenario-derived data in quantifying the tree cooling effect of transpiration and shading.

3.4. Respective Cooling Effects of Tree Transpiration and Canopy Shading

Based on the non-transpiring and lawn scenarios established in ENVI-met, we derived 1718 hourly transpiration and shade cooling effects. Figure 6 illustrates the annual and diurnal dynamics of partitioned cooling processes, revealing mean annual cooling intensities of 0.30 ± 0.13 °C for transpiration and 0.17 ± 0.27 °C for shading, corresponding to relative contributions of 44.7 ± 26.3% and 21.2 ± 51.6%, respectively. ΔT_{Sh} followed a unimodal annual pattern (Figure 6a), peaking in July with a maximum daily mean of 0.57 °C, and showing significantly reduced or even negative effects during the winter months. Specifically, during winter (December–February), ΔT_{Sh} exhibited net thermal accumulation, reaching a minimum value of −0.52 °C.

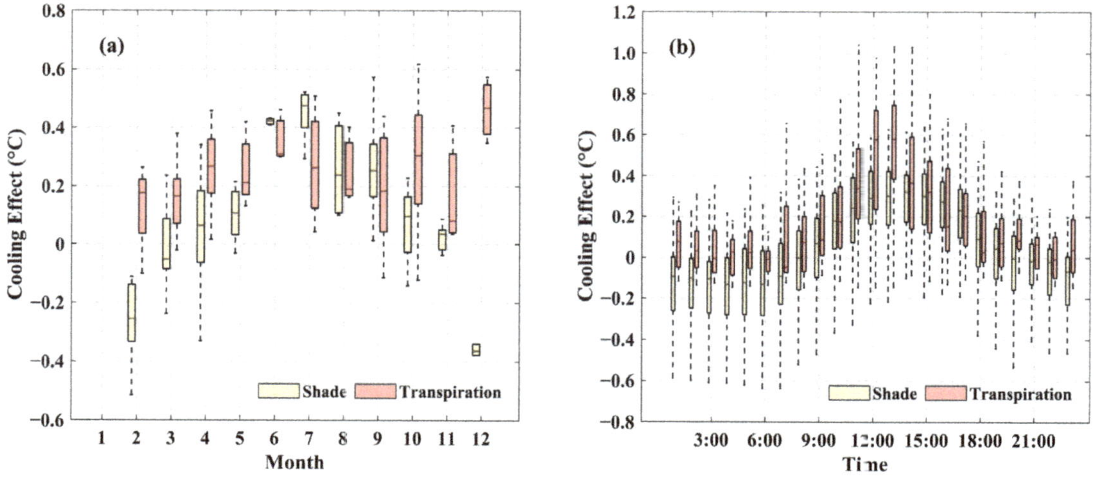

Figure 6. Shading and transpiration cooling effect of the individual tree. (**a**) Annual variations in the tree cooling effect on a daily scale. (**b**) Intraday variations in the tree cooling effect on an hourly scale.

ΔT_{Tr} displayed larger intra-annual variability, with daily means ranging from −0.12 °C to 0.82 °C. Positive cooling effects occurred in over 94% of cases, whereas negative anomalies were concentrated in February, September, and October. These instances were likely driven by increased evapotranspiration from lawn surfaces compared to tree transpiration, thereby amplifying the $(T_{l,no\_tr} - T_{l,m})$ differentials in Equation (3).

Diurnal patterns of both ΔT_{Sh} and ΔT_{Tr} exhibited sinusoidal patterns, with maximum cooling intensities observed at 13:00 (ΔT_{Tr}: 1.03 °C; ΔT_{Sh}: 0.63 °C), followed by gradual declines to nocturnal minima around 06:00. Notably, transpiration consistently provided stronger cooling than shading throughout daytime hours, underscoring its dominant role in tree-mediated cooling.

4. Discussion

4.1. Tree Cooling Effects and Thermal Response Patterns

Traditionally, the cooling effect of trees is considered directly linked to background air temperature [50]. Our observations reveal a unimodal variation pattern in tree-induced cooling across both annual and diurnal cycles, peaking during summer (July) and midday (13:00), consistent with air temperature trends. Similar patterns have been reported in studies on green space cooling [51–53]. Figure 7a delineates the nonlinear thermal response function derived from 3246 valid hourly observations, revealing accelerated cooling intensification with rising ambient temperatures. The tree cooling effect intensifies with rising temperatures, particularly above 30 °C, where the average cooling reaches 0.47 °C.

To further elucidate the impact of background temperature on the cooling effect of trees, air temperature data were stratified into deciles, and the median cooling effect of tree transpiration and shading was evaluated within each bin (Figure 7b). The shade cooling effect exhibits quasi-linear enhancement with temperature. At lower temperatures (<23.1 °C, representing the bottom 30th percentile and encompassing 84.7% winter or nighttime data), shading produced net warming, with temperature differences ranging from −0.23 to −0.02 °C, contributing negatively (−51.3% to −15.7%) to combined cooling effects (Figure 7c). These results provide evidence for canopy-induced thermal inversions under low-temperature conditions. Above 31.1 °C, shading dominates with a median cooling of 0.56 °C with a cooling contribution of 56.9%. In contrast, transpiration cooling follows a near-exponential pattern, consistent with its regulation by stomatal behavior and vapor pressure deficit [54]. In the initial 50th percentile, transpiration cooling was minimal, with a median effect of only 0.1 °C. As temperatures increased, cooling via transpiration intensified, with peaking at 0.6 °C. However, above 31.1 °C, a slight decline was observed, with cooling stabilizing around 0.4 °C, likely due to partial stomatal closure under extreme heat stress, which limits water loss and reduces transpiration efficiency.

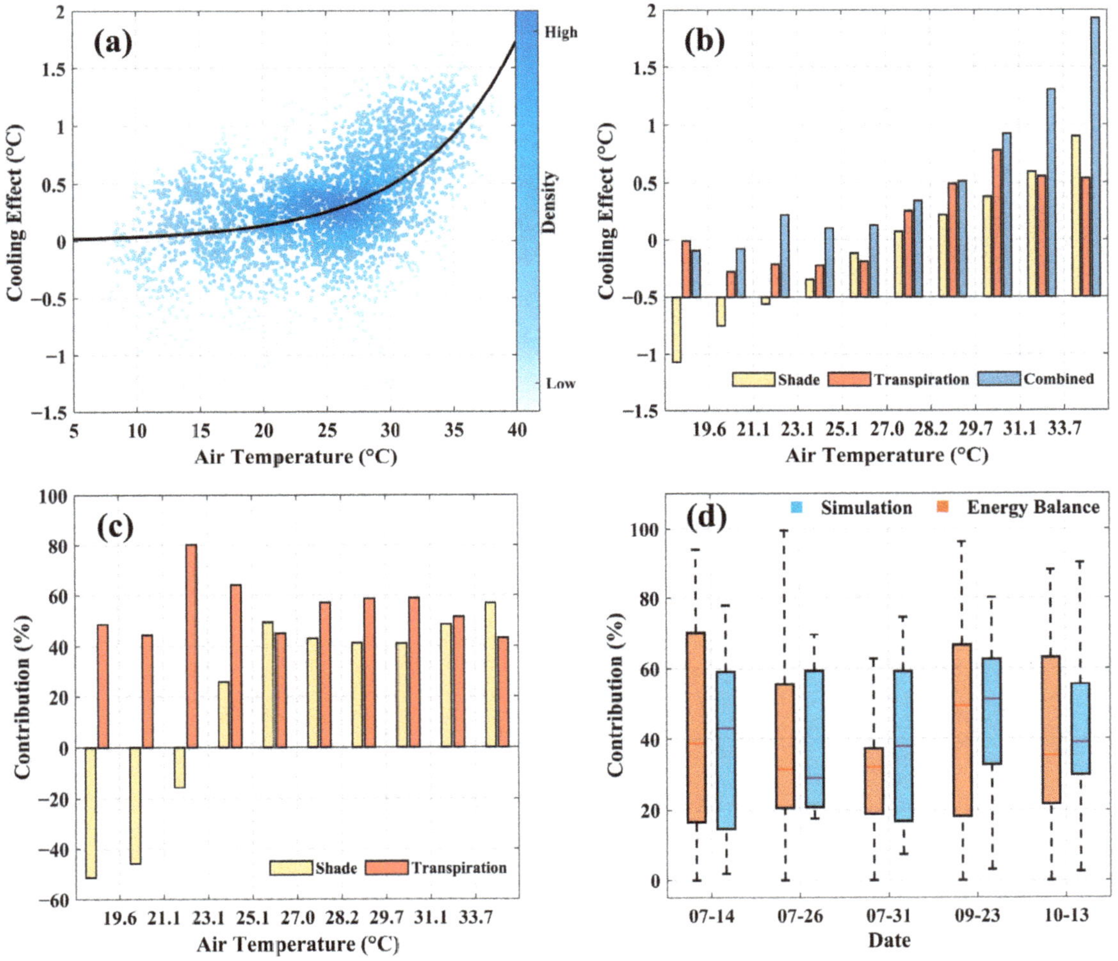

Figure 7. (**a**) Scatter plot of the combined cooling effect and air temperature, the black line represents the regression line. (**b**) Cooling effect of transpiration and shading under different air temperature ranges. (**c**) Cooling contribution of transpiration and shading. (**d**) Comparisons of transpiration cooling contribution between the ENVI-met simulation and flux measured data.

Correlational analysis revealed critical radiative drivers of arboreal cooling efficacy. The combined cooling effect demonstrated a significant positive correlation with air temperature (r = 0.56, $p < 0.05$), while solar radiation over open lawns exhibited the strongest association (r = 0.68, $p < 0.05$). Notably, a negative correlation emerged between the tree's combined cooling effect and longwave radiation in shaded areas (r = −0.39, $p < 0.05$), revealing a trade-off in the surface energy budget. Specifically, canopy cover reduces daytime surface heating by attenuating shortwave radiation but concurrently results in nocturnal warming through longwave radiation trapping [55,56]. This dynamic alters the surface–atmosphere radiative exchange, especially within shaded microclimates. These findings quantitatively resolve the paradox of canopy thermal impacts, demonstrating how tree shade transitions from a cooling asset to a thermal liability depending on radiative regime and phenological state.

For transpiration cooling, the simulation results demonstrated a significant positive correlation with measured latent heat (r = 0.77, $p < 0.01$). The model-based estimates of its cooling contribution agreed well with flux-derived values, with median differences in less than 10% (Figure 7d). This consistency between methods partly underscores the robustness of the quantification approach. However, the contribution calculated from flux data exhibited greater variability. This discrepancy is likely attributable to the low and unstable nighttime energy fluxes, whereas the simulated temperatures remained relatively stable.

4.2. Dynamic Shifts in the Dominant Cooling Mechanism

Over the course of the one-year observation experiment, the daily average cooling effects of trees exhibited remarkable variability (Figure 4). Based on the daily mean incoming clearness index, 215 observation days were categorized into 104 sunny days and 111 cloudy days, with sunny days defined as those having a daily average clearness index surpassing 0.80. Figure 8a compares the distributions of tree-induced cooling effects on both sunny and cloudy days. Overall, *F. concinna* provided stronger cooling under sunny conditions, with mean shading, transpiration and combined cooling effects reaching 0.25 ± 0.24 °C, 0.32 ± 0.31 °C, and 0.65 ± 0.38 °C, respectively. Under cloudy conditions, they decreased to 0.18 ± 0.24 °C, 0.29 ± 0.28 °C, and 0.62 ± 0.29 °C, respectively. The reduction in solar radiation not only limited the potential for shading-induced temperature regulation but also decreased the available energy for transpiration, thereby weakening both cooling pathways. Figure 8b further demonstrates the relative contributions of shading and transpiration to combined cooling across different weather scenarios. The median contribution of shading decreased from 60.0% in sunny days to 40.2% under cloudy conditions, while that of transpiration increased from 44.4% to 50.0%.

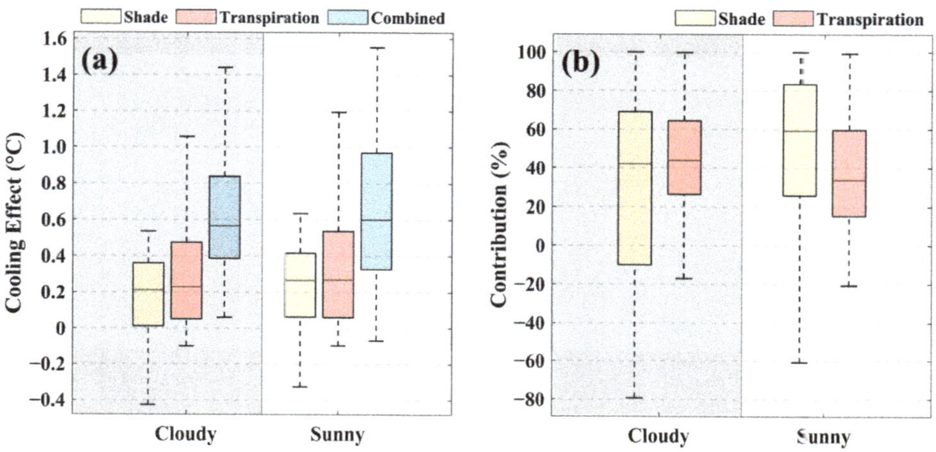

Figure 8. (**a**) The shading, transpiration and combined cooling effect of a tree under different weather conditions. (**b**) The cooling contribution of shading and transpiration under different weather conditions.

The dual roles of tree shading and transpiration in environmental cooling have long been central to urban climate research. In previous studies, shading has been considered the dominant cooling mechanism. For instance, a study conducted in Givatayim, Israel, found that direct shading accounted for approximately 80% of the cooling effect beneath the canopy, with transpiration contributing to a lesser extent [57]. Similarly, research conducted in Shanghai, China, reported that during heat waves, shading remained the principal mechanism for urban cooling at midday [58]. However, findings from the present study challenge this conventional view, highlighting the substantial and often dominant role of transpiration—particularly during nighttime periods. The observed higher contribution of transpiration in this study is likely attributable to the high transpiration capacity of *F. concinna*. It was demonstrated that *F. concinna* was among the highest transpiration rates across fifteen urban tree species in Los Angeles [59]. In subtropical climates, ample precipitation ensures consistent soil moisture, while elevated air temperatures increase the vapor pressure deficit—together enhancing the transpiration-driven cooling effect.

Trees regulate urban thermal environments through two primary biophysical mechanisms: passive radiative shading and active evaporative cooling via transpiration. *F. concinna* shows distinct responses of these processes to environmental drivers. Shading reduces solar heat gain through canopy interception, exhibiting a near-linear relationship with ambient temperature and solar irradiance. In contrast, transpiration operates as an active physiological process, governed by more complex interactions among air temperature, humidity, wind speed, and vapor pressure deficit, all of which influence stomatal conductance and latent heat flux [26,60]. The synergistic interaction between these mechanisms creates temperature-dependent compensatory effects. During periods of extreme heat, stomatal closure can limit transpiration, thereby amplifying the relative importance of shading. Conversely, at night or on cloudy days, when shading becomes less effective or may even induce warming through longwave radiation trapping, transpiration can continue to provide cooling. This temporal and phenological variation in thermal regulation underscores the necessity of species-specific and site-specific strategies in urban greening. To maximize thermal mitigation, urban design should integrate both canopy structure and physiological function: ensuring continuous shading coverage while also selecting species with high transpiration potential. Adopting the "right tree, right place" principle, planners should prioritize transpiration-efficient species in areas prone to thermal accumulation, and shade-dominant species (e.g., large tree canopy) in highly irradiated zones. These findings provide actionable insights for enhancing the resilience of urban environments under intensifying climate extremes such as heatwaves and urban heat islands.

4.3. Limitations and Future Research

This study has several limitations that should be acknowledged. The most significant limitation is its focus on a single tree species at one location. While this allowed for a controlled, year-long analysis, the findings' generalizability to other species with different physiological traits (e.g., deciduous habits, varying stomatal regulation) or in different urban contexts requires further investigation. Future studies could encompass a broader range of species and urban settings to develop more universally applicable guidelines. Secondly, the results of this study, such as the individual cooling effects and contributions of transpiration and shading, were derived exclusively from ENVI-met simulations. Although sap flow and radiation flux data were collected during the observation period, their utility for direct comparison and validation of the model results was limited due to the amount of valid data. Future research should prioritize obtaining a more comprehensive flux observation dataset to enable robust comparison and mutual validation of these two methodologies. Finally, this study focused on the biophysical mechanisms of shading and transpiration. Other factors influencing urban tree cooling, such as the impact of particulate matter deposition on leaf albedo and physiological function [61,62], were not considered. Investigating the interplay between these additional factors and the fundamental cooling mechanisms quantified here presents a promising avenue for future work.

5. Conclusions

This study proposed a quantitative framework to determine tree transpiration and canopy shading-induced cooling effects through high-frequency microclimate monitoring and scenario-based modeling. The key findings are as follows:

(1) The annual combined cooling effect of *F. concinna* is 0.31 ± 0.46 °C. This combined cooling effect has a clear unimodal variation pattern both seasonally and diurnally. The combined cooling effect exponentially increases as air temperatures rise and notably increases under higher temperature conditions (such as during heat waves).

(2) The shade cooling effect of the *F. concinna* is highly sensitive to solar radiation. Canopy shading could substantially reduce the amount of shortwave radiation reaching the ground surface during the daytime, especially on clear summer days when peak cooling could reach up to 0.63 °C. During the night, however, the canopy attenuates longwave radiation loss and results in local warming.

(3) *F. concinna* has a significant transpiration cooling effect, and it has a strong positive correlation with air temperature, with a maximum cooling effect of 1.03 °C.

Annually, the average contributions of shading and transpiration to combined cooling were 21.2 ± 51.6% and 44.7 ± 26.3%, respectively. The high transpiration rate of *F. concinna* and the favorable climatic conditions of the study area (i.e., warm temperatures and abundant soil moisture) facilitated dominant transpiration-driven cooling. Notably, during heatwave events, shading emerged as the predominant mechanism due to the decline in stomatal conductance under extreme stress.

Author Contributions: Conceptualization, G.Y.Q. and Z.S.; Funding acquisition, C.Y and G.Y.Q.; Investigation, Z.S. and Z.L.; Methodology, Z.S.; Visualization, Z.S. and W.H.; Writing—original draft, Z.S.; Writing—review and editing, G.Y.Q., C.Y. and W.H. All authors have read and agreed to the published version of the manuscript.

Funding: This research was funded by the Science, Technology and Innovation Commission of Shenzhen Municipality, grant number GXWD20201231165807007-20200827105738001, the Chinese Ministry of Science and Technology Projects, grant number 2017FY100206-03, and the National Natural Science Foundation of China, grant number 42571024.

Data Availability Statement: Data is contained within the article. Further inquiries can be directed to G.-Y.Q. (qiugy@pkusz.edu.cn) and Z.S. (shiz@pku.edu.cn).

Conflicts of Interest: The authors declare no conflicts of interest.

References

1. Yuan, Y.; Santamouris, M.; Xu, D.; Geng, X.; Li, C.; Cheng, W.; Su, L.; Xiong, P.; Fan, Z.; Wang, X.; et al. Surface urban heat island effects intensify more rapidly in lower income countries. *npj Urban Sustain.* **2025**, *5*, 11. [CrossRef]
2. Portela, C.I.; Massi, K.G.; Rodrigues, T.; Alcântara, E. Impact of urban and industrial features on land surface temperature: Evidences from satellite thermal indices. *Sust. Cities Soc.* **2020**, *56*, 102100. [CrossRef]
3. Rajagopal, P.; Priya, R.S.; Senthil, R. A review of recent developments in the impact of environmental measures on urban heat island. *Sust. Cities Soc.* **2023**, *88*, 104279. [CrossRef]
4. Wang, J.; Zhou, W.; Pickett, S.T.A.; Yu, W.; Li, W. A multiscale analysis of urbanization effects on ecosystem services supply in an urban megaregion. *Sci. Total Environ.* **2019**, *662*, 824–833. [CrossRef]
5. Guerreiro, S.B.; Dawson, R.J.; Kilsby, C.; Lewis, E.; Ford, A. Future heat-waves, droughts and floods in 571 European cities. *Environ. Res. Lett.* **2018**, *13*, 34009. [CrossRef]

6. IPCC. *Climate Change 2021: The Physical Science Basis. Contribution of Working Group I to the Sixth Assessment Report of the Intergovernmental Panel on Climate Change*; Cambridge University Press: Cambridge, UK; New York, NY, USA, 2021.
7. Kim, S.W.; Brown, R.D. Urban heat island (UHI) intensity and magnitude estimations: A systematic literature review. *Sci. Total Environ.* **2021**, *779*, 146389. [CrossRef]
8. Buchin, O.; Hoelscher, M.; Meier, F.; Nehls, T.; Ziegler, F. Evaluation of the health-risk reduction potential of countermeasures to urban heat islands. *Energy Build.* **2016**, *114*, 27–37. [CrossRef]
9. Scherer, D.; Fehrenbach, U.; Lakes, T.; Lauf, S.; Meier, F.; Schuster, C. Quantification of heat-stress related mortality hazard, vulnerability and risk in Berlin, Germany. *Die Erde* **2014**, *144*, 238–259.
10. Schinasi, L.H.; Benmarhnia, T.; De Roos, A.J. Modification of the association between high ambient temperature and health by urban microclimate indicators: A systematic review and meta-analysis. *Environ. Res.* **2018**, *161*, 168–180. [CrossRef] [PubMed]
11. Kumar, P.; Sharma, A. Study on importance, procedure, and scope of outdoor thermal comfort—A review. *Sust. Cities Soc.* **2020**, *61*, 102297. [CrossRef]
12. Deilami, K.; Kamruzzaman, M.; Liu, Y. Urban heat island effect: A systematic review of spatio-temporal factors, data, methods, and mitigation measures. *Int. J. Appl. Earth Obs. Geoinf.* **2018**, *67*, 30–42. [CrossRef]
13. Georgescu, M.; Morefield, P.E.; Bierwagen, B.G.; Weaver, C.P. Urban adaptation can roll back warming of emerging megapolitan regions. *Proc. Natl. Acad. Sci. USA* **2014**, *111*, 2909–2914. [CrossRef] [PubMed]
14. Lai, D.; Lian, Z.; Liu, W.; Guo, C.; Liu, W.; Liu, K.; Chen, Q. A comprehensive review of thermal comfort studies in urban open spaces. *Sci. Total Environ.* **2020**, *742*, 140092. [CrossRef] [PubMed]
15. Sadeghi, M.; Chaston, T.; Hanigan, I.; de Dear, R.; Santamouris, M.; Jalaludin, B.; Morgan, G.G. The health benefits of greening strategies to cool urban environments–A heat health impact method. *Build. Environ.* **2022**, *207*, 108546. [CrossRef]
16. Paschalis, A.; Chakraborty, T.C.; Fatichi, S.; Meili, N.; Manoli, G. Urban forests as main regulator of the evaporative cooling effect in cities. *Agu Adv.* **2021**, *2*, e2020AV000303. [CrossRef]
17. Schwaab, J.; Meier, R.; Mussetti, G.; Seneviratne, S.; Bürgi, C.; Davin, E.L. The role of urban trees in reducing land surface temperatures in European cities. *Nat. Commun.* **2021**, *12*, 6763. [CrossRef]
18. Ziter, C.D.; Pedersen, E.J.; Kucharik, C.J.; Turner, M.G. Scale-dependent interactions between tree canopy cover and impervious surfaces reduce daytime urban heat during summer. *Proc. Natl. Acad. Sci. USA* **2019**, *116*, 7575–7580. [CrossRef]
19. Bowler, D.E.; Buyung-Ali, L.; Knight, T.M.; Pullin, A.S. Urban greening to cool towns and cities: A systematic review of the empirical evidence. *Landsc. Urban Plan.* **2010**, *97*, 147–155. [CrossRef]
20. Sabrin, S.; Karimi, M.; Nazari, R.; Pratt, J.; Bryk, J. Effects of Different Urban-Vegetation Morphology on the Canopy-level Thermal Comfort and the Cooling Benefits of Shade Trees: Case-study in Philadelphia. *Sustain. Cities Soc.* **2021**, *66*, 102684. [CrossRef]
21. Wang, C.; Wang, Z.H.; Yang, J. Cooling effect of urban trees on the built environment of contiguous United States. *Earth's Future* **2018**, *6*, 1066–1081. [CrossRef]
22. Qiu, G.Y.; Yu, X.; Wen, H.; Yan, C. An advanced approach for measuring the transpiration rate of individual urban trees by the 3D three-temperature model and thermal infrared remote sensing. *J. Hydrol. (Amst.)* **2020**, *587*, 125034.
23. Winbourne, J.B.; Jones, T.S.; Garvey, S.M.; Harrison, J.L.; Wang, L.; Li, D.; Templer, P.H.; Hutyra, L.R. Tree transpiration and urban temperatures: Current understanding, implications, and future research directions. *Bioscience* **2020**, *70*, 576–588. [CrossRef]
24. Massetti, L.; Petralli, M.; Napoli, M.; Brandani, G.; Orlandini, S.; Pearlmutter, D. Effects of deciduous shade trees on surface temperature and pedestrian thermal stress during summer and autumn. *Int. J. Biometeorol.* **2019**, *63*, 467–479. [CrossRef]
25. Rahman, M.A.; Moser, A.; Rötzer, T.; Pauleit, S. Microclimatic differences and their influence on transpirational cooling of Tilia cordata in two contrasting street canyons in Munich, Germany. *Agric. For. Meteorol.* **2017**, *232*, 443–456. [CrossRef]
26. Rahman, M.A.; Stratopoulos, L.M.F.; Moser-Reischl, A.; Zölch, T.; Häberle, K.; Rötzer, T.; Pretzsch, H.; Pauleit, S. Traits of trees for cooling urban heat islands: A meta-analysis. *Build. Environ.* **2020**, *170*, 106606. [CrossRef]
27. Chen, X.; Zhao, P.; Hu, Y.; Ouyang, L.; Zhu, L.; Ni, G. Canopy transpiration and its cooling effect of three urban tree species in a subtropical city-Guangzhou, China. *Urban For. Urban Green.* **2019**, *43*, 126368. [CrossRef]
28. Lindén, J.; Fonti, P.; Esper, J. Temporal variations in microclimate cooling induced by urban trees in Mainz, Germany. *Urban For. Urban Green.* **2016**, *20*, 198–209. [CrossRef]
29. Shashua-Bar, L.; Hoffman, M.E. Vegetation as a climatic component in the design of an urban street: An empirical model for predicting the cooling effect of urban green areas with trees. *Energy Build.* **2000**, *31*, 221–235. [CrossRef]
30. Kim, J.; Khouakhi, A.; Corstanje, R.; Johnston, A.S.A. Greater local cooling effects of trees across globally distributed urban green spaces. *Sci. Total Environ.* **2024**, *911*, 168494. [CrossRef] [PubMed]
31. Li, H.; Zhao, Y.; Wang, C.; Ürge-Vorsatz, D.; Carmeliet, J.; Bardhan, R. Cooling efficacy of trees across cities is determined by background climate, urban morphology, and tree trait. *Commun. Earth Environ.* **2024**, *5*, 754. [CrossRef]
32. Jiao, M.; Zhou, W.; Zheng, Z.; Wang, J.; Qian, Y. Patch size of trees affects its cooling effectiveness: A perspective from shading and transpiration processes. *Agric. For. Meteorol.* **2017**, *247*, 293–299. [CrossRef]

33. Wu, Z.; Shi, Y.; Ren, L.; Hang, J. Scaled outdoor experiments to assess impacts of tree evapotranspiration and shading on microclimates and energy fluxes in 2D street canyons. *Sust. Cities Soc.* **2024**, *108*, 105486. [CrossRef]
34. Tan, P.Y.; Wong, N.H.; Tan, C.L.; Jusuf, S.K.; Chang, M.F.; Chiam, Z.Q. A method to partition the relative effects of evaporative cooling and shading on air temperature within vegetation canopy. *J. Urban Ecol.* **2018**, *4*, 1–11. [CrossRef]
35. Pace, R.; De Fino, F.; Rahman, M.A.; Pauleit, S.; Nowak, D.J.; Grote, R. A single tree model to consistently simulate cooling, shading, and pollution uptake of urban trees. *Int. J. Biometeorol.* **2021**, *65*, 277–289. [CrossRef] [PubMed]
36. Manickathan, L.; Defraeye, T.; Allegrini, J.; Derome, D.; Carmeliet, J. Parametric study of the influence of environmental factors and tree properties on the transpirative cooling effect of trees. *Agric. For. Meteorol.* **2018**, *248*, 259–274. [CrossRef]
37. Yu, H.; Zhang, T.; Fukuda, H.; Ma, X. The effect of landscape configuration on outdoor thermal environment: A case of urban Plaza in Xi'an, China. *Build. Environ.* **2023**, *231*, 110027. [CrossRef]
38. Zheng, X.; Chen, L.; Yang, J. Simulation framework for early design guidance of urban streets to improve outdoor thermal comfort and building energy efficiency in summer. *Build. Environ.* **2023**, *228*, 109815. [CrossRef]
39. Marchin, R.M.; Backes, D.; Ossola, A.; Leishman, M.R.; Tjoelker, M.G.; Ellsworth, D.S. Extreme heat increases stomatal conductance and drought-induced mortality risk in vulnerable plant species. *Glob. Change Biol.* **2022**, *28*, 1133–1146. [CrossRef]
40. Hayat, M.; Xiang, J.; Yan, C.; Xiong, B.; Wang, B.; Qin, L.; Saeed, S.; Hussain, A.; Zou, Z.; Qiu, G.Y. Environmental control on transpiration and its cooling effect of Ficus concinna in a subtropical city Shenzhen, southern China. *Agric. For. Meteorol.* **2022**, *312*, 108715. [CrossRef]
41. Zhang, Y.; Chiew, F.H.S.; Peña-Arancibia, J.; Sun, F.; Li, H.; Leuning, R. Global variation of transpiration and soil evaporation and the role of their major climate drivers. *J. Geophys. Res. Atmos.* **2017**, *122*, 6868–6881. [CrossRef]
42. Toparlar, Y.; Blocken, B.; Maiheu, B.; van Heijst, G.J.F. A review on the CFD analysis of urban microclimate. *Renew. Sustain. Energy Rev.* **2017**, *80*, 1613–1640. [CrossRef]
43. Salata, F.; Golasi, I.; de Lieto Vollaro, R.; de Lieto Vollaro, A. Urban microclimate and outdoor thermal comfort. A proper procedure to fit ENVI-met simulation outputs to experimental data. *Sust. Cities Soc.* **2016**, *26*, 318–343. [CrossRef]
44. Le, A.; Chan, Y. Comparison between ENVI-met and ANSYS-fluent when used for microclimate simulation. *J. Phys. Conf. Ser.* **2023**, *2600*, 82019. [CrossRef]
45. Simon, H.; Lindén, J.; Hoffmann, D.; Braun, P.; Bruse, M.; Esper, J. Modeling transpiration and leaf temperature of urban trees–A case study evaluating the microclimate model ENVI-met against measurement data. *Landsc. Urban Plan.* **2018**, *174*, 33–40. [CrossRef]
46. Liu, Z.; Cheng, W.; Jim, C.Y.; Morakinyo, T.E.; Shi, Y.; Ng, E. Heat mitigation benefits of urban green and blue infrastructures: A systematic review of modeling techniques, validation and scenario simulation in ENVI-met V4. *Build. Environ.* **2021**, *200*, 107939. [CrossRef]
47. Wang, B.; Yan, C.; Shi, Z.; Ding, J.; Zhang, T.; Qin, L.; Qiu, G.Y. Seasonal variation in water uptake patterns of two greening species and their responses to rainfall events in a subtropical megacity of China. *J. Hydrol.* **2023**, *618*, 129262. [CrossRef]
48. Cheng, K.Y.; Lau, K.; Shek, Y.T.; Liu, Z.; Ng, E. Evaluation on the performance of tree view factor in a high-density subtropical city: A case study in Hong Kong. *Build. Environ.* **2023**, *239*, 110431. [CrossRef]
49. Bruse, M. ENVI-Met Implementation of the Jacobs A−gs Model to Calculate the Stomata Conductance. Available online: http://www.envi-met.net/documents/new_a_gs.pdf (accessed on 16 February 2004).
50. Cheng, X.; Peng, J.; Dong, J.; Liu, Y.; Wang, Y. Non-linear effects of meteorological variables on cooling efficiency of African urban trees. *Environ. Int.* **2022**, *169*, 107489. [CrossRef] [PubMed]
51. de Abreu-Harbich, L.V.; Labaki, L.C.; Matzarakis, A. Effect of tree planting design and tree species on human thermal comfort in the tropics. *Landsc. Urban Plan.* **2015**, *138*, 99–109. [CrossRef]
52. Kong, F.; Yan, W.; Zheng, G.; Yin, H.; Cavan, G.; Zhan, W.; Zhang, N.; Cheng, L. Retrieval of three-dimensional tree canopy and shade using terrestrial laser scanning (TLS) data to analyze the cooling effect of vegetation. *Agric. For. Meteorol.* **2016**, *217*, 22–34. [CrossRef]
53. Zhang, L.; Zhan, Q.; Lan, Y. Effects of the tree distribution and species on outdoor environment conditions in a hot summer and cold winter zone: A case study in Wuhan residential quarters. *Build. Environ.* **2018**, *130*, 27–39. [CrossRef]
54. Lambers, H.; Chapin, F.S.; Pons, T.L. Plant Water Relations. In *Plant Physiological Ecology*; Lambers, H., Chapin, F.S., Pons, T.L., Eds.; Springer: New York, NY, USA, 2008; pp. 163–223.
55. Morakinyo, T.E.; Kong, L.; Lau, K.K.; Yuan, C.; Ng, E. A study on the impact of shadow-cast and tree species on in-canyon and neighborhood's thermal comfort. *Build. Environ.* **2017**, *115*, 1–17. [CrossRef]
56. Wujeska-Klause, A.; Pfautsch, S. The best urban trees for daytime cooling leave nights slightly warmer. *Forests.* **2020**, *11*, 945. [CrossRef]
57. Shashua-Bar, L.; Pearlmutter, D.; Erell, E. The cooling efficiency of urban landscape strategies in a hot dry climate. *Landsc. Urban Plan.* **2009**, *92*, 179–186. [CrossRef]
58. Yu, Z.; Chen, J.; Chen, J.; Zhan, W.; Wang, C.; Ma, W.; Yao, X.; Zhou, S.; Zhu, K.; Sun, R. Enhanced observations from an optimized soil-canopy-photosynthesis and energy flux model revealed evapotranspiration-shading cooling dynamics of urban vegetation during extreme heat. *Remote Sens. Environ.* **2024**, *305*, 114098. [CrossRef]

59. Pataki, D.E.; McCarthy, H.R.; Litvak, E.; Pincetl, S. Transpiration of urban forests in the Los Angeles metropolitan area. *Ecol. Appl.* **2011**, *21*, 661–677. [CrossRef] [PubMed]
60. Zhu, Y.; Cheng, Z.; Feng, K.; Chen, Z.; Cao, C.; Huang, J.; Ye, H.; Gao, Y. Influencing factors for transpiration rate: A numerical simulation of an individual leaf system. *Therm. Sci. Eng. Prog.* **2022**, *27*, 101110. [CrossRef]
61. Rai, P.K.; Chutia, B.M.; Patil, S.K. Monitoring of spatial variations of particulate matter (PM) pollution through bio-magnetic aspects of roadside plant leaves in an Indo-Burma hot spot region. *Urban For. Urban Green.* **2014**, *13*, 761–770. [CrossRef]
62. Rai, P.K. Impacts of particulate matter pollution on plants: Implications for environmental biomonitoring. *Ecotoxicol. Environ. Saf.* **2016**, *129*, 120–136. [CrossRef]

forests MDPI

Article

Response of Common Garden Plant Leaf Traits to Air Pollution in Urban Parks of Suzhou City (China)

Zhiyu Yang [1], Xing Zhang [1,2,*], Yanting Qu [2], Fei Gao [1] and Yutong Li [1]

1 School of Architecture and Urban Planning, Suzhou University of Science and Technology, Suzhou 215129, China; yzyworkin@163.com (Z.Y.); gaofei@usts.edu.cn (F.G.); zl_y98@foxmail.com (Y.L.)
2 Institute of Nature and Ecology, Heilongjiang Academy of Sciences, Harbin 150040, China; quyanting1976@126.com
* Correspondence: 2605@usts.edu.cn

Abstract: In this study, to comprehensively investigate the impact of garden plants on air quality, we measured the leaves of 18 common garden plants in three different areas, namely, Suzhou industrial parks (clean air area (CAA)), Xiangcheng district parks (lightly polluted area (LPA)), and Huqiu district parks (highly polluted area (HPA)). We also measured the leaf functional traits of different life-types of plants. To explore the trade-off strategies of the leaf traits of common garden plants in response to air pollution and to assess the adaptive capacity of different life types of plants to air pollution. The results show that plants in the polluted area had higher leaf dry matter content (LDMC) and leaf nitrogen content per unit mass (N_{mass}), and a lower specific leaf area (SLA), maximum net photosynthetic rate per unit area (A_{area}), transpiration rate (Tr), stomatal conductance (Gs), and chlorophyll value (SPAD). Pearson correlation analysis showed that SLA was significantly positively correlated with N_{mass}, Tr, photosynthetic use efficiency (PNUE), and SPAD, and significantly negatively correlated with LDMC, while A_{area} was significantly positively correlated with chlorophyll value. Redundancy analysis revealed that the correlation between each leaf functional trait and atmospheric pollution factors was as follows: LDMC > N_{mass} > SLA > LA > A_{area} > Tr > PNUE > SPAD. The results suggest that different plant types have varying levels of adaptability to environmental conditions. Trees were found to be the most adaptable, followed by shrubs, herbs, and lianas. Additionally, under the stress of air pollution, herbs and lianas exhibited characteristics of "fast investment-return" on the leaf economic spectrum, meaning they were able to quickly allocate resources to maximize their return. However, trees and shrubs displayed traits of "slow investment-return", indicating a more conservative approach to resource allocation. These results provide valuable insights into the leaf trade-off strategies of plants in Suzhou Park under air pollution stress and can guide the selection of suitable plant species in similar environments.

Keywords: garden plants; leaf economic spectrum; air pollution; leaf functional traits; trade-off strategies

1. Introduction

Since 1978, China has undergone rapid development, resulting in the significant emission of air pollutants that pose a threat to human respiratory health and physiological function, hinder urban ecological construction, and impede the healthy development of cities [1]. According to air pollution data from 2018, only one-third of the 338 monitored cities in China met the air quality standards [2,3]. Landscape plants play a crucial role in urban vegetation communities, providing diverse forms and species that facilitate the creation of aesthetically pleasing urban environments [4]. In addition to their ornamental function, plants have ecological functions, including carbon sequestration, oxygen release [5–7], dust retention, and noise reduction [8–10]. The leaves of plants serve as the primary site for energy conversion and material exchange with the external environment, and their functional traits can respond sensitively to environmental changes. Wright's Leaf economics spectrum (LES) theory, first proposed in 2004 [11], suggests that plants adopt a strategy of acquisition or conservatism, depending on the availability of resources, and trade-off resource allocation to enhance their adaptive capacity under environmental stress [12,13]. The LES theory offers a quantitative framework for predicting plant resource utilization efficiency and trade-off strategies [14]. However, it is crucial to validate the accuracy of the LES theory at the local scale [15]. Extensive research has demonstrated that leaf economic spectra are primarily influenced by environmental factors such as light, temperature, moisture, and notably, atmospheric pollutants including CO_2, CO, SO_2, NO_2, and O_3. Atmospheric pollution emerges as a significant factor impacting plant growth and its application in landscaping. The response of the leaf economic spectrum to atmospheric pollution predominantly centers on leaf structural and chemical traits. Osnas et al. have highlighted that specific leaf area (SLA) represents one of the trade-offs adopted by plants when faced with limited environmental resources [16]. In the presence of atmospheric pollution, plants may modify their resource allocation strategy, directing more resources towards growth and production capacity. This adjustment aims to fortify defensive tissues and structures, thereby enhancing vitality and resistance [17,18]. Chen and his team conducted a study on the leaf functional traits of 89 plant species in the eastern part of Guangdong Province, China [19]. Their findings revealed significant differences in specific leaf area (SLA) and leaf dry matter content (LDMC) among the different species. Under the stress of atmospheric pollution, the plants exhibited a tendency towards low SLA and high LDMC [20]. The response of herbaceous plants to O_3 was studied by Evans et al., who found no significant differences in leaf thickness, fenestrated tissue thickness, and spongy tissue thickness [21]. Other studies have reported an increase in the thickness of epidermal tissues and a decrease in the thickness of chloroplastic tissues in acacia leaves contaminated with NO_2 and SO_2 [22]. Hang and other scholars demonstrated that environmental stresses, such as vehicle exhaust and soot, resulted in a decreasing trend in leaf nitrogen content of garden shrubs in South China [23]. Additionally, Lai conducted a study on the leaf functional traits of garden plants under different concentrations of NO_2 fumigation, and the results showed significant differences in the nitrogen and phosphorus contents of the leaves of different plants in response to varying concentrations of NO_2.

Chlorophyll is an important pigment for photosynthesis in green plants, which not only absorbs and converts light energy, but also its content directly affects the photosynthetic capacity of green plants [24]. At the same time, chlorophyll is also one of the most important indicators of the degree of external stress on plants [25,26]. Different types of pollution in the ecological environment will affect the chlorophyll content of green plants, thus reflecting the level of plant resistance and sensitivity. Therefore, the change of chlorophyll content can be used as one of the important indicators to assess the ecological adaptability of plants. Gao Chuanyou et al. [27] showed that when plants are exposed to pollution stress for a long period of time, their photosynthetic efficiency decreases, resulting in a lack of basic substances required for growth and development. This is due to the effect of pollution on plants, resulting in the decomposition of chlorophyll a and b, which leads to a decrease in the total chlorophyll content [28]. Plant photosynthesis is also extremely sensitive to air pollution, and studies have shown that the exposure of plants to bursts of high SO_2 concentrations in the vicinity of plants can lead to a cessation of photosynthesis. Different species of plants respond differently to SO_2 concentrations. In a study by Samuel B M et al. [29], pine, spruce, larch and lime were fumigated with SO_2 for one hour. The results showed that photosynthesis was reduced in all four species. For crops such as soybean, wheat, rice and potato, Wu Liying et al. [30] conducted SO_2 fumigation tests and found that their photosynthesis was inhibited to different degrees.

In addition, there are many studies focusing on the effect of plant photosynthetic traits on NO_2. Okano K [31] showed that the net photosynthetic rate of gerbera increased after two weeks of NO_2 fumigation at a concentration of 0.2 μL/L, but NO_2 at concentrations of 0.5 μL/L and 1.0 μL/L inhibited the net photosynthetic rate of the plant to varying degrees. Sabaratnam and Gupta [32,33] found that the net photosynthetic rate of soybean was significantly reduced at 0.4 μL/L of NO_2 and increased at 0.2 μL/L of NO_2. The current study mainly focuses on the effects of atmospheric pollutants on the photosynthetic characteristics of plants, mainly related to SO_2, NO_2 and other aspects, but the study of the comprehensive effects of atmospheric pollutants is still limited. These studies highlight the sensitivity of leaf functional traits to atmospheric pollution and the potential for plants to adapt to environmental stress through adjustments in their leaf anatomy and nutrient allocation strategies [34].

In recent years, Chinese scholars have conducted extensive research on the Leaf economics spectrum (LES) theory, investigating various factors such as altitude [35], thermal environment [36], light intensity [37], and moisture gradients [38]. Specifically, studies conducted in China have explored the response of leaf functional traits to atmospheric pollution. Zhu [39] and Li [40] conducted experiments on greening tree species in Beijing and street trees in Suzhou, respectively, and observed the presence of a global leaf economic spectrum even in the presence of atmospheric pollution.

The Leaf economic spectrum (LES) theory posits that plants employ diverse resource allocation strategies in response to resource availability, thereby enhancing their ability to adapt to environmental stress. Leaf functional traits, including leaf dry matter content (LDMC), specific leaf area (SLA), leaf nitrogen content per unit mass (N_{mass}), and maximum net photosynthetic rate per unit area (A_{area}), play a vital role in this trade-off strategy. In this paper, the following leaf functional traits were selected as indicators for analysis: leaf dry matter content (LDMC); specific leaf area (SLA); the net photosynthetic rate per unit area (A_{area}); stomatal conductance (Gs); transpiration rate (Tr); the nitrogen content of leaves per unit mass (N_{mass}); the calculation of photosynthetic nitrogen utilization (PNUE); and chlorophyll values (SPAD).

This study aims to investigate the key issues in the planning and design of urban park plant landscapes. Based on the leaf economic spectrum traits, the study aims to scientifically evaluate the resource trade-off strategies of different life types and species of plants under atmospheric changes, and to investigate the responses of common garden plants under different environmental gradients, so as to provide new ideas and reference bases for the design of landscape plants.

2. Materials and Methods

2.1. Study Area

Suzhou, Jiangsu Province (119°55′~121°20′ E, 30°47′~32°02′ N) is located on the eastern coast of mainland China, with a subtropical monsoon maritime climate, four distinct seasons, and an annual rainfall of about 1100 mm. The average annual temperature is 15.7 °C. Suzhou is located in the intersection of northern subtropical and central subtropical zones. The bioclimatic characteristics, affected by the southeast monsoon, presenting a warm and humid climate. The natural vegetation belongs to the northern subtropical deciduous, evergreen broad-leaved mixed forest zone. Suzhou, located in the Yangtze River Delta city cluster, boasts a warm and humid climate along with abundant rainfall. It is renowned for its historical and cultural significance, showcasing precious classical gardens and landscapes. The city has a rich tradition of carefully selecting and matching landscape vegetation.

The city's economic structure is dominated by industry, and the industrial structure is dominated by heavy industry and mostly located in the upwind direction of the city. The number of key industrial enterprises in Suzhou reaches 10,233, contributing to 96.7% of the SO_2, 59.9% of the NO_x, and 49.7% of the $PM_{2.5}$.

In terms of time distribution, the $PM_{2.5}$ exceedance pollution is most serious in autumn and winter, and the $PM_{2.5}$ exceedance pollution is least serious in July to September. $PM_{2.5}$ mainly comes from the local area of Suzhou, mainly from motor vehicle exhaust, which reaches 33.6%, followed by coal combustion with 26.0%, industrial process with 12.8%, dust with 12.7%, biomass combustion with 5.9%, and other emission sources with 8.1%.The emissions of other pollutant are as follows: SO_2—98,400 tonnes; No_x—21,200 tonnes; $PM_{2.5}$—72,700 tonnes; and PM_{10}—156,600 tonnes. The main sources of SO_2 are electricity, iron and steel, and textile; the main sources of No_x are electricity, iron and steel, and motor vehicles; and the main sources of $PM_{2.5}$ and PM_{10} are iron and steel, road dust, construction dust, and electricity.

According to monitoring data requested by the research team from the Meteorological Service of the City, three administrative districts in Suzhou with different air pollution levels were selected (Figure 1), namely, Xiangcheng district (lightly polluted area (LPA)), Hi-tech district (highly polluted area (HPA)) in the heavily polluted area and Suzhou Industrial Park (clean air area (CAA)) in the clean air area as the control, and three parks in each area were selected for sampling experiments. The air quality data in this study were obtained submitting a request to the official website of Suzhou Bureau of Ecology and Environment (suzhou.gov.cn), and the data included the daily average values of SO_2, NO_2, PM_{10}, $PM_{2.5}$, and the daily average value of the air quality index (AQI) from 1 October to 30 November 2021.

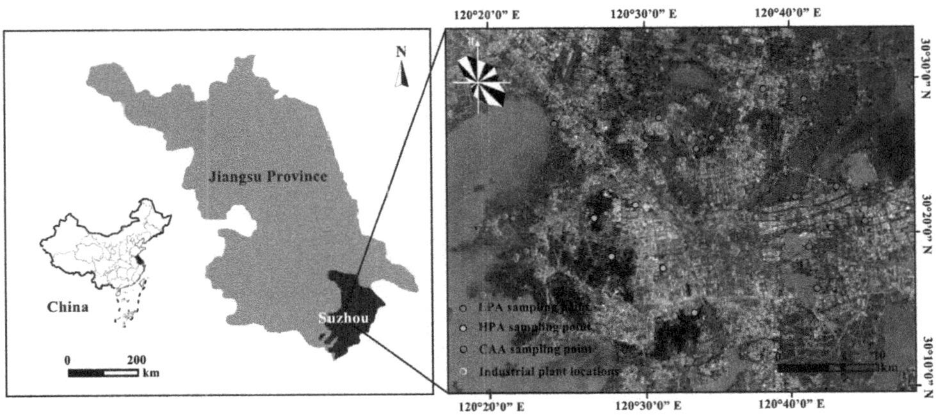

Figure 1. Study area and gradient zone delineation.

2.2. *Collection and Measurement of Samples*

The sampling areas in this paper are all urban parks and green spaces with good soil homogeneity, which ensures the relative consistency of tree species and management conditions. Sampling selected 18 species of similar age, good growth, and that are representative of the common garden tree species in Suzhou City according to the life type and are divided into four categories: trees, shrubs, herbs and lianas. The study of the sub-four categories of tree species life types in the different air quality environment, response, and trade-off mechanisms is shown in Table 1.

2.3. *Sample Collection and Measurement*

Plants with high rainfall in spring and summer have a higher capacity to settle and adsorb pollutants due to their high metabolic activity [41], and the pollutants accumulated by the plants reach their peak in autumn [42,43].

Therefore, the present study was conducted in autumn (early October to late November), with a total of 32 days of sampling, from 9:00 a.m. to 15:00 p.m., and no less than 30 plants of each tree species were collected, with 20 leaves per plant. The sampling sites were clusters of trees, and the park management office was consulted before sampling to ensure that the plants were of similar age and in good condition. We selected trees of similar diameter at breast height for sampling. Mature fresh leaves were collected to reflect the growth of the plant, free from pests and diseases, and then placed in Ziplock bags and refrigerated in the laboratory for later use.

Table 1. Eighteen experimental plant species in Suzhou.

Number	Plant-Life	Species	Familia	Genus
1	Tree	*Cinnamomum camphora* (L.) J.Pres	*Lauraceae*	*Camphora*
2		*Osmanthus fragrans* (Thunb.) Lour	*Oleaceae*	*Osmanthus*
3		*Magnolia grandiflora* L.	*Magnoliaceae*	*Magnolia*
4		*Koelreuteria paniculata* Laxm.	*Sapindaceae*	*Koelreuteria*
5		*Ginkgo biloba* L.	*Ginkgoaceae*	*Ginkgo*
6		*Sapindus saponaria* L.	*Sapindaceae*	*Sapindus*
7	Shrub	*Loropetalum chinense* (R. Br.) Oliver	*Hamamelidaceae*	*Loropetalum*
8		*Viburnum odoratissimum* Ker Gawl.	*Viburnaceae*	*Viburnum*
9		*Pittosporum tobira* (Thunb.) W. T. Aiton	*Pittosporaceae*	*Pittosporum*
10		*Hibiscus mutabilis* L.	*Malvaceae*	*Hibiscus*
11		*Buxus sinica* (Rehder & E. H. Wilson) M. Cheng	*Buxaceae*	*Buxus*
12		*Lagerstroemia indica* L.	*Lythraceae*	*Lagerstroemia*
13	Herb	*Ophiopogon bodinieri* H. Lév.	*Asparagaceae*	*Ophiopogon*
14		*Oxalis corniculata* L.	*Oxalidaceae*	*Oxalis*
15		*Ophiopogon japonicus* (L. f.) Ker Gawl.	*Asparagaceae*	*Ophiopogon*
16	Liane	*Jasminum mesnyi* Hance	*Oleaceae*	*Jasminum*
17		*Parthenocissus tricuspidata* (Siebold & Zucc.) Planch.	*Vitaceae*	*Parthenocissus*
18		*Trachelospermum jasminoides* (Lindl.) Lem.	*Apocynaceae*	*Trachelospermum*

2.3.1. Measurement of Leaf Structural Traits

The cleaned leaves were placed on an analytical balance to weigh the leaf fresh weight (LFW) and vernier calipers to measure the leaf thickness (LT), with 30 slices measured for each species. The leaves were first placed in an oven at 80 °C for 3 h and then set to dry at 55 °C for 48 h to a constant weight. After removal, the leaves corresponding to the leaf fresh weight number were weighed again for leaf dry weight (LDW), and leaf dry matter content (LDMC) was calculated with the following Formula (1):

$$\mathrm{LDMC}\left(\mathrm{g}\cdot\mathrm{g}^{-1}\right) = \mathrm{LDW}(\mathrm{g})/\mathrm{LFW}(\mathrm{g}) \quad (1)$$

Leaf area measurement (LA), leaf length (LL), and leaf width (LW) were determined using a leaf area scanner (MICROTEK ScanMaker i800plus). Leaf mass per area (LMA) was calculated according to Formula (2), and specific leaf area (SLA) was calculated according to Formula (3).

$$\mathrm{LMA}\left(\mathrm{g}\cdot\mathrm{cm}^{-2}\right) = \mathrm{LDW}(\mathrm{g})/\mathrm{LA}\left(\mathrm{cm}^{2}\right) \quad (2)$$

$$\mathrm{SLA}\left(\mathrm{cm}^{2}\cdot\mathrm{g}^{-1}\right) = \mathrm{LA}\left(\mathrm{cm}^{2}\right)/\mathrm{LDW}(\mathrm{g}) \quad (3)$$

Dried samples were ground and crushed through a sieve, and 1.0–15.0 mg of each sample was weighed using an analytical balance (XPE105, METTLER TOLEDO). Samples were wrapped in tin capsules and placed in an elemental analyzer (EURO EA3000) for the determination of the nitrogen content of leaves per unit mass (N_{mass}), and the mean value of the same species was taken as the value of the assay.

2.3.2. Measurement of Leaf Physiological Traits

Chlorophyll values (SPAD) of mature and undamaged plant leaves were determined using a portable chlorophyll meter (TYS-B) in an outdoor environment with natural light. The 3rd to 5th mature leaves of the plant at the front of the branches in the mid-height part of the sunny side were measured using a photosynthesizer (PPsystem, Li-Cor6400XT, Beijing, China), and the values were obtained in a standard leaf chamber. The instrument was set with LED red and blue light sources, light intensity of 1500 μmol $m^{-2}\cdot s^{-1}$, CO_2 concentration of 400 μmol L^{-1}, and leaf chamber temperature of 30 °C, and the average value was taken after three repetitions of measurement. The net photosynthetic rate per unit area (A_{area}), transpiration rate (Tr), and stomatal conductance (Gs) were measured. The calculation of photosynthetic nitrogen utilization (PNUE) was performed using Formula (4):

$$PNUE\left(\mu mol \cdot g^{-1} N \cdot s^{-1}\right) = A_{area}\left(\mu mol CO_2 m^2 \cdot s^{-1}\right)/N_{area}\left(g \cdot cm^{-2}\right) \quad (4)$$

2.4. Data Processing and Statistics

For leaf functional traits, SPSS 22.0 software was used, and the degree of variation in each of the leaf economic spectrum traits was assessed using the coefficient of variation. The calculation of the coefficient of variation (C.V) was performed using Formula (5):

$$C.V = (SD/MN) \cdot 100\% \quad (5)$$

where SD is the standard deviation, and MN is the mean.

One-way analysis of variance (ANOVA) was adopted to investigate the response of structural and physiological traits to different gradients of atmospheric pollution; Pearson correlation analysis (PCA) was used to investigate the correlation of leaf traits between different types of landscape plants and different gradients of atmospheric pollution; Canoco 5.0 software was used to carry out the redundancy analysis (RDA) was performed using Canoco 5.0 software to investigate the correlation between leaf traits of different landscape plants and atmospheric pollutants, and to screen the leaf traits related to air pollution.

To calculate the value of fuzzy membership function analysis (FMFA), when leaf traits are positively correlated with thermal environment, Formula (6) was used and is as follows:

$$R(X_i) = (X_i - X_{min})/(X_{max} - X_{min}) \quad (6)$$

If negatively correlated, Formula (6) is as follows:

$$R(X_i)_{inverse} = 1 - (X_i - X_{min})/(X_{max} - X_{min}) \quad (7)$$

where X_i is the measured value of the i-th indicator, X_{max} indicates the maximum value of the measured indicator, and X_{min} is the minimum value of the measured indicator.

3. Results

3.1. Air Pollution Gradient Analysis

From the data of the Suzhou Eco-Environmental Bureau, the air pollution level in the three gradient zones follows the trend HPA > LPA > CAA. Figure 2 shows that compared to CAA, the content of SO_2, NO_2, $PM_{2.5}$, and PM_{10} in HPA and LPA increased significantly, and that the urban air quality composite index AQI was the most distinctive among the three zones. The degree of air quality differentiation among the three gradient zones was significant. The range of contaminants is shown in Table 2.

Table 2. Atmospheric pollution concentration.

	Area	**SO_2 ($\mu g/m^3$)**	**NO_2 ($\mu g/m^3$)**	**PM_{10} ($\mu g/m^3$)**	**$PM_{2.5}$ ($\mu g/m^3$)**	**AQI**
CAA	Average value	4.87	27.00	54.00	11.00	46.00
	Min	1.00	5.00	8.00	3.00	18.00
	Max	17.00	84.00	94.00	63.00	170.00
LPA	Average value	7.93	28.90	62.20	25.00	79.00
	Min	3.00	4.00	10.00	4.00	18.00
	Max	15.00	105.00	99.00	66.00	169.00
HPA	Average value	9.37	43.80	65.53	27.37	126.00
	Min	6.00	6.00	7.00	5.00	15.00
	Max	32.00	70.00	164.00	77.00	500.00

Note: The pollutant values in the table are 24 h average concentrations.

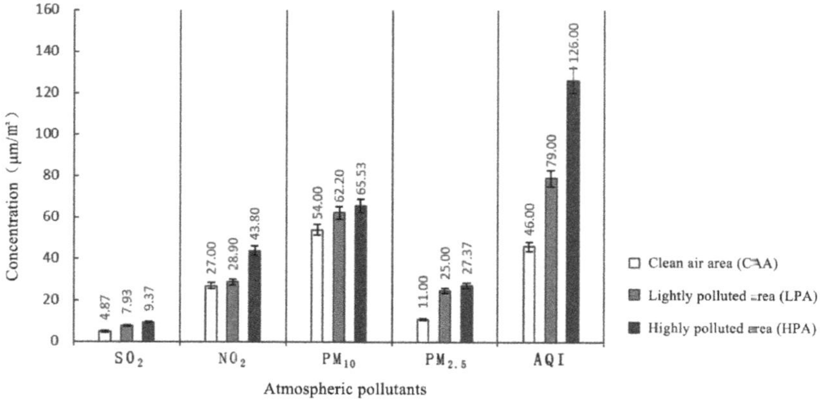

Figure 2. Air pollution.

3.2. Relationships between Air Pollutants and Leaf Functional Traits

In this study, redundancy analysis was used in order to investigate the relationship between air pollution factors and different leaf functional traits of 18 garden plants in Suzhou. Redundancy analyses were performed with plant leaf functional traits as species variable groups and each pollutant in the air environment as environmental variable groups. Each arrow points to the direction of the sharpest increase in the value of the corresponding environmental variable. The angle between the arrows indicates the correlation between the environmental variable and the leaf trait: a positive correlation is present when the angle is less than 90°, and a negative correlation is present when the angle is greater than 90°. The length of the arrows indicates the fit of the environmental variables.

The environmental correlations between each trait and air pollution were as follows: LDMC > N_{mass} > SLA > LA > A_{area} > TR > PNUE > SPAD. Each pollutant was positively correlated with the LDMC and N_{mass} of trees, and negatively correlated with SLA, Gs, Tr, SPAD, A_{area}, and PNUE, with Axis 1 and Axis 2 explaining 76.17% and 4.82% of all the information, respectively, cumulatively explaining 80.99% of the information (Figure 3a). Each pollution factor was positively correlated with shrub LDMC, N_{mass}, and SLA, and negatively correlated with Gs, Tr, SPAD, A_{area}, and PNUE, with Axis 1 explaining 61.02% of all the information and Axis 2 explaining 1.48%, cumulatively explaining 62.5% of the information (Figure 3b). Vine LDMC, N_{mass}, and LA were positively correlated with each pollutant, and other leaf functional traits were negatively correlated with each pollutant factor, with Axis 1 explaining 82.96% of all the information and Axis 2 explaining 6.19%, with a cumulative total of 89.15% (Figure 3c). Herbaceous LDMC, N_{mass}, and LA were positively correlated with each air pollutant, and all other functional traits were negatively correlated, with Axis 1 and Axis 2 explaining 79.44% and 3% of all the information, respectively, and cumulatively explaining 82.44% of the information (Figure 3d). It can be seen that the first two axes clearly reflect the relationship between plant leaf functional traits and air pollutants, and are mainly determined by Axis 1, indicating that the degree of linear combination between the sorting axes and atmospheric environmental factors can better reflect the correlation between the environment and plant leaf functional traits, and the sorting results are reliable.

3.3. Leaf Functional Traits

ANOVA analysis was conducted on 18 garden tree species in Suzhou City, and it was concluded that the differences between the traits of different garden plant life type in the three air quality gradient zones in Suzhou City were significant (Table 3).

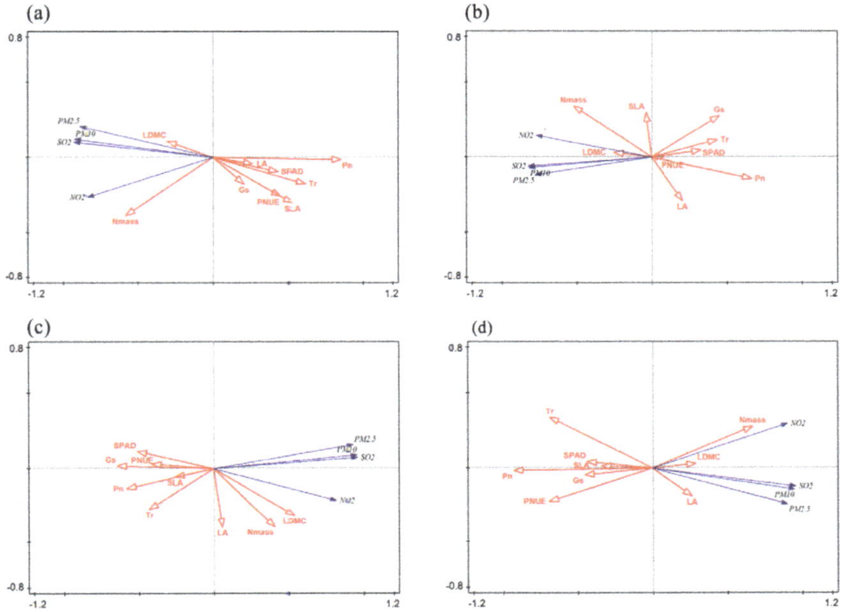

Figure 3. (**a**) Trees. (**b**) Shrubs. (**c**) Lianas. (**d**) Herbs.

In CAA and LPA, the LDMC was trees > shrubs > lianas > herbs, and in HPA, lianas > trees > shrubs > herbs; herb SLA was the largest in the three gradient zones, which was herbs > lianas > trees > shrubs in CAA and LPA, and herbs > lianas > shrubs > trees in HPA, and herbs were significantly different from trees, shrubs, and lianas. The A_{area} of shrubs were the largest in all three gradient zones and differed significantly from other life types; Gs of different life types showed herbs > shrubs > lianas > trees in the clean zone and LPA, and shrubs > herbs > lianas > trees in the HPA; shrub Tr was the highest in all three gradient zones, and showed shrubs > trees > herbs > lianas in the CAA, shrubs > herbs > lianas > trees in the LPA, and shrubs > herbs > lianas > trees in the HPA. N_{mass} was herbs > lianas > trees > shrubs in the CAA, lianas > herbs > trees > shrubs in the LPA, and lianas > herbs > shrubs > trees in the HPA; herb PNUE was the highest in all the three gradient zones, herbs > lianas > shrubs > trees in the CAA, and herbs > shrubs > lianas > trees in the remaining two zones; the SPAD of each plant life type was the highest in the CAA. In the CAA, the SPAD of each life type was lianas > shrubs > herbs > trees; in the LPA, it was lianas > shrubs > trees > herbs; and in the HPA, it was lianas > shrubs > tree s> herbs, and it formed a significant difference among the four life types.

Table 3. Plant leaf functional traits in gradient zones.

Area	Plant-Life	LDMC	SLA	A_{area}	Gs	Tr	N_{mass}	PNUE	SPAD
CAA	Tree	0.433 ± 0.071 a	10,013.975 ± 2606.067 b	16.48 ± 3.705 b	0.006 ± 0.004 a	0.62 ± 0.952 a	2.586 ± 0.395 b	2.437 ± 1.148 b	47.023 ± 9.302 a
	Shrub	0.375 ± 0.047 a	6659.786 ± 2683.895 b	20.669 ± 2.898 a	0.013 ± 0.008 ab	0.762 ± 0.86 a	2.058 ± 0.583 c	8.012 ± 6.319 b	51.42 ± 12.868 a
	Herb	0.296 ± 0.092 b	12,045.93 ± 7313.533 b	18.767 ± 3.335 ab	0.008 ± 0.002 bc	0.27 ± 0.104 a	2.68 ± 0.449 b	17.039 ± 3.079 b	52.994 ± 8.389 a
	Vine	0.248 ± 0.103 b	41,495.877 ± 45,318.948 a	17.148 ± 4.436 b	0.015 ± 0.014 a	0.443 ± 0.329 a	3.105 ± 0.251 a	35.425 ± 41.661 a	51.068 ± 8.449 a
LPA	Tree	0.512 ± 0.105 a	7254.972 ± 1575.539 b	14.734 ± 2.957 b	0.005 ± 0.003 b	0.162 ± 0.056 b	2.293 ± 0.239 c	1.918 ± 1.084 c	45.018 ± 6.056 bc
	Shrub	0.385 ± 0.053 b	5740.056 ± 3100.166 b	17.884 ± 2.592 a	0.009 ± 0.005 ab	0.577 ± 0.614 a	2.111 ± 0.476 c	8.562 ± 8.643 b	48.891 ± 11.567 ab
	Herb	0.381 ± 0.038 b	8010.565 ± 5156.448 b	12.901 ± 1.469 b	0.006 ± 0.003 b	0.177 ± 0.076 b	3.361 ± 0.106 a	6.29 ± 3.896 bc	53.133 ± 2.204 a
	Vine	0.253 ± 0.084 c	27,374.194 ± 27,628.919 a	12.934 ± 1.397 b	0.011 ± 0.01 a	0.201 ± 0.089 b	2.781 ± 0.781 b	17.138 ± 11.502 a	40.67 ± 4.988 c
HPA	Tree	0.491 ± 0.151 a	6822.454 ± 2559.228 b	12.926 ± 3.033 b	0.006 ± 0.003 b	0.113 ± 0.067 b	2.466 ± 0.227 c	1.896 ± 1.17 c	42.611 ± 9.565 ab
	Shrub	0.381 ± 0.136 b	7111.725 ± 8816.375 b	18.764 ± 5.07 a	0.013 ± 0.013 a	0.449 ± 0.425 a	2.697 ± 1.01 bc	9.002 ± 12.428 a	47.066 ± 12.162 a
	Herb	0.495 ± 0.092 a	7959.507 ± 4706.202 b	13.042 ± 2.754 b	0.008 ± 0.005 ab	0.123 ± 0.047 b	3.868 ± 0.749 a	4.121 ± 3.28 ab	49.276 ± 1.217 a
	Vine	0.279 ± 0.109 b	20,180.6 ± 14,521.167 a	12.046 ± 1.263 b	0.008 ± 0.005 ab	0.179 ± 0.135 b	3.222 ± 0.825 b	10.265 ± 7.399 a	35.589 ± 6.299 b

(a–c) indicate significant differences in functional traits ($p < 0.05$). Note: leaf dry matter content (LDMC); specific leaf area (SLA); the net photosynthetic rate per unit area (A_{area}); stomatal conductance (Gs); transpiration rate (Tr); the nitrogen content of leaves per unit mass (N_{mass}); calculation of photosynthetic nitrogen utilization (PNUE); chlorophyll values (SPAD).

3.4. Characteristics of Variation in Plant Leaf Traits in Different Gradient Zones

There were interspecific differences in leaf traits among trees, shrubs, lianas, and herbs under different air pollution environment, and the variation of each trait among the 18 garden plants ranged from 8.1% to 153.5% for CAA, 4.1% to 106.5% for LPA, and 2.5% to 138.1% for HPA. The leaf trait with the highest coefficient of variation for both CAA and LPA was Tr, and the leaf trait with the highest coefficient of variation for HPA was PNUE. The coefficients of variation for LDMC, N_{mass}, and SPAD were all smaller in the three different air environments. The smallest coefficients of variation for trees in the three different gradients were 15.3%, 10.4%, and 9.2% for N_{mass}, respectively. The smallest coefficients of variation for shrubs in CAA and LPA were LDMC, and the largest was Tr. The smallest coefficient of variation for shrubs in HPA was SPAD, and the largest was PNUE. The smallest coefficient of variation for lianas in the three different gradients was SPAD, and the largest coefficient of variation for lianas was PNUE. The minimum value was SPAD in all three groups; the maximum values were SLA in CAA and LPA, and PNUE in HPA. The minimum value of the coefficient of variation for herbs was N_{mass} in CAA, and the minimum value was A_{area} in both contaminated areas (Table 4).

Table 4. Characteristics of variation in plant leaf functional traits in different gradient zones.

Area	Plant-Life	LDMC	SLA	A_{area}	Gs	Tr	N_{mass}	PNUE	SPAD
CAA	Tree	0.164	0.260	0.225	0.667	1.535	0.153	0.471	0.198
	Shrub	0.125	0.403	0.140	0.615	1.129	0.283	0.789	0.250
	Herb	0.311	0.607	0.178	0.250	0.385	0.168	0.181	0.158
	Liane	0.415	1.092	0.259	0.933	0.743	0.081	1.176	0.165
LPA	Tree	0.206	0.217	0.201	0.620	0.344	0.104	0.565	0.135
	Shrub	0.137	0.540	0.145	0.604	1.065	0.225	1.009	0.237
	Herb	0.099	0.644	0.114	0.417	0.431	0.032	0.619	0.041
	Liane	0.332	1.009	0.108	0.853	0.444	0.281	0.671	0.123
HPA	Tree	0.308	0.375	0.235	0.500	0.593	0.092	0.617	0.224
	Shrub	0.357	1.240	0.270	1.000	0.947	0.374	1.381	0.258
	Herb	0.186	0.591	0.211	0.625	0.382	0.194	0.796	0.025
	Liane	0.391	0.720	0.105	0.625	0.754	0.256	0.721	0.177

Note: leaf dry matter content (LDMC); specific leaf area (SLA); the net photosynthetic rate per unit area (A_{area}); stomatal conductance (Gs); transpiration rate (Tr); the nitrogen content of leaves per unit mass (N_{mass}); calculation of photosynthetic nitrogen utilization (PNUE); chlorophyll values (SPAD).

3.5. Correlation Analysis between Leaf Functional Traits

3.5.1. Leaf Traits of Plants

As can be seen in Figure 4, among leaf structural traits, LDMC was highly significantly negatively correlated with SLA ($p < 0.01$). Among leaf physiological traits, N_{mass} was highly significantly negatively correlated with A_{area} and significantly negatively correlated with Tr ($p < 0.05$). A_{area} was highly significantly positively correlated with Gs and Tr, and significantly positively correlated with PNUE and SPAD. Gs was highly significantly positively correlated with PNUE, and significantly negatively correlated with SPAD. SPAD was highly significantly negatively correlated with both Tr and PNUE.

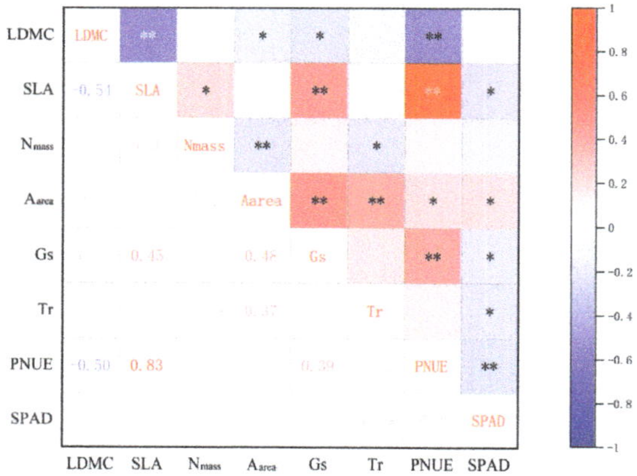

Figure 4. Comprehensive correlation analysis of plant leaf functional traits. * ≤ 0.05; ** ≤ 0.01.

Among the structural and physiological traits of leaves, LDMC was significantly negatively correlated with A_{area} and Gs, and highly significantly negatively correlated with PNUE; SLA was significantly positively correlated with N_{mass}, highly significantly positively correlated with Gs and PNUE, and significantly negatively correlated with SPAD.

3.5.2. Leaf Traits of Plants of Different Life Forms

As can be seen from Figure 5, between leaf structural traits, LDMC and SLA were significantly negatively correlated in tree plant leaves. Between leaf structural and physiological traits, LDMC was highly significantly negatively correlated with A_{area}, and SLA was significantly positively correlated with Gs and PNUE. Between leaf physiological traits, N_{mass} was highly significantly positively correlated with Tr, A_{area} was significantly positively correlated with Tr, and Gs was significantly positively correlated with PNUE. SPAD was highly significantly negatively correlated with PNUE and Tr.

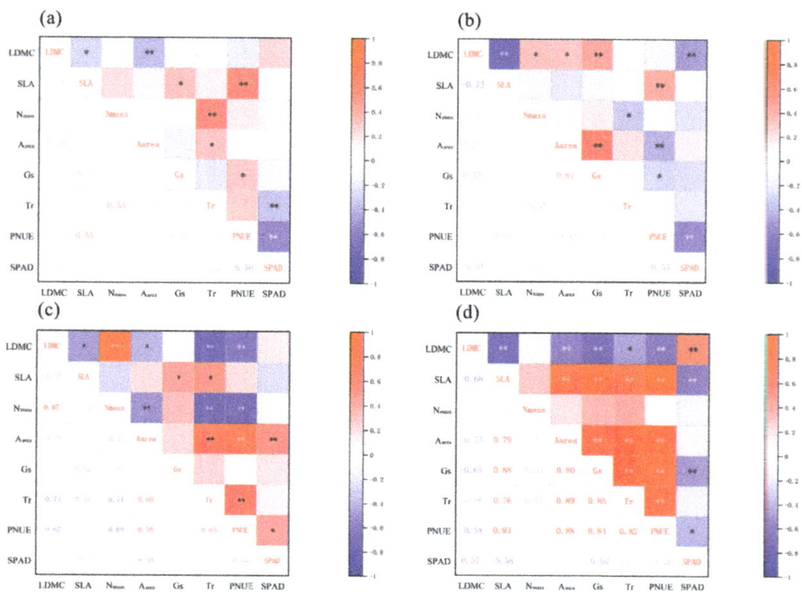

Figure 5. Correlation analysis of leaf traits. In the figure, (**a**) trees, (**b**) shrubs, (**c**) lianas, (**d**) herbs. * ≤ 0.05; ** ≤ 0.01.

Between leaf structural traits, shrub plant leaf LDMC was highly significantly negatively correlated with SLA. Between leaf structural traits and physiological traits, LDMC was significantly positively correlated with N_{mass} and A_{area}, highly significantly negatively correlated with SPAD, and highly significantly negatively correlated with SLA and PNUE. Between leaf physiological traits, N_{mass} was significantly negatively correlated with Tr, and A_{area} was highly significantly positively correlated with Gs.Gs was highly significant positively correlated with PNUE, Gs was highly significant negatively correlated with PNUE, and PNUE was highly significant negatively correlated with SPAD.

Between the structural traits of leaves, LDMC of lianas was significantly negatively correlated with SLA. between the structural and physiological traits of leaves, LDMC was highly significantly positively correlated with N_{mass}, highly significantly negatively correlated with A_{area}, highly significantly negatively correlated with Tr and PNUE, and significantly positively correlated with Gs and Tr. Between the physiological traits of leaves, N_{mass} was highly significantly negatively correlated with A_{area}, Tr and PNUE were highly significant negatively correlated, A_{area} was highly significant positively correlated with Tr, PNUE and SPAD, Tr was highly significant positively correlated with PNUE, and PNUE was significantly positively correlated with SPAD.

In herbaceous plants, the structural trait LDMC was highly significantly negatively correlated with SLA; between structural shape and physiological traits, LDMC was highly significantly negatively correlated with A_{area}, Gs, and PNUE, highly significantly negatively correlated with Tr, and highly significantly positively correlated with SPAD, and SLA was highly significantly positively correlated with A_{area}, Gs, Tr, and PNUE, and highly significantly negatively correlated with SPAD; and among the leaf physiological traits, A_{area} was significantly positively correlated with Gs, Tr, and PNUE; Gs was significantly positively correlated with Tr and PNUE and significantly negatively correlated with SPAD; Tr was significantly positively correlated with PNUE; and PNUE was significantly negatively correlated with SPAD.

3.6. Fuzzy Affiliation Function Analysis

Among the four life types, the order of adaptability to the environment was trees (0.480) > shrubs (0.418) > herbs (0.374) > lianas (0.367) (Figure 6). Among the trees, the order of adaptability from strong to weak was *Magnolia grandiflora*, *Koelreuteria paniculata*, *Camphora officinarum*, *Osmanthus fragrans*, *Ginkgo biloba*, and *Sapindus saponaria*, respectively. Shrubs were ranked from strongest to weakest in order of adaptability as *Hibiscus mutabilis*, *Pittosporum tobira*, *Loropetalum chinense*, *Viburnum odoratissimum*, *Buxus sinica*, and *Lagerstroemia indica*. The adaptability of herbaceous plants was ranked as *Oxalis corniculata* > *Ophiopogon japonicus* > *Ophiopogon bodinieri*, and that of lianas was ranked as *Parthenocissus tricuspidata* > *Jasminum mesnyi* > *Trachelospermum jasminoides*. The adaptability of plants with different life types in CAA was greater than that in HPA, except for *Buxus sinica*. In HPA, the most adaptable tree was *Magnolia grandiflora* (0.538), and the weakest was *Sapindus Saponaria* (0.359); the most adaptable shrub was *Hibiscus mutabilis* (0.447), and the weakest was *Lagerstroemia indica* (0.339). The strongest herb was *Oxalis corniculata* (0.423), and the weakest was *Ophiopogon bodinieri* (0.316); among lianas, the strongest was *Parthenocissus tricuspidata* (0.405), and the weakest was *Trachelospermum jasminoides* (0.320).

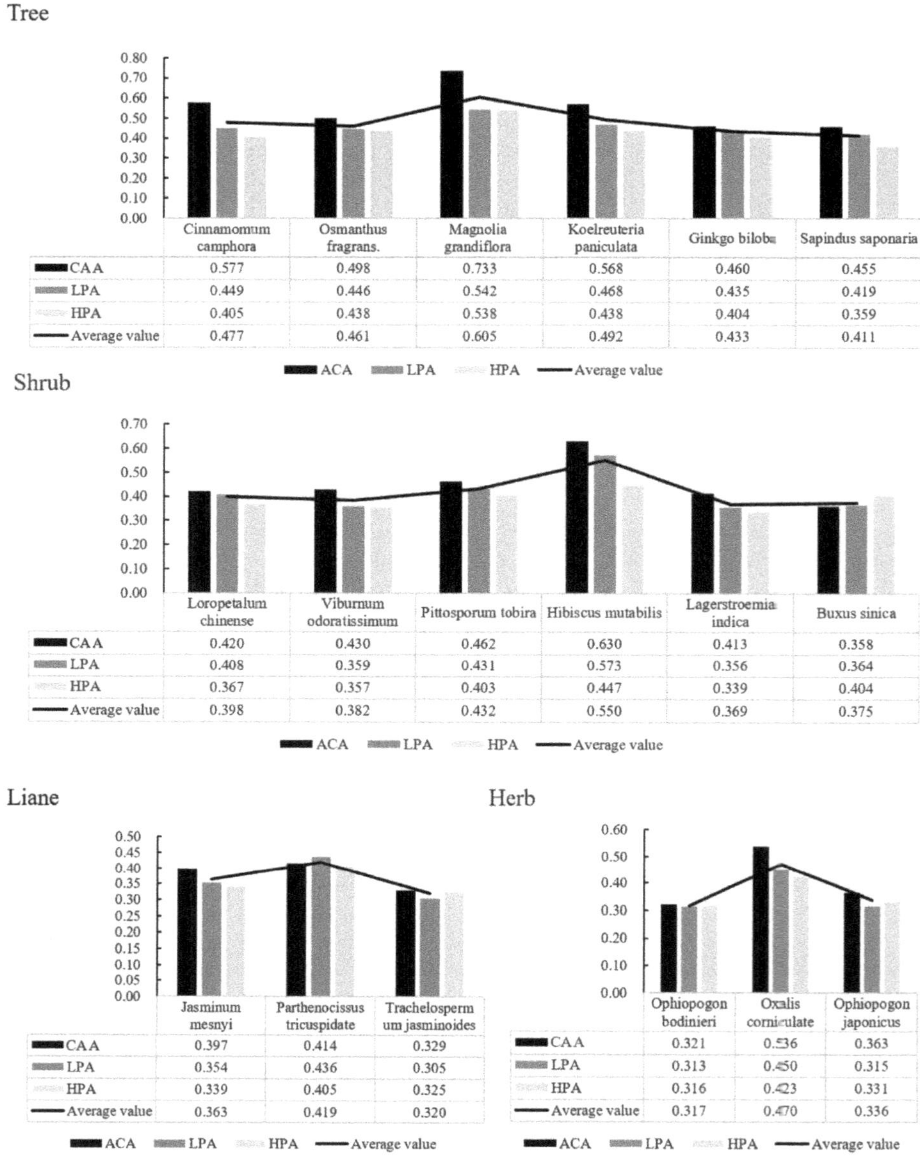

Figure 6. Fuzzy affiliation function analysis of plant environmental adaptation.

4. Discussion

4.1. Leaf Functional Traits in Relation to Air Quality

In this study, we observed a positive correlation between the leaf functional traits specific leaf area (SLA) and leaf nitrogen content (N_{mass}), which aligns with the findings of previous research conducted on Jinhua Beishan plants [44]. Furthermore, N_{mass} exhibited a highly significant positive correlation with leaf net assimilation rate (A_{area}), which in turn showed positive correlations with stomatal conductance (Gs), transpiration rate (Tr), photosynthetic nitrogen use efficiency (PNUE), and chlorophyll content (SPAD). The positive correlation between SLA and N_{mass} suggests that plants have the ability to modify their leaf structure to enhance nitrogen accumulation, thereby improving photosynthetic efficiency per unit area and enhancing leaf transpiration rate. This adaptive response aids in mitigating the detrimental effects of air pollution. However, under high-pollution conditions (HPA), we observed a decrease in Tr across all plant life types. This decline may be attributed to the plants' regulation of stomatal [45,46] closure or contraction upon reaching a certain threshold of air pollution. This adaptive mechanism serves to prevent the intrusion of pollutants, consequently reducing transpiration rates.

The size of the SPAD value reflects the chlorophyll content in leaves, and it can be considered an indicator of leaf development. The correlation between A_{area} and SPAD revealed a decrease in both variables under HPA conditions. This decline can be attributed to the inadequate accumulation of organic matter in plant leaves under environmental stress, leading to premature leaf senescence and chlorophyll decomposition, ultimately resulting in a decrease in SPAD values. Du's research presents the same results, with his team finding that chlorophyll a and b are damaged to varying degrees in atmospheric pollution environments, and that total chlorophyll content decreases [47].

The results of redundancy analyses showed that LDMC and N_{mass} were leaf functional traits with consistent and strong correlation with the direction of change of each pollution factor among all plant types, which indicated that LDMC and N_{mass} could be used as good leaf traits to reflect urban air quality. In the face of air pollution stress, plants improved their physiology by increasing LDMC and at the same time regulated the rate of leaf nitrogen building to potentially affect their photosynthetic utilization efficiency and chlorophyll value, improving their adaptability to air pollution environments. This is consistent with the findings of Juanxia Li [48] and Jiyou Zhu [39].

4.2. Relationships between Leaf Traits of Different Life Types

Differences in functional traits among species serve as the foundation for species coexistence in natural ecosystems, while intraspecific trait variation also contributes significantly to species coexistence and distribution [49]. Only by considering both intraspecific and interspecific trait variations can we accurately assess the response of species to environmental changes and resource competition during community development [50]. Leaf dry matter content (LDMC) serves as a critical indicator of nutrient resource conservation in plants [51]. High LDMC values indicate the ability of plants to thrive in resource-limited environments, enabling them to develop growth responses and maintain nutrient conservation efficiency [52]. In this study, the mean LDMC value for trees, shrubs, lianas, and herbs in Suzhou were found to be 0.479 $g \cdot g^{-1}$, 0.380 $g \cdot g^{-1}$, 0.371 $g \cdot g^{-1}$, and 0.26 $g \cdot g^{-1}$, respectively. The higher LDMC values observed in trees and shrubs suggest their enhanced resistance to physical damage. Moreover, the significant increase in LDMC among all plant types in areas with higher pollution levels reflects the leaf's efficient nutrient stabilization mechanism, enabling survival in more polluted air environments.

Research has demonstrated that plants with lower specific leaf area (SLA) values exhibit efficient nutrient conservation, while those with higher SLA values are better at capturing light and have relatively higher growth rates [53]. We investigated the variations in SLA among plant life forms in different air pollution gradients. The results show significant differences in SLA between herbaceous plants and other plant types across the air quality gradients. Additionally, SLA values for all plant types were negatively correlated with air pollution levels. Trees and shrubs exhibited lower SLA values compared to herbs and lianas, which occupy the middle and upper layers of the community and are influenced by factors such as strong light. These tree and shrub species allocate more biomass and nitrogen to build cell walls, enabling them to accumulate photosynthesis products necessary for overwintering and enhancing leaf toughness [54]. In contrast, herbaceous plants tend to allocate more nitrogen to thylakoids and carboxylases involved in respiration, resulting in higher photosynthetic capacity, as evidenced by their larger leaf area [55], lower leaf thickness, and higher SLA values [56].

We aimed to investigate the variations in A_{area}, Tr, and Gs among different plant life forms. The mean A_{area} values were 14.713 μmolm^{-2}·s^{-1} for trees, 19.116 μmolm^{-2}·s^{-1} for shrubs, 14.903 μmolm^{-2}·s^{-1} for lianas, and 14.043 μmolm^{-2}·s^{-1} for herbs. The A_{area} of shrubs was significantly higher than that of other plant types, indicating their greater ability to absorb CO_2 and nitrogen from the air and use them for photosynthesis, ultimately improving plant growth efficiency and productivity. Furthermore, A_{area} was significantly correlated with the Tr and Gs of different plant types, which is consistent with the findings of Shao et al.'s study on the main greening tree species in Shanghai [57].

We also examined the SLA, N_{mass}, and PNUE of different plant types. The results reveal that herbaceous plants had significantly higher SLA, N_{mass}, and PNUE values compared to trees, shrubs, and lianas. This finding is similar to the results of Song He et al.'s study on the leaf economic spectrum of Beijing Botanical Gardens, which suggests that the specific combination of structural and physiological traits of herbaceous plants, such as thin leaf blades and high photosynthetic capacity, gives them a competitive advantage in their habitat [58].

4.3. Analysis of Plants in the Leaf Economic Spectrum

The leaf economic spectrum represents a quantitative manifestation of the variation in plant leaf functional traits, as assessed through a series of trait indicators (Figure 7). In this study, we observed significantly higher values of SLA, N_{mass}, and PNUE in herbs and lianas compared to shrubs and trees, with the lowest values found in trees. This suggests that herbs and lianas occupy the "fast investment-return" end of the leaf economic spectrum, characterized by high nutrient concentrations, rapid photosynthetic rates, fast respiration rates, and short lifespans. On the other hand, shrubs and trees occupy the "slow investment-return" end of the spectrum, characterized by low nutrient concentrations, high photosynthetic rates, slow respiration rates, and long lifespans.

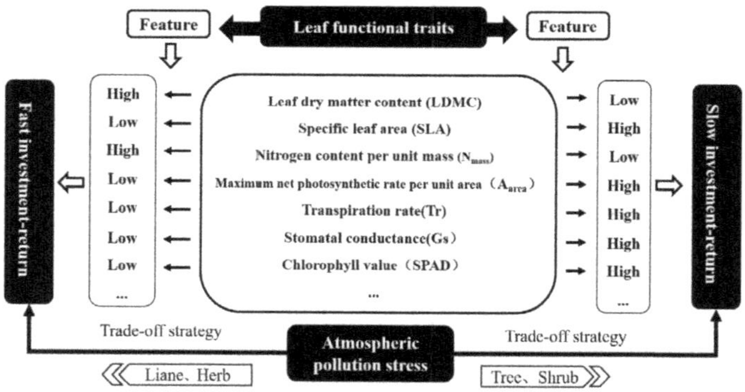

Figure 7. Leaf economic spectrum under the stress of atmospheric pollution [59–61].

However, the position of different life types on the leaf economic spectrum is also shifted under different atmospheric pollution environments. We observed that trees under LPA and HPA were biased towards the "fast investment-return" end of the spectrum, whereas plants in CAA were biased towards the "slow investment-return end"; however, we did not observe any significant positional differences for shrubs. The largest deviation in the leaf economic spectrum was for lianas, with very significant shifts to both ends of the spectrum for both plants in CAA and HPA; herbs were second only to lianas, but there was not much difference between the shifts for plants in the CAA and LPA environments. Overall, plants in CAA were more oriented towards the "slow investment-return end", while plants in LPA and HPA were more oriented towards the "fast investment-return" end, which confirms that plants will make different levels of resource This confirms that plants make different levels of resource trade-offs in air pollution environments.

4.4. Research Limitations

In this study, the relationship between leaf economic spectral traits and the effect of air pollution on 18 plant species in Suzhou City was analyzed only in a single season of a single year, while environmental factors such as temperature, wind speed, potential evapotranspiration, and water [52,63] have seasonal differences and may lead to variations in the trade-off strategies of the plants and cause bias in the experimental results. In addition, this may also introduce a bias in the relationship between leaves to air pollutants and the overall plant response to the environment are different. In future studies, we can continue to expand the research level by combining the functional traits of the whole plant, or set up experimental sample plots by considering the effects of leaf litter and roots on air pollution, and conduct research from a community perspective by combining leaf area index, plant height and crown size, and try to monitor the experiments for many years in order to ensure a more comprehensive experiment.

5. Conclusions

The results of this study on garden plants in Suzhou City show correlations among leaf economic spectrum traits that are consistent with global patterns. In this study, we investigated the response of plant leaf traits under urban air pollution, enriched the small-scale study of leaf economic spectrum theory, and proposed a plant-screening model under different air pollution environments so as to make the plant configuration more in line with the objective quantitative standards, and to provide a model reference for the construction of scientific, reasonable and ecologically balanced urban parks and green spaces under the air pollution environment.

Author Contributions: Z.Y., experimentation, data analysis, writing and editing; X.Z. and F.G., review and editing; Y.Q., project administration; Y.L., experimentation. All authors have read and agreed to the published version of the manuscript.

Funding: This research was funded by "Heilongjiang academy of sciences shuangtiyanzhen project (grant NO. STYZ2022ZR01)"; "Jiangsu postgraduate research and innovation programme (grant No. KYCX22_3310)"; and "14th Five-Year Plan" Jiangsu Province Key Discipline Construction Project Funding (Landscape Architecture) (grant No. 082240006/002/002).

Data Availability Statement: Data are contained within the article.

Conflicts of Interest: The authors declare no conflict of interest.

References

1. Cao, D.; Ramirez, C.D. Air pollution, government pollution regulation, and industrial production in China. *J. Syst. Sci. Complex.* **2022**, *33*, 1064–1079. [CrossRef]
2. Vennemo, H.; Aunan, K.; Lindhjem, H.; Seip, H.M. Environmental pollution in China: Status and trends. *Rev. Environ. Econ. Policy* **2009**, *3*, 209–230. [CrossRef]
3. Zheng, S.; Kahn, M. Understanding China's urban pollution dynamics. *J. Econ.* **2013**, *51*, 731–772. [CrossRef]
4. Costa, M. Vegetation Dispersion, Interspersion, and Landscape Preference. *Front. Psychol.* **2022**, *13*, 771543. [CrossRef]
5. Temjen, W.; Singh, M.R.; Ajungla, T. Carbon sequestration potential of two Musa cultivars from Mokokchung, Nagaland, North East India along an altitudinal gradient. *Curr. Sci.* **2022**, *123*, 925–927. [CrossRef]
6. Jin, S.; Zhang, E.; Guo, H.; Hu, C.; Zhang, Y.; Yan, D. Comprehensive evaluation of carbon sequestration potential of landscape tree species and its influencing factors analysis: Implications for urban green space management. *Carbon Balance Manag.* **2023**, *18*, 17. [CrossRef]
7. Parr, J.; Sullivan, L.; Chen, B.; Ye, G.; Zheng, W. Carbon bio-sequestration within the phytoliths of economic bamboo species. *Glob. Chang. Biol.* **2010**, *16*, 2661–2667. [CrossRef]
8. Zhao, Y.; Lei, S.; Yang, X.; Gong, C.; Wang, C.; Cheng, W.; Li, H.; She, C. Study on spectral response and estimation of grassland plants dust retention based on hyperspectral data. *Remote Sens.* **2020**, *12*, 2019. [CrossRef]
9. Li, H.; Zhu, X.; Kong, W.; Zheng, M.; Guo, X.; Wang, T. Physiological response of urban greening shrubs to atmospheric particulate matter pollution: An integral view of ecosystem service and plant function. *Environ. Exp. Bot.* **2023**, *213*, 105439. [CrossRef]
10. Xinru, H.E.; Zhang, Y.; Bing, S.U.N.; Pujie, W.E.I.; Die, H.U. Study on leaf epidermis structure and dust-retention ability of five Machilus species. *Not. Bot. Horti Agrobot. Cluj-Napoca* **2019**, *47*, 1224–1229.

11. Wright, I.J.; Reich, P.B.; Westoby, M.; Ackerly, D.D.; Baruch, Z.; Bongers, F.; Cavender-Bares, J.; Chapin, T.; Cornelissen, J.H Diemer, M.; et al. The worldwide leaf economics spectrum. *Nature* **2004**, *428*, 821–827. [CrossRef] [PubMed]
12. Reich, P.B.; Walters, M.B.; Ellsworth, D.S. From tropics to tundra: Global convergence in plant functioning. *Proc. Natl. Acad. Sc USA* **1997**, *94*, 13730–13734. [CrossRef] [PubMed]
13. Shipley, B.; De Bello, F.; Cornelissen, J.H.; Laliberté, E.; Laughlin, D.C.; Reich, P.B. Reinforcing loose foundation stones in trait-based plant ecology. *Oecologia* **2016**, *180*, 923–931. [CrossRef] [PubMed]
14. Westoby, M.; Falster, D.S.; Moles, A.T.; Vesk, P.A.; Wright, I.J. Plant ecological strategies: Some leading dimensions of variation between species. *Annu. Rev. Ecol. Syst.* **2002**, *33*, 125–159. [CrossRef]
15. Anderegg, L.D.; Berner, L.T.; Badgley, G.; Sethi, M.L.; Law, B.E.; HilleRisLambers, J. Within-species patterns challenge ou understanding of the leaf economics spectrum. *Ecol. Lett.* **2018**, *21*, 734–744. [CrossRef]
16. Osnas, J.L.; Lichstein, J.W.; Reich, P.B.; Pacala, S.W. Global leaf trait relationships: Mass, area, and the leaf economics spectrum *Science* **2013**, *340*, 741–744. [CrossRef]
17. Le, Y. Comparative analysis of different software PCA for plant trait integration. *J. Cent. South For. Univ. Sci. Technol.* **2015**, *35* 59–64.
18. Bucci, S.J.; Goldstein, G.; Meinzer, F.C.; Scholz, F.G.; Franco, A.C.; Bustamante, M. Functional convergence in hydraulic architecture and water relations of tropical savanna trees: From leaf to whole plant. *Tree Physiol.* **2004**, *24*, 891–899. [CrossRef]
19. Chen, W. Characteristics of variation in leaf functional traits of 89 common plant species in Guangdong. *J. Ecol.* **2016**, *35* 2101–2109.
20. Zhu, J. Spectral characteristics of large-leaved boxwood and its leaf functional traits in response to foliar dustfall. *Spectrosc. Spectr Anal.* **2020**, *40*, 1620–1625.
21. Evans, L.S.; Albury, K.; Jennings, N. Relationships between anatomical characteristics and ozone sensitivity of leaves of several herbaceous dicotyledonous plant species at Great Smoky Mountains National Park. *Environ. Exp. Bot.* **1996**, *36*, 413–420 [CrossRef]
22. Zhang, T.; Shan, L.; Xue, Q. Effect of nitrogen and phosphorus nutrition on water relations in wheat. *J. Plant Nutr. Fertil.* **2000**, *6* 147–151.
23. Hang, X.; Weng, S.; Yuan, Z. Study on leaf trait characteristics of five garden shrubs in South China and their response to environment. *J. Northwest For. Coll.* **2014**, *29*, 243–247.
24. Zhang, Q. *Experimental Study on the Introduction of Raspberries and Blackberries in Ya'an, Sichuan*; Sichuan Agricultural University Ya'an, China, 2004.
25. Su, X.; Hu, D.; Lin, Z. Effects of atmospheric pollution on chlorophyll fluorescence characteristics of two green plants in Guangzhou city. *J. Plant Ecol.* **2002**, *26*, 599–604.
26. Li, J.; Liu, N.; Ren, H.; Shen, W.J.; Jian, S.G. Ecological adaptations of seven plant species to tropical coral island environments *Ecol. Environ. J.* **2016**, *25*, 790–794.
27. Gao, C. Research on dust retention effect and physiological characteristics of typical garden plants in Nanning. *Res. Soil Water Conserv.* **2016**, *23*, 187–192.
28. Cai, Y. *Study on the Dust Retention Effect of Urban Keystone Tree Species and Their Photosynthetic Characteristics*; Fujian University of Agriculture and Forestry: Fuzhou, China, 2010.
29. Mclaughlin, S.B.; Mcconathy, R.K. Effects of SO_2 and O_3 on allocation of 14C-labeled photosynthate in Phaseolus vulgaris. *Plant Physiol.* **1983**, *73*, 630–635. [CrossRef]
30. Wu, L.-Y. Effects of sulphur dioxide on photosynthetic and respiratory intensities of crops. *Agric. Environ. Prot.* **1989**, *8*, 9–12.
31. Okano, K.; Totsuka, T.; Fukuzawa, T.; Tazaki, T. Growth responses of plants to various concentrations of nitrogen dioxide. *Environ Pollut. Ser. A Ecol. Biol.* **1985**, *38*, 361–373. [CrossRef]
32. Sabaratnam, S.; GuPat, G.; Mulehi, C. Effects of nitrogen dioxide on leaf chloroPhyll and Nitrogen content of soybean. *Environ. Pollut.* **1988**, *51*, 113–120. [CrossRef]
33. Sabaratnam, S.; Gupat, G. Effeets of nitrogen dioxide on bioehemieal and Physiologieal Charaeteristies of soybean. *Environ. Pollut.* **1988**, *55*, 149–158. [CrossRef] [PubMed]
34. Lai, X.; Wang, H.; Zhong, Y. Response of leaf functional traits of nine garden plants to short-term NO_2 pollution under artificially controlled conditions. *J. Ecol.* **2019**, *39*, 8058–8067.
35. Liu, M.; Yu, R.; Mu, R. Photosynthetic characteristics of three typical greening tree species at different altitudes in Beishan, Lanzhou. *J. Ecol. Environ.* **2021**, *30*, 1943–1951.
36. Zhu, J.; Yu, Q.; Liu, Y. Plant functional traits and their leaf economic profiles in response to urban thermal environment. *J. Beijing For. Univ.* **2018**, *40*, 72–81.
37. Niinemets, Ü.; Keenan, T.F.; Hallik, L. A worldwide analysis of within-canopy variations in leaf structural, chemical and physiological traits across plant functional types. *New Phytol.* **2015**, *205*, 973–993. [CrossRef]
38. Chen, S.; Zhang, X.; Qu, Y. *LES Traits of Garden Plants in Shihu Lake in Response to Water Environment*; Northeast Agricultural University: Suzhou, China, 2021.
39. Zhu, J.; Xu, C.; Qin, G. Response of leaf functional traits of three typical greening plants to air pollution and their leaf economic spectra analyses—A case study in Beijing. *J. Cent. South For. Univ. Sci. Technol.* **2019**, *39*, 91–98.

40. Li, Y.; Zhang, X.; Qu, Y. Response strategies of leaf functional traits of common street trees to air quality in Suzhou. *J. Northeast Agric. Univ.* **2022**, *53*, 16–27.
41. Wang, A.-Y. *Particulate Matter Retention Effects and Physiological Responses of Plants under Different Pollution Levels in Zhengzhou City*; Henan Agricultural University: Zhengzhou, China, 2023.
42. Liu, C. Characterisation of volatile organic pollutants in Suqian City during early autumn. *Jiangxi Chem. Ind.* **2023**, *39*, 97–100.
43. Xie, X.; Zhao, Z.; Li, S. Study on pollution characteristics and potential sources of O_3 and $PM_{2.5}$ in Akhdara. *J. Environ. Sci.* **2023**, 1–13.
44. Yuan, Q. *Functional Traits of Plant Leaves and Their Relationship with the Environment in Beishan, Jinhua*; Zhejiang Normal University: Jinhua, China, 2020.
45. He, C.; Li, J.; Guo, M. Changes in photosynthetic properties and water use efficiency of four tree leaves with tree height. *J. Ecol.* **2008**, *28*, 3008–3016.
46. Cheng, J.-F.; Jiang, K.; Shen, Y.-G. Effects of pruning and stemming on the photosynthetic characteristics of leaves of Dicotyledonus pendula. *Chin. J. Ecol. Agric.* **2009**, *17*, 469–473. [CrossRef]
47. Du, M.; Zhang, N. Effects of atmospheric pollution on chlorophyll content of urban green plants. *China Environ. Monit.* **2007**, *23*, 86–88.
48. Li, J.; He, S.; Jing, Y. Physiological and ecological responses of 10 garden plants to air pollution and ecological responses of 10 garden plant species to atmospheric pollution. *J. Ecol. Environ.* **2020**, *29*, 1205–1214.
49. Li, H.; Lv, G.; Jiang, L.; Wang, J. Scale Change and Correlation of Plant Functional Characteristics in the Desert Community of Ebinur Lake. *Sustainability* **2021**, *13*, 4983. [CrossRef]
50. Tang, Q. *Variation of Plant Functional T Raits in Subtropical Evergreen Deciduous Broad-Leaved Mixed Forest*; China Academy of Forestry Sciences: Beijing, China, 2016.
51. Garnier, E.; Shipley, B.; Roumet, C.; Laurent, G. A standardized protocol for the determination of specific leaf area and leaf dry matter content. *Funct. Ecol.* **2001**, *15*, 688–695. [CrossRef]
52. Li, S.; Su, P.; Zhang, H. Characteristics of water and functional traits of desert plant leaves and their interrelationships. *J. Plant Physiol.* **2013**, *49*, 153–160.
53. Wang, R.; Ji, X.; Liu, X. Diversity analysis of leaf and photosynthetic characteristics among millets germplasm in different ecological zones. *J. Shanxi Agric. Univ. Nat. Sci. Ed.* **2014**, *34*, 97–102.
54. Warren, C.R.; Adams, M.A. Evergreen trees do not maximize instantaneous photosynthesis. *Trends Plant Sci.* **2004**, *9*, 270–274. [CrossRef]
55. Poorter, L.; Wright, S.J.; Paz, H.; Ackerly, D.D.; Condit, R.; Ibarra-Manríquez, G.; Harms, K.E.; Licona, J.C.; Martínez-Ramos, M.; Mazer, S.J. Are functional traits good predictors of demographic rates? Evidence from five neotropical forests. *Ecology* **2008**, *89*, 1908–1920. [CrossRef]
56. Takashima, T.; Hikosaka, K.; Hirose, T. Photosynthesis or persistence: Nitrogen allocation in leaves of evergreen and deciduous Quercus species. *Plant Cell Environ.* **2004**, *27*, 1047–1054. [CrossRef]
57. Shao, Y.; Zhang, J.; Wang, B. Study on summer photosynthetic characteristics and carbon sequestration and oxygen release capacity of major greening tree species in Shanghai. *J. Anhui Agric. Univ.* **2016**, *43*, 94–101.
58. Song, H.; Yu, H.; Chen, Y. Leaf economic spectrum of different functional plants in Beijing Botanical Garden. *J. Appl. Ecol.* **2016**, *27*, 1861–1869.
59. Chen, Y.; Xu, Z. Progress in the study of economic spectrum of plant leaves. *J. Plant Ecol.* **2014**, *38*, 1135–1153.
60. Pérez-Ramos, I.M.; Roumet, C.; Cruz, P.; Blanchard, A.; Autran, P.; Garnier, E. Evidence for a 'plant community economics spectrum' driven by nutrient and water limitations in a Mediterranean rangeland of southern France. *J. Ecol.* **2012**, *100*, 1315–1327. [CrossRef]
61. Onoda, Y.; Westoby, M.; Adler, P.B.; Choong, A.M.F.; Clissold, F.J.; Cornelissen, J.H.C.; Díaz, S.; Dominy, N.J.; Elgart, A.; Enrico, L. Global patterns of leaf mechanical properties. *Ecol. Lett.* **2011**, *14*, 301–312. [CrossRef]
62. Peco, B.; Navarro, E.; Carmona, C.P.; Medina, N.G.; Marques, M.J. Effects of grazing abandonment on soil multifunctionality: The role of plant functional traits. *Agric. Ecosyst. Environ.* **2017**, *249*, 215–225. [CrossRef]
63. Medvigy, D.; Jeong, S.J.; Clark, K.L.; Skowronski, N.S.; Schäfer, K.V. Effects of seasonal variation of photosynthetic capacity on the carbon fluxes of a temperate deciduous forest. *J. Geophys. Res. Biogeosci.* **2013**, *118*, 1703–1714. [CrossRef]

MDPI AG
Grosspeteranlage 5
4052 Basel
Switzerland
Tel.: +41 61 683 77 34

Forests Editorial Office
E-mail: forests@mdpi.com
www.mdpi.com/journal/forests

www.ingramcontent.com/pod-product-compliance
Lightning Source LLC
Chambersburg PA
CBHW061447210926
51850CB00008B/1636

* 9 7 8 3 7 2 5 8 6 7 6 4 6 *